Ma
21

Springer Proceedings in Mathematics & Statistics

Volume 351

This book series features volumes composed of selected contributions from workshops and conferences in all areas of current research in mathematics and statistics, including operation research and optimization. In addition to an overall evaluation of the interest, scientific quality, and timeliness of each proposal at the hands of the publisher, individual contributions are all refereed to the high quality standards of leading journals in the field. Thus, this series provides the research community with well-edited, authoritative reports on developments in the most exciting areas of mathematical and statistical research today.

More information about this series at http://www.springer.com/series/10533

Allaberen Ashyralyev · Tynysbek Sh. Kalmenov ·
Michael V. Ruzhansky · Makhmud A. Sadybekov ·
Durvudkhan Suragan
Editors

Functional Analysis in Interdisciplinary Applications—II

ICAAM, Lefkosa, Cyprus, September 6–9, 2018

Springer

Editors
Allaberen Ashyralyev
Department of Mathematics
Near East University
Mersin, Turkey

Tynysbek Sh. Kalmenov
Institute of Mathematics
and Mathematical Modeling
Almaty, Kazakhstan

Michael V. Ruzhansky
Department of Mathematics:
Analysis, Logic and Discrete Mathematics
Ghent University
Ghent, Belgium

School of Mathematical Sciences
Queen Mary University of London
London, UK

Makhmud A. Sadybekov
Institute of Mathematics
and Mathematical Modeling
Almaty, Kazakhstan

Durvudkhan Suragan
Department of Mathematics
Nazarbayev University
Astana, Kazakhstan

ISSN 2194-1009 ISSN 2194-1017 (electronic)
Springer Proceedings in Mathematics & Statistics
ISBN 978-3-030-69291-9 ISBN 978-3-030-69292-6 (eBook)
https://doi.org/10.1007/978-3-030-69292-6

Mathematics Subject Classification: 35N25, 47A58, 58J32, 65M12, 92D30

This Springer imprint is published by the registered company Springer Nature Switzerland AG
The registered company address is: Gewerbestrasse 11, 6330 Cham, Switzerland

Preface

Functional analysis is an important branch of mathematical analysis which deals with the transformations of functions and their algebraic and topological properties. Its applications are encountered in numerous areas, such as the theories of dynamic and stochastic differential equations, difference and integral equations, numerical analysis, computational analysis, fractional calculus, fourier analysis, mathematical and theoretical physics, control and optimization theories, probability theory, and mathematical statistics.

Motivated by their large applicability for real-life problems, applications of functional analysis have been the aim of an intensive study effort in the last decades, yielding significant progress in the theory of functions and functional spaces, differential and difference equations and boundary value problems, differential and integral operators and spectral theory, and mathematical methods in physical and engineering sciences. The present volume is devoted to these investigations. The publication of this collection of papers is based on the materials of the Mini-symposium *Functional Analysis in Interdisciplinary Applications* organized in the framework of the Fourth International Conference on Analysis and Applied Mathematics (ICAAM, September 6–9, 2018 in Cyprus, Mersin 10, Turkey). The aim of the Mini-symposium is to unite mathematicians working in the areas of functional analysis and its interdisciplinary applications to share new trends of applications of the functional analysis.

The conference ICAAM is organized biannually. Previous conferences were held in Gumushane, Turkey in 2012; in Shymkent, Kazakhstan in 2014; in Almaty, Kazakhstan in 2016. The main goal of the conference ICAAM is to bring together the mathematicians working in the area of analysis and applied mathematics to share new trends of applications of mathematics. The developments in the field of applied mathematics open new research areas in analysis and vice versa. That is why, we organize the conference series to provide a forum for researches and scientists to communicate their recent developments and to present their original results in various fields of analysis and applied mathematics.

This issue presents 21 papers by authors from different countries: Turkey, Kazakhstan, USA, Russian Federation, Iraq, Turkmenistan. Especially, we are pleased with the fact that many articles are written by co-authors who work in different universities in the world. We are confident that such international integration provides an opportunity for a significant increase in the quality and quantity of scientific publications. Contains numerous new results and all them went through a refereeing process. This volume contains four different parts. The first part contains the contributed papers focusing on various aspects of the theory of functions and functional spaces, including investigations of finite-difference analogue of the integral geometry problem with a weight function, results concerning the cone rectangular metric spaces over Banach algebras and fixed point results of T-contraction mappings and ternary semigroups of continuous mappings, results of representation variety of Abelian groups and Reidemeister torsion, study of structure of fractional spaces generated by a two-dimensional difference Neutron transport operator and its applications, boundary conditions of volume hyperbolic potential in a domain with curvilinear boundary. The second part is devoted to the research on difference and differential equations and boundary value problems. Correct and ill-posed problems for partial differential equations, construction problems and properties of their solutions are presented. The third part contains the results of studies on differential and integral operators and on the spectral theory. Uncertainty type principles for radial derivatives, basic theory of impulsive quaternion-valued linear systems, Lyapunov-type inequality for fractional sub-Laplacians are considered. The fourth part is focused on the simulation of problems arising in real-world applications of applied sciences, such as attractors for weak solutions of a regularized model of viscoelastic mediums motion with memory in non-autonomous case, unified numerical method for solving system of nonlinear wave equations, trends and risk of HIV/AIDS in Turkey, comparison of the rate of induced intrinsic pathway of apoptosis on COLO-320 and COLO-741.

Mersin, Turkey — Allaberen Ashyralyev
Almaty, Kazakhstan — Tynysbek Sh. Kalmenov
Ghent, Belgium — Michael V. Ruzhansky
Almaty, Kazakhstan — Makhmud A. Sadybekov
Nur-Sultan, Kazakhstan — Durvudkhan Suragan

Contents

Theory of Functions and Functional Spaces

Attractors for Weak Solutions of a Regularized Model of Viscoelastic Mediums Motion with Memory in Non-autonomous Case

Aleksandr Boldyrev and Victor Zvyagin

Abstract We study existence of attractors for weak solutions of the regularized model for viscoelastic medium motion with memory in non-autonomous case. We apply the theory of trajectory attractors for non-invariant trajectory spaces and prove the existence of trajectory attractor, global attractor, uniform trajectory attractor, and uniform global attractor for this system. For the proof of the existence theorems the approximating-topological method is used. This method was introduced by V. G. Zvyagin and was developed in his papers and papers of his colleagues.

Keywords Regularized model · Viscoelastic medium with memory · Weak solution · Trajectory attractor · Global attractor · Uniform attractor

1 Introduction

Let Ω be a bounded domain with a smooth boundary Γ in the space $\mathbb{R}^n$, $n = 2, 3$. We look at the following initial-boundary value problem for the regularized model of the motion of fluid media with memory:

A. Boldyrev (✉) · V. Zvyagin
Laboratory of mathematical fluid dynamics of Research Institute of Mathematics, Voronezh State University, Universitetskaya pl. 1, 394018 Voronezh, Russian Federation
e-mail: al-boldyrev@mail.ru

V. Zvyagin
e-mail: zvg_vsu@mail.ru

A. Ashyralyev et al. (eds.), *Functional Analysis in Interdisciplinary Applications—II*, Springer Proceedings in Mathematics & Statistics 351,
https://doi.org/10.1007/978-3-030-69292-6_1

$$\frac{\partial v}{\partial t}+\sum_{i=1}^{n} v_i\frac{\partial v}{\partial x_i}-\mu_1 \mathrm{Div}\int_0^t e^{-\frac{t-s}{\lambda}}\mathcal{E}(v)(s,Z_\delta(s;t,x))\,ds$$

$$-\mu_0\mathrm{Div}\mathcal{E}(v)=-\mathrm{grad}\,p+f(t,x),\quad (t,x)\in[0,+\infty)\times\Omega, \tag{1}$$

$$\mathrm{div}\,v=0,\quad (t,x)\in[0,+\infty)\times\Omega, \tag{2}$$

$$v\Big|_{\Gamma}=0,\quad v(0,x)=v^0(x),\ x\in\Omega,\quad \int_{\Omega} p\,dx=0, \tag{3}$$

where $v(t,x)=(v_1(t,x),\dots,v_n(t,x))$ is the vector-valued velocity function of particles in the fluid, $p(t,x)$ is the pressure, $f(t,x)$ is the density of the external forces ($(t,x)\in[0,+\infty)\times\Omega$), $\mu_0>0$, $\mu_1\geq 0$ are some constants, $\mathcal{E}(v)=(\mathcal{E}_{ij}(v))$ is the rate of deformation tensor, with components defined by

$$\mathcal{E}_{ij}=\frac{1}{2}\left(\frac{\partial v_i}{\partial x_j}+\frac{\partial v_j}{\partial x_i}\right),\quad (i,j=1\dots n),$$

$\mathrm{Div}A$ is the divergence of the $n\times n$-matrix $A=(A_{ij})$, that is,

$$\mathrm{Div}A=\left(\sum_{i=1}^{n}\frac{\partial A_{i1}}{\partial x_i},\dots,\sum_{i=1}^{n}\frac{\partial A_{in}}{\partial x_i}\right);$$

We shall explain the notation $Z_\delta(s;t,x)$. To do this we introduce the spaces V and H which are standard in mathematical problems in hydrodynamics.

Let $\mathcal{D}(\Omega)^n$ be the space of C^∞-functions from Ω into $\mathbb{R}^n$ which have compact support in Ω. We set

$$\mathcal{V}=\left\{v=(v_1,\dots,v_n)\colon v_i\in\mathcal{D}(\Omega)^n,\ i=1,\dots,n;\ \mathrm{div}\,v=0\right\}.$$

Let H be the closure of $\mathcal{V}$ in the norm of $L_2(\Omega)^n$ and let V be the closure of $\mathcal{V}$ in the norm of $W_2^1(\Omega)^n$. The space V is a Hilbert space with the inner product $(v,u)_V=\sum_{i,j=1}^{n}\int_\Omega \mathcal{E}_{ij}(u)\cdot\mathcal{E}_{ij}(v)\,dx$ and the corresponding norm $\|v\|_V$.

We look at the trajectory defined by the equation

$$z(\tau;t,x)=x+\int_t^\tau S_\delta v(s,z(s;t,x))\,ds,\quad \tau\in[0,T],\ (t,x)\in(0,T)\times\Omega. \tag{4}$$

In (4) we use the regularization operator $S_\delta\colon H\to C^1(\Omega)^n\cap V$ ($\delta>0$). It has the following properties: $S_\delta(v)\to v$ in H as $\delta\to 0$ and the map generated by this operator is continuous ($S_\delta\colon L_2(0,T;H)\to L_2(0,T;C^1(\overline{\Omega})^n\cap V)$). One construction of an operator of this type can be found in [2]. We must introduce a regularization operator because for problems in hydrodynamics the velocity vector $v\in L_2(0,T;V)$, so a trajectory $z(\tau; t, x)$ can only be defined for a regularized velocity field (smoother than the original velocity field).

Thus for each $v \in L_2(0, T; V)$ Eq. (4) has a unique solution $Z_\delta(v)$. If $v \in L_2^{\mathrm{loc}}(\mathbb{R}_+; V)$ then we can assume that the function $Z_\delta(\tau; t, x)$ is defined for $\tau \geq 0$ and $t > 0$. This is the function involved in (1).

Note also that when the constant μ_1 vanishes, system (1)–(2) is transformed into the Navier–Stokes system.

It was established in [7, 11] that the initial-boundary value problem (1)–(3) is solvable in the weak sense. A description of the large-time behaviour of solutions is usually considered to be next in importance. This description (the so-called limiting regimes) also characterizes the process as a whole. These questions are answered using the theory of attractors of the corresponding systems.

The aim of this paper is to investigate uniform attractors for problem (1)–(3) in non-autonomous case. Existence attractors for problem (1)–(3) in autonomous case was established in [9].

2 Basic Concepts and Facts in the Theory of Attractors

Let E and E_0 be Banach spaces, $E \subset E_0$ (the embedding is assumed to be continuous); we also assume that E is reflexive.

Let $L_\infty(\mathbb{R}_+; E)$ be the Banach space of essentially bounded functions on $\mathbb{R}_+$ taking values in E.

The linear space $C(\mathbb{R}_+; E_0)$ consists of continuous function on $\mathbb{R}_+$ taking values in E_0. Convergence in this space is treated as uniform convergence on each interval $[0, T]$, $T > 0$.

Now we look at the shift operators $\mathrm{T}(h)$, to each function g this operator assigns the function $\mathrm{T}(h)g$ such that

$$\mathrm{T}(h)g(t) = g(t + h).$$

Let us consider a certain normalized space $\mathscr{X}$ and assume that with each σ from a certain nonempty set $\Sigma \subset \mathscr{X}$ a nonempty set $\mathscr{H}_\sigma^+ \subset C(\mathbb{R}_+; E_0) \cap L_\infty(\mathbb{R}_+; E)$ is assigned, which is said to be a trajectory space. In this case one says that a family of trajectory spaces $\mathscr{H}_\sigma^+ : \sigma \in \Sigma$ is given.

A set Σ is said to be a set of symbols, and its elements are symbols. A set $\mathscr{H}_\Sigma^+ = \bigcup\limits_{\sigma \in \Sigma} \mathscr{H}_\sigma^+$ is said to be a unified space of trajectories of family $\mathscr{H}_\sigma^+ : \sigma \in \Sigma$.

Definition 1 A set $P \subset C(\mathbb{R}_+; E_0) \cap L_\infty(\mathbb{R}_+; E)$ is called a uniformly (with respect to $\sigma \in \Sigma$) attracting set (for the family of trajectory spaces $\mathscr{H}_\sigma^+ : \sigma \in \Sigma$), if for any nonempty set $B \subset \mathscr{H}_\Sigma^+$, bounded with respect to the norm of $L_\infty(\mathbb{R}_+; E)$, we have $\lim\limits_{h \to \infty} \sup\limits_{u \in B} \inf\limits_{v \in P} \|\mathrm{T}(h)u - v\|_{C(\mathbb{R}_+; E_0)} = 0$.

Definition 2 A set $P \subset C(\mathbb{R}_+; E_0) \cap L_\infty(\mathbb{R}_+; E)$ is called a uniformly (with respect to $\sigma \in \Sigma$) absorbing (for the family of trajectory spaces $\mathcal{H}_\sigma^+ : \sigma \in \Sigma$), if for any set $B \subset \mathcal{H}_\Sigma^+$, bounded with respect to the norm of $L_\infty(\mathbb{R}_+; E)$, there exists $h \geq 0$ such that $\mathrm{T}(t)B \subset P$ whenever $t \geq h$.

It follows from these definitions that each uniformly absorbing set is uniformly attracting.

Definition 3 A set $\mathcal{U} \subset C(\mathbb{R}_+; E_0) \cap L_\infty(\mathbb{R}_+; E)$ is called a uniform trajectory semi-attractor of the family of trajectory spaces $\mathcal{H}_\sigma^+ : \sigma \in \Sigma$, if the following conditions hold:

(i) $\mathcal{U}$ is compact in $C(\mathbb{R}_+; E_0)$ and bounded in $L_\infty(\mathbb{R}_+; E)$;
(ii) $\mathrm{T}(t)\mathcal{U} \subset \mathcal{U}$ for all $t \geq 0$;
(iii) the set $\mathcal{U}$ is uniformly attracting.

Definition 4 A set $\mathcal{U} \subset C(\mathbb{R}_+; E_0) \cap L_\infty(\mathbb{R}_+; E)$ is called a uniform trajectory semi-attractor of the family of trajectory spaces $\mathcal{H}_\sigma^+ : \sigma \in \Sigma$, if it satisfies the conditions (i), (iii) of Definition 3, and it satisfies the following condition:

(ii') $\mathrm{T}(t)P = P$ for all $t \geq 0$.

Definition 5 The minimal uniform trajectory attractor of the family of trajectory spaces $\mathcal{H}_\sigma^+ : \sigma \in \Sigma$ is the least with respect to inclusion uniform trajectory attractor, i.e., the uniform trajectory attractor that is contained in any uniform trajectory attractor.

Definition 6 A nonempty set $\mathcal{A} \subset E$ is called the uniform global attractor (in E_0) of the family of trajectory spaces $\mathcal{H}_\sigma^+ : \sigma \in \Sigma$, if the following conditions hold:

(i) $\mathcal{A}$ is compact in E_0 and bounded in E;
(ii) for any set $B \subset \mathcal{H}_\Sigma^+$ bounded in $L_\infty(\mathbb{R}_+; E)$ the attraction condition holds: $\sup\limits_{u \in B} \inf\limits_{y \in \mathcal{A}} \|u(t) - y\|_{E_0} \to 0 \, (t \to \infty)$;
(iii) $\mathcal{A}$ is contained in any nonempty set that satisfies (i) and (ii).

Obviously, minimal uniform trajectory attractor and uniform global attractor are unique if they exist.

We have the following theorem for the existence: (see [6, 11])

Theorem 1 *Assume that there exists a uniform trajectory semi-attractor P of the family of trajectory spaces $\mathcal{H}_\sigma^+ : \sigma \in \Sigma$. Then there exists a minimal uniform trajectory attractor $\mathcal{U} \subset P$ of $\mathcal{H}_\sigma^+ : \sigma \in \Sigma$.*

Theorem 2 *If there exists a minimal uniform trajectory attractor $\mathcal{U}$ of the family of trajectory spaces $\mathcal{H}_\sigma^+ : \sigma \in \Sigma$, then there is a uniform global attractor $\mathcal{A}$ of $\mathcal{H}_\sigma^+ : \sigma \in \Sigma$ and for all $t \geq 0$ one has*

$$\mathcal{A} = \mathcal{U}(t).$$

To describe the corresponding trajectory space we need the concept of a weak solution of this problem. To introduce it we translate (1)–(3) in operator form.

We denote the dual spaces to H and V by H^* and V^*, respectively. By Riesz's theorem H can be identified canonically with H^*. Bearing in mind this identification we have the chain of embeddings

$$V \subset H \equiv H^* \subset V^*, \tag{5}$$

where both embeddings are dense and compact.

We set $CG[0,T] = C([0,T] \times [0,T], C^1 D(\overline{\Omega}))$, where $C^1 D(\overline{\Omega})$ is the class of continuous bijective maps $z\colon \overline{\Omega} \to \overline{\Omega}$ which coincide with the identity map on Γ and have continuous first order partial derivatives such that $\det\left(\frac{\partial z}{\partial x}\right) = 1$ at each point in $\overline{\Omega}$ (the topology in $C^1 D(\overline{\Omega})$ is induced from $C(\overline{\Omega})^n$).

Let $\langle \varphi, v \rangle$ denote the action on the vector v of the linear functional φ in the dual space. Introduce the maps:
(1) a linear operator $A\colon V \to V^*$, $\langle A(u), h \rangle = \mu_0(\mathcal{E}(u), \mathcal{E}(h))_{L_2(\Omega)^{n^2}}$, $u, h \in V$;
(2) a map $K : V \to V^*$, $\langle K(u), h \rangle = \sum\limits_{i,j=1}^{n} \left(u_i u_j, \frac{\partial h_i}{\partial x_j}\right)_{L_2(\Omega)}$, $u, h \in V$;
($V \subset L_4(\Omega)^n$ then $u_i u_j \in L_2(\Omega)$ and $\frac{\partial h_i}{\partial x_j} \in L_2(\Omega)$ therefore the inner product is well-posed);
(3) if $v \in L_2(0,T;V)$ and $z \in CG[0,T]$, then for each fixed $t \in (0,T)$ we introduce the functional on V

$$\langle C(v,z)(t), h \rangle = \mu_1 \left(\int\limits_0^t e^{-\frac{t-s}{\lambda}} \mathcal{E}(v)(s, z(s;t,x))\, ds, \mathcal{E}(h) \right)_{L_2(\Omega)^{n^2}}.$$

Thus we have defined an operator $C : L_2(0,T;V) \times CG[0,T] \to L_2(0,T;V^*)$ and $L_2^{\mathrm{loc}}(\mathbb{R}_+;V) \times CG \to L_2^{\mathrm{loc}}(\mathbb{R}_+;V^*)$, where $L_2^{\mathrm{loc}}(\mathbb{R}_+;V) = \{v : v|_{[0,T]} \in L_2(0,T;V)\ \forall T > 0\}$ and $CG = C(\mathbb{R}_+ \times \mathbb{R}_+, C^1 D(\overline{\Omega}))$.

Let

$$W_1(0,T) = \left\{v\colon v \in L_2(0,T;V),\ v' \in L_1(0,T;V^*)\right\}.$$

Definition 7 The weak solution of problem (1)–(3) on $[0,T]$ with $v^0 \in H$, $f \in L_2(0,T;V^*)$ is a function $v \in W_1$, if it satisfies the identity

$$v' + A(v) - K(v) + C(v, Z_\delta(v)) = f, \tag{6}$$

$$v(0) = v^0. \tag{7}$$

Since for $n = 2, 3$ we have $W_1(0,T) \subset C([0,T], V^*)$, condition (7) makes sense for $v^0 \in H$ in view of (5).

That problem (1)–(3) has weak solutions was shown in [7].

Note that the continuous embedding $V \subset H$ means that there exists a constant K_0, depending only on Ω, such that

$$\|u\|_H \leq K_0 \|u\|_V \quad (u \in V). \tag{8}$$

As a pair of Banach spaces, which are need for introducing trajectories spaces, we choose

$$E = H, \quad E_0 = V_\theta^*,$$

where V_θ^* ($\theta \in (0, 1)$) is the conjugate space for V_θ, which is a closure of the set $\mathscr{V}$ in the norm of the Sobolev space $H^\theta(\Omega)^n$.

In problem (1)–(3) the function f depends on time, therefore it can be considered as a symbol. Hence, it is natural to consider the set of symbols in the space $\mathscr{X} = L_2^{\text{loc}}(\mathbb{R}_+, V^*)$. The space $\mathscr{X}$ is a Banach one with the norm

$$\|\varphi\|_{\mathscr{X}} = \sup_{t \geq 0} \|\varphi\|_{L_2(t,t+1;V^*)}.$$

We can take $\Sigma = \{f\}$ as a space of symbols, in this case we will investigate trajectory and global attractors of problem (1)–(3). However, since the solutions of non-autonomous problems depend on the initial moments of time, then we can obtain different attractors for two different initial moments of time. In this connection, it is interesting to consider a uniform attractor, i.e., attractor that does not depend on the initial moment of time (see [1, 3]). For that it is necessary to extend the set of symbols, namely, for f to take $\Sigma = \{T(h)f, \forall h \geq 0\}$ and together with the initial problem to consider a family of problems, where f is replaced with $\sigma \in \Sigma$. We note that in this case for any $\sigma \in \Sigma$ we have

$$\|\sigma\|_{\mathscr{X}} = \sup_{t \geq 0} \|T(h)f\|_{L_2(t,t+1;V^*)} = \sup_{t \geq 0} \|f\|_{L_2(t+h,t+h+1;V^*)} \leq \|f\|_{\mathscr{X}}. \tag{9}$$

Definition 8 As a space of trajectories $\mathscr{H}_\sigma^+$ of problem (6)–(7), corresponding to the symbol $\sigma \in \Sigma$ we take a space of functions $v \in L_\infty(\mathbb{R}_+; H) \cap L_2^{\text{loc}}(\mathbb{R}_+; V)$ with $v' \in L_1^{\text{loc}}(\mathbb{R}_+; V^*)$ such that a constraint of v to any segment $[0, T]$ a solution to the problem

$$v' + A(v) - K(v) + C(v, Z_\delta(v)) = \sigma \tag{10}$$

with certain $v^0 \in H$, and for any $t \geq 0$ the inequalities are fulfilled

$$\|v\|_{L_\infty(t,t+1;H)} \leq C_1 \left(1 + \|v\|^2_{L_\infty(\mathbb{R}_+,H)} e^{-2\gamma t}\right)^{1/2}, \tag{11}$$

$$\int_0^t e^{-2\gamma(t-s)} \|v(s)\|_V^2 \, ds \leq C_1^2 \left(1 + \|v\|^2_{L_\infty(\mathbb{R}_+,H)} e^{-2\gamma t}\right) \tag{12}$$

with certain constants $C_1 > 0$ and $0 < \gamma < \lambda^{-1}$, depending on μ_0, μ_1, $\|f\|_{\mathscr{X}}$ and independent of v, v^0.

It will be further proved that this space lies in the class $L_\infty(\mathbb{R}_+; H) \cap C(\mathbb{R}_+; V_\theta^*)$ and is nonempty.

The main results of this paper are the following two existence theorems for attractors.

Theorem 3 *Let* $f \in L_2^{\text{loc}}(\mathbb{R}_+, V^*)$, $\Sigma = \{T(h)f, \forall h \geq 0\}$ *and the parameters* μ_0, μ_1, λ *in (1)–(3) are related by* $\mu_0 - \mu_1\lambda > 0$. *Then there exists the minimal uniform trajectory attractor* $\mathcal{U}$ *of the family of trajectory spaces* $\mathcal{H}_\sigma^+ : \sigma \in \Sigma$ *of problem (1)–(3).*

Theorem 4 *Let* $f \in L_2^{\text{loc}}(\mathbb{R}_+, V^*)$, $\Sigma = \{T(h)f, \forall h \geq 0\}$ *and the parameters* μ_0, μ_1, λ *in (1)–(3) are related by* $\mu_0 - \mu_1\lambda > 0$. *Then there exists the minimal uniform trajectory attractor* $\mathcal{U}$ *and the uniform global attractor* $\mathcal{A}$ *of the family of trajectory spaces* $\mathcal{H}_\sigma^+ : \sigma \in \Sigma$ *of problem (1)–(3).*

To prove this theorems we need some auxiliary results and properties of operators, which we present in the next section.

3 Properties of the Operators and Estimates for Solutions

In this paper we use the topological approximation method for analyzing equations which was presented in [8] and further developed in [11]. Bearing this method in mind, we introduce the approximation equations which will be used in problem (1)–(3). To do this we modify Eq. (6) to fit all the terms in the space $L_2(0, T; V^*)$. We consider the operator

$$K_\varepsilon : V \to V^*, \ \langle K_\varepsilon(u), h \rangle = \sum_{i,j=1}^{n} \left(\frac{u_i u_j}{1 + \varepsilon |u|^2}, \frac{\partial h_i}{\partial x_j} \right)_{L_2(\Omega)} \qquad (\varepsilon \geq 0).$$

Note, that for $\varepsilon = 0$ the operator K_ε and K are the same.

Look at the initial problem

$$v' + A(v) - K_\varepsilon(v) + C(v, Z_\delta(v)) = f \quad (\varepsilon > 0); \tag{13}$$

$$v(0) = v^0, \tag{14}$$

in the space

$$W(0, T) = \left\{ v : v \in L_2(0, T; V), \ v' \in L_2(0, T; V^*) \right\}. \tag{15}$$

Assume that $W(0, T)$ is endowed with the norm

$$\|v\|_{W(0,T)} = \|v\|_{L_2(0,T;V)} + \|v'\|_{L_2(0,T;V^*)} \quad (v \in W(0, T)).$$

The space $W(0, T)$ is a Banach space and we know ([5, chapter III, Lemma 1.2]) that $W(0, T) \subset C([0, T], H)$, thus the initial condition (14) makes sense for $v^0 \in H$.

First we point out the properties of the operators A, K_ε and $C(v, z)$ which are of interest to us.

Lemma 1 *The operator $A\colon V \to V^*$ is bounded.*

Proof Let $u, h \in V$, then by the definition of A

$$|\langle A(u), h\rangle| = \mu_0 |(\mathscr{E}(u), \mathscr{E}(h))_{L_2(\Omega)^{n^2}}|$$
$$\leq \mu_0 \|\mathscr{E}(u)\|_{L_2(\Omega)^{n^2}} \|\mathscr{E}(h)\|_{L_2(\Omega)^{n^2}} = \mu_0 \|u\|_V \|h\|_V,$$

and since h is arbitrary, it follows that $\|A(u)\|_{V^*} \leq \mu_0 \|u\|_V$, which shows that A is continuous. □

Lemma 2 *The operator $K_\varepsilon\colon V \to V^*$ ($\varepsilon \geq 0$) is continuous and*

$$\|K_\varepsilon(u)\|_{V^*} \leq c_0 \|u\|_H^{1/2} \|u\|_V^{3/2} \tag{16}$$

where the constant c_0 is independent of u.

The proof of Lemma 2 can be found in [10] for $\varepsilon = 0$ and in [8] for $\varepsilon > 0$.

For matrices $A = (a_{ij})$ and $B = (b_{ij})$ of order n we set

$$A : B = \sum_{i,j=1}^{n} a_{ij} b_{ij}.$$

Lemma 3 *Let $v \in L_2(0, T; V)$, $z \in CG[0, T]$ and $h \in V$. Then for $t \in [0, T]$*

$$|\langle C(v, z)(t), h\rangle| \leq \mu_1 \left(\int_0^t e^{-\frac{t-s}{\lambda}} \|v(s)\|_V \, ds \right) \|h\|_V. \tag{17}$$

Proof

$$
\begin{aligned}
|\langle C(v,z)(t),h\rangle| &= \left|\mu_1\left(\int_0^t e^{-(t-s)/\lambda}\mathscr{E}(v)(s,z(s;t,x))\,ds,\mathscr{E}(h)\right)\right|\\
&= \mu_1\left|\int_\Omega\left(\int_0^t e^{-(t-s)/\lambda}\mathscr{E}(v)(s,z(s;t,x))\,ds\right):\mathscr{E}(h)\,dx\right|\\
&= \mu_1\left|\int_\Omega\int_0^t e^{-(t-s)/\lambda}\mathscr{E}(v)(s,z(s;t,x)):\mathscr{E}(h)\,ds\,dx\right|\\
&= \mu_1\left|\int_0^t e^{-(t-s)/\lambda}\int_\Omega \mathscr{E}(v)(s,z(s;t,x)):\mathscr{E}(h)\,dx\,ds\right|\\
&\le \mu_1\int_0^t e^{-(t-s)/\lambda}\left|\int_\Omega \mathscr{E}(v)(s,z(s;t,x)):\mathscr{E}(h)\,dx\right|ds\\
&\le \mu_1\int_0^t e^{-(t-s)/\lambda}\left(\int_\Omega|\mathscr{E}(v)(s,z(s;t,x))|^2\,dx\right)^{\frac{1}{2}}\left(\int_\Omega|\mathscr{E}(h)|^2\,dx\right)^{\frac{1}{2}}ds.
\end{aligned}
$$

In the integral

$$
\int_\Omega|\mathscr{E}(v)(s,z(s;t,x))|^2\,dx
$$

we can make the change of variables $z = z(s;t,x)$; since for fixed s and t we have $\det(\partial z/\partial x) = 1$, it follows that

$$
\int_\Omega|\mathscr{E}(v)(s,z(s;t,x))|^2\,dx = \int_\Omega|\mathscr{E}(v)(s,z)|^2\,dz.
$$

Then we obtain

$$
\begin{aligned}
|\langle C(v,z)(t),h\rangle| &\le \mu_1\int_0^t e^{-(t-s)/\lambda}\left(\int_\Omega|\mathscr{E}(v)(s,z)|^2\,dz\right)^{\frac{1}{2}}\left(\int_\Omega|\mathscr{E}(h)|^2\,dx\right)^{\frac{1}{2}}ds\\
&= \mu_1\int_0^t e^{-(t-s)/\lambda}\|v(s)\|_V\|h\|_V\,ds = \mu_1\left(\int_0^t e^{-(t-s)/\lambda}\|v(s)\|_V\,ds\right)\|h\|_V.
\end{aligned}
$$

The proof of (17) is complete. □

Lemma 4 *Let φ is the summable function on $[0,t]$ ($t > 1$), and let $a > 1$ is the real number. Then*

$$
\int_0^t a^s\varphi(s)ds \le \frac{2a^{t+1}}{a-1}\sup_{s\in[0,t-1]}\int_s^{s+1}|\varphi(s)|ds. \tag{18}
$$

Proof Let $[t]$ is the integer part t. Then

$$\int_0^t a^s\varphi(s)ds \le \sum_{i=0}^{[t]-1}\int_i^{i+1} a^s|\varphi(s)|ds + \int_{[t]}^t a^s|\varphi(s)|ds$$

$$\le \sum_{i=0}^{[t]-1}\int_i^{i+1} a^{i+1}|\varphi(s)|ds + \int_{[t]}^t a^t|\varphi(s)|ds \le \sum_{i=0}^{[t]-1} a^{i+1}\int_i^{i+1}|\varphi(s)|ds + a^t\int_{t-1}^t|\varphi(s)|ds$$

$$\le (a + a^2 + \ldots + a^{[t]} + a^t)\sup_{s\in[0,t-1]}\int_s^{s+1}|\varphi(s)|ds$$

$$= \left(\frac{a^{[t]+1}-a}{a-1} + a^t\right)\sup_{s\in[0,t-1]}\int_s^{s+1}|\varphi(s)|ds \le \frac{2a^{t+1}}{a-1}\sup_{s\in[0,t-1]}\int_s^{s+1}|\varphi(s)|ds.$$

Now we turn to the required estimates.

Lemma 5 *Assume that* $\mu_0 - \mu_1\lambda > 0$ *and let* v *be a solution of (13), (14) on an interval* $[0, T]$, $T > 1$. *Then*

$$\|v\|_{L_\infty(t,t+1;H)} \le C_2\left(\|v^0\|_H^2 e^{-2\gamma t} + \sup_{\tau\in[0,T-1]}\|f\|^2_{L_2(\tau,\tau+1;V^*)}\right)^{1/2}, \tag{19}$$

$$\int_0^t e^{-2\gamma(t-s)}\|v(s)\|_V^2\,ds \le C_2^2\left(\|v^0\|_H^2 e^{-2\gamma t} + \sup_{\tau\in[0,T-1]}\|f\|^2_{L_2(\tau,\tau+1;V^*)}\right), \tag{20}$$

for $t \in [0, T-1]$ *and* $t \in [0, T]$ *respectively, where the constants* $0 < \gamma < \lambda^{-1}$ *and* $C_2 > 0$ *depend on* μ_0, μ_1, *but are independent of* v^0 *and* v.

Proof Applying the left parts and right parts of the Eq. (13) belonging to $L_2(0, T; V^*)$ to the function $v \in L_2(0, T; V)$ we obtain the equality

$$\langle v'(t), v(t)\rangle + \langle Av(t), v(t)\rangle - \langle K_\varepsilon(v(t)), v(t)\rangle + \langle C(v, Z_\delta(v))(t), v(t)\rangle = \langle f(t), v(t)\rangle \tag{21}$$

which holds for almost all $t \in (0, T)$. We consider each term on the left-hand side separately. Since $v' \in L_2(0, T; V^*)$ we have

$$\langle v'(t), v(t)\rangle = \frac{1}{2}\frac{d}{dt}\|v(t)\|_H^2 \tag{22}$$

(see [8, p. 62]). By the definition of the operator A

$$\langle A(v(t)), v(t)\rangle = \mu_0(\mathcal{E}(v(t)), \mathcal{E}(v(t)))_{L_2(\Omega)^{n^2}} = \mu_0\|v(t)\|_V^2. \tag{23}$$

By the properties of K_ε for $\varepsilon > 0$ (see [8, p. 45–56])

$$\langle K_\varepsilon(v(t)), v(t)\rangle = 0. \tag{24}$$

The last term on the left-hand side of (21) can be estimated with the help of (17):

$$|\langle C(v, Z_\delta(v))(t), v(t)\rangle| \le \mu_1\Big(\int_0^t e^{-\frac{t-s}{\lambda}}\|v(s)\|_V\,ds\Big)\|v(t)\|_V. \tag{25}$$

Using relations (22)–(25), from (21) we obtain

$$\frac{1}{2}\frac{d}{dt}\|v(t)\|_H^2 + \mu_0\|v(t)\|_V^2 - \mu_1\Big(\int_0^t e^{-\frac{t-s}{\lambda}}\|v(s)\|_V\,ds\Big)\|v(t)\|_V \le \langle f(t), v(t)\rangle.$$

We use Cauchy's inequality to estimate the right-hand side. Let α and β be some numbers to be specified such that

$$\alpha > 0, \quad \beta > 0, \quad \alpha + \beta < 1. \tag{26}$$

Then we have

$$|\langle f(t), v(t)\rangle| \le \|f(t)\|_{V^*}\|v(t)\|_V \le (1-\alpha-\beta)\mu_0\|v(t)\|_V^2 + \frac{\|f(t)\|_{V^*}^2}{4(1-\alpha-\beta)\mu_0}.$$

Setting $R_1 = 1/(4(1-\alpha-\beta)\mu_0)$ we obtain

$$\frac{1}{2}\frac{d}{dt}\|v(t)\|_H^2 + \mu_0\|v(t)\|_V^2 - \mu_1\Big(\int_0^t e^{-\frac{t-s}{\lambda}}\|v(s)\|_V\,ds\Big)\|v(t)\|_V$$
$$\le (1-\alpha-\beta)\mu_0\|v(t)\|_V^2 + R_1\|f(t)\|_{V^*}^2.$$

Carrying the first term on the right-hand side over to the left and multiplying both sides by 2 we obtain

$$\frac{d}{dt}\|v(t)\|_H^2 + 2\alpha\mu_0\|v(t)\|_V^2$$
$$+\Big\{2\beta\mu_0\|v(t)\|_V^2 - 2\mu_1\|v(t)\|_V\int_0^t e^{-\frac{t-s}{\lambda}}\|v(s)\|_V\,ds\Big\} \le 2R_1\|f(t)\|_{V^*}^2.$$

Now we multiply both sides of this inequality by $e^{2\gamma t}$, where $\gamma > 0$ will be defined later and make the substitution

$$e^{2\gamma t}\frac{d}{dt}\|v(t)\|_H^2 = \frac{d}{dt}\left(e^{2\gamma t}\|v(t)\|_H^2\right) - 2\gamma e^{2\gamma t}\|v(t)\|_H^2,$$

on the left-hand side. For the moment we denote the expression in the curly brackets by M; then we have

$$\frac{d}{dt}\left(e^{2\gamma t}\|v(t)\|_H^2\right) + 2\alpha\mu_0 e^{2\gamma t}\|v(t)\|_V^2 + e^{2\gamma t}M - 2\gamma e^{2\gamma t}\|v(t)\|_H^2 \leq 2R_1 e^{2\gamma t}\|f(t)\|_{V*}^2.$$

We find an estimate for the norm $\|v(t)\|_H$ using (8):

$$\frac{d}{dt}\left(e^{2\gamma t}\|v(t)\|_H^2\right) + 2\alpha\mu_0 e^{2\gamma t}\|v(t)\|_V^2 + e^{2\gamma t}\left(M - 2\gamma K_0^2\|v(t)\|_V^2\right) \leq 2R_1 e^{2\gamma t}\|f(t)\|_{V*}^2.$$

The last identity holds for almost all $t \in (0, T)$. We integrate it from 0 to τ, then for all $\tau \in [0, T]$ we obtain

$$e^{2\gamma\tau}\|v(\tau)\|_H^2 + 2\alpha\mu_0\int_0^\tau e^{2\gamma t}\|v(t)\|_V^2\,dt + \int_0^\tau e^{2\gamma t}\left(M - 2\gamma K_0^2\|v(t)\|_V^2\right)dt \leq \|v^0\|_H^2 + 2R_1\int_0^\tau e^{2\gamma t}\|f(t)\|_{V*}^2\,dt.$$

Multiplying both sides by $e^{-2\gamma\tau}$ we substitute in the expression corresponding to M; this yields

$$\|v(\tau)\|_H^2 + 2\alpha\mu_0\int_0^\tau e^{-2\gamma(\tau-t)}\|v(t)\|_V^2\,dt$$

$$+\int_0^\tau e^{-2\gamma(\tau-t)}\left(2\beta\mu_0\|v(t)\|_V^2 - 2\mu_1\|v(t)\|_V\int_0^t e^{-\frac{t-s}{\lambda}}\|v(s)\|_V\,ds - \right.$$

$$\left. 2\gamma K_0^2\|v(t)\|_V^2\right)dt \leq \|v^0\|_H^2 e^{-2\gamma\tau} + 2R_1\int_0^\tau e^{-2\gamma(\tau-t)}\|f(t)\|_{V*}^2\,dt.$$

Now we interchange t and τ.

$$\|v(t)\|_H^2 + 2\alpha\mu_0 \int_0^t e^{-2\gamma(t-\tau)} \|v(\tau)\|_V^2 \, d\tau + 2 \int_0^t e^{-2\gamma(t-\tau)}$$
$$\cdot \left(\left(\beta\mu_0 - \gamma K_0^2\right) \|v(\tau)\|_V^2 - \mu_1 \|v(\tau)\|_V \int_0^\tau e^{-\frac{\tau-s}{\lambda}} \|v(s)\|_V \, ds \right) d\tau$$
$$\leq \|v^0\|_H^2 e^{-2\gamma t} + 2R_1 \int_0^t e^{-2\gamma(t-\tau)} \|f(\tau)\|_{V^*}^2 \, d\tau. \qquad (27)$$

Let

$$I = \int_0^t e^{-2\gamma(t-\tau)} \left(\left(\beta\mu_0 - \gamma K_0^2\right) \|v(\tau)\|_V^2 - \mu_1 \|v(\tau)\|_V \int_0^\tau e^{-\frac{\tau-s}{\lambda}} \|v(s)\|_V \, ds \right) d\tau$$

denote the integral on the left-hand side of (27). We are going to show that $I \geq 0$ for a suitable choice of α, β and γ. We look at the functions

$$j_1(\tau) = \|v(\tau)\|_V,$$
$$j_2(\tau) = \int_0^\tau e^{-(\tau-s)/\lambda} \|v(s)\|_V \, ds = \int_0^\tau e^{-(\tau-s)/\lambda} j_1(s) \, ds.$$

The function j_1 is square integrable, while j_2 is absolutely continuous on $[0, T]$. Differentiating the integral with respect to the variable upper limit we obtain

$$j_2'(\tau) = j_1(\tau) - \frac{1}{\lambda} \int_0^\tau e^{-(\tau-s)/\lambda} j_1(s) \, ds = j_1(\tau) - \frac{1}{\lambda} j_2(\tau).$$

Thus

$$j_2'(\tau) + \frac{1}{\lambda} j_2(\tau) = j_1(\tau) \qquad (28)$$

for almost all $\tau \in [0, T]$. Furthermore, from the definition of j_2 we obviously deduce the equality

$$j_2(0) = 0 \qquad (29)$$

In the second factor under the integral sign in I we use the new notation and transform this factor with the help of (28):

$$
\begin{aligned}
\left(\beta\mu_0 - \gamma K_0^2\right) \|v(\tau)\|_V^2 - \mu_1 \|v(\tau)\|_V \int_0^\tau e^{-(\tau-s)/\lambda} \|v(s)\|_V \, ds& \\
= \left(\beta\mu_0 - \gamma K_0^2\right) j_1^2 - \mu_1 j_1 j_2& \\
= \left(\beta\mu_0 - \gamma K_0^2\right) \left(j_2'(\tau) + \frac{1}{\lambda} j_2(\tau)\right)^2 - \mu_1 \left(j_2'(\tau) + \frac{1}{\lambda} j_2(\tau)\right) j_2& \\
= \left(\beta\mu_0 - \gamma K_0^2\right) \left(j_2'\right)^2 + \left(\frac{2\left(\beta\mu_0 - \gamma K_0^2\right)}{\lambda} - \mu_1\right) j_2 j_2'& \\
+ \left(\frac{\beta\mu_0 - \gamma K_0^2}{\lambda} - \frac{\mu_1}{\lambda}\right) j_2^2 = A\left(j_2'\right)^2 + 2B j_2 j_2' + C j_2^2,&
\end{aligned}
$$

where

$$
\left.
\begin{aligned}
A &= \beta\mu_0 - \gamma K_0^2, \\
B &= \frac{1}{2}\left(\frac{2(\beta\mu_0 - \gamma K_0^2)}{\lambda} - \mu_1\right), \\
C &= \frac{\beta\mu_0 - \gamma K_0^2}{\lambda} - \frac{\mu_1}{\lambda}.
\end{aligned}
\right\}
\tag{30}
$$

We claim that we can choose α, β and γ such that

$$
\frac{\beta\mu_0 - \gamma K_0^2}{\lambda} - \mu_1 > 0, \tag{31}
$$

$$
C - 2\gamma B \geq 0. \tag{32}
$$

In fact, we have the inequality

$$
\frac{\mu_0}{\lambda} - \mu_1 > 0,
$$

and therefore it is clear that we can find β sufficiently close to 1 such that

$$
\frac{\beta\mu_0}{\lambda} - \mu_1 > 0. \tag{33}
$$

Now, as

$$
\begin{aligned}
C - 2\gamma B &= \frac{\beta\mu_0 - \gamma K_0^2}{\lambda^2} - \frac{\mu_1}{\lambda} - \gamma\left(\frac{2\beta\mu_0 - \gamma K_0^2}{\lambda} - \mu_1\right) \\
&\geq \frac{1}{\lambda}\left(\frac{\beta\mu_0}{\lambda} - \mu_1\right) - \gamma\left(\frac{K_0^2}{\lambda^2} + \frac{2\mu_0}{\lambda} - \mu_1\right),
\end{aligned}
$$

taking (33) into account we see that we can choose γ so that (32) holds; in addition, γ can be taken sufficiently small so that (31) is satisfied, because

$$
\frac{\beta\mu_0 - \gamma K_0^2}{\lambda} - \mu_1 = \left(\frac{\beta\mu_0}{\lambda} - \mu_1\right) - \gamma\frac{K_0^2}{\lambda},
$$

and the expression in the brackets is positive by (33). Finally, we choose α to satisfy conditions (26). In what follows we assume that we have taken α, β and γ such that (31) and (32) holds.

Note that as follows from (31), the coefficients A and B defined by (30) are positive.

Now we go over to the integral I itself. Using the new notation

$$I = \int_0^t e^{-2\gamma(t-\tau)} \left(A \left(j_2'(\tau) \right)^2 + 2B j_2(\tau) j_2'(\tau) + C(j_2(\tau))^2 \right) d\tau.$$

Before we transform the expression on the right, we note that by integration by parts

$$\int\limits_0^t 2e^{-2\gamma(t-\tau)} j_2(\tau) j_2'(\tau)\, d\tau = e^{-2\gamma(t-\tau)} (j_2(\tau))^2 \Big|_0^t - 2\gamma \int_0^t e^{-2\gamma(t-\tau)} (j_2(\tau))^2 \, d\tau$$

$$= (j_2(t))^2 - 2\gamma \int_0^t e^{-2\gamma(t-\tau)} (j_2(\tau))^2 \, d\tau$$

(we have used (29)). Then we obtain

$$I = A \int_0^t e^{-2\gamma(t-\tau)} \left(j_2'(\tau) \right)^2 d\tau + B \int_0^t 2e^{-2\gamma(t-\tau)} j_2(\tau) j_2'(\tau)\, d\tau$$
$$+C \int_0^t e^{-2\gamma(t-\tau)} (j_2(\tau))^2 \, d\tau = A \int_0^t e^{-2\gamma(t-\tau)} \left(j_2'(\tau) \right)^2 d\tau + B \left(j_2(t) \right)^2$$
$$+(C - 2\gamma B) \int_0^t e^{-2\gamma(t-\tau)} (j_2(\tau))^2 \, d\tau \geq 0.$$

We have shown that I is nonnegative, so it follows from (27) that

$$\|v(t)\|_H^2 + 2\alpha\mu_0 \int\limits_0^t e^{-2\gamma(t-\tau)} \|v(\tau)\|_V^2 \, d\tau$$
$$\leq \|v^0\|_H^2 e^{-2\gamma t} + 2R_1 \int\limits_0^t e^{-2\gamma(t-\tau)} \|f(\tau)\|_{V^*}^2 \, d\tau, \tag{34}$$

for all $t \in [0, T]$.

The integral on the right-hand side of (34) can be estimated with the help of (18):

$$e^{-2\gamma t}\int\limits_0^t e^{2\gamma\tau}\|f(\tau)\|_{V^*}^2\,d\tau \le e^{-2\gamma t}\frac{2e^{2\gamma(t+1)}}{e^{2\gamma}-1}\sup_{\tau\in[0,t-1]}\int\limits_{\tau}^{\tau+1}\|f(s)\|_{V^*}^2\,ds$$
$$\le \frac{2e^{2\gamma}}{(e^{2\gamma}-1)}\sup_{\tau\in[0,T-1]}\|f\|_{L_2(\tau,\tau+1;V^*)}^2.$$

Then

$$\|v(t)\|_H^2 + 2\alpha\mu_0\int\limits_0^t e^{-2\gamma(t-\tau)}\|v(\tau)\|_V^2\,d\tau$$
$$\le \|v^0\|_H^2 e^{-2\gamma t} + \frac{4R_1 e^{2\gamma}}{(e^{2\gamma}-1)}\sup_{\tau\in[0,T-1]}\|f\|_{L_2(\tau,\tau+1;V^*)}^2. \tag{35}$$

Let $t\in[0,T-1]$. From (35) we see that for $t\le s\le t+1$

$$\|v(s)\|_H^2 \le \|v^0\|_H^2 e^{-2\gamma s} + \frac{4R_1 e^{2\gamma}}{(e^{2\gamma}-1)}\sup_{\tau\in[0,T-1]}\|f\|_{L_2(\tau,\tau+1;V^*)}^2$$
$$\le \|v^0\|_H^2 e^{-2\gamma t} + \frac{4R_1 e^{2\gamma}}{(e^{2\gamma}-1)}\sup_{\tau\in[0,T-1]}\|f\|_{L_2(\tau,\tau+1;V^*)}^2,$$

Passing to the essential maximum for $s\in[t,t+1]$ we obtain

$$\|v\|_{L_\infty(t,t+1;H)}^2 \le \|v^0\|_H^2 e^{-2\gamma t} + \frac{4R_1 e^{2\gamma}}{(e^{2\gamma}-1)}\sup_{\tau\in[0,T-1]}\|f\|_{L_2(\tau,\tau+1;V^*)}^2,$$

which yields (19).

Now it follows from (35) that for each $t\in[0,T]$

$$2\alpha\mu_0\int\limits_0^t e^{-2\gamma(t-\tau)}\|v(\tau)\|_V^2\,d\tau \le \|v^0\|_H^2 e^{-2\gamma t} + \frac{4R_1 e^{2\gamma}}{(e^{2\gamma}-1)}\sup_{\tau\in[0,T-1]}\|f\|_{L_2(\tau,\tau+1;V^*)}^2,$$

which gives us (20).

We can assume without loss of generality that C_2 in (19), (20) is the same constant. □

Lemma 6 *Let v be a solution of (6), (7) or (13), (14) on the interval $[0,T]$, $T>1$ and assume that $v\in L_\infty(0,T;H)$. Then the derivative v' belongs to $L_{4/3}(0,T;V^*)$ and*

$$\|v'\|_{L_{4/3}(t,t+1;V^*)} \le \mathscr{M}(I_1(t), I_2(t)), \tag{36}$$

for $0\le t\le T-1$, where

$$I_1(t) = \text{vrai} \max_{s\in[t,t+1]} \|v(s)\|_H,$$

$$I_2(t) = \max_{\tau\in[t,t+1]} \int_0^\tau e^{-2\gamma(\tau-s)} \|v(s)\|_V^2\, ds,$$

and $\mathcal{M}$ is a continuous function of two nonnegative real variables such that $\mathcal{M}(I_1, I_2)$ is nondecreasing in I_1 for fixed I_2 and in I_2 for fixed I_1.

Proof We start with an estimate for the norm $\|v\|_{L_2(t,t+1;V)}$ in terms of I_2. For $s \geq t$ we have $e^{-2\gamma(t-s)} \geq 1$, so that

$$\|v\|^2_{L_2(t,t+1;V)} = \int_t^{t+1} \|v(s)\|_V^2\, ds \leq \int_t^{t+1} e^{-2\gamma(t-s)} \|v(s)\|_V^2\, ds$$

$$= e^{2\gamma} \int_t^{t+1} e^{-2\gamma(t+1-s)} \|v(s)\|_V^2\, ds \leq e^{2\gamma} \int_0^{t+1} e^{-2\gamma(t+1-s)} \|v(s)\|_V^2\, ds$$

$$\leq e^{2\gamma} \max_{\tau\in[t,t+1]} \int_0^\tau e^{-2\gamma(\tau-s)} \|v(s)\|_V^2\, ds = e^{2\gamma} I_2. \tag{37}$$

Now we turn to an estimate for the derivative. Since v satisfies (6) or (13), expressing the derivative from (6) or (13), we obtain

$$v' = -Av + K_\varepsilon + C(v, Z_\delta(v)) + f. \tag{38}$$

where $\varepsilon \geq 0$

We claim that all terms on the right-hand side belong to $L_{4/3}(0, T; V^*)$ and will estimate their norms in $L_{4/3}(t, t+1; V^*)$, $0 \leq t \leq T-1$.

Since $A\colon V \to V^*$ is a bounded operator (Lemma 1) it follows that

$$\|Av(s)\|_{V^*} \leq \mu_0 \|v(s)\|_V,$$

so that $\|Av\|_{V^*}$ is square integrable on $[0, T]$ because $\|v\|_V$ is, and therefore it is also integrable to the power 4/3. Thus

$$\|Av\|_{L_{4/3}(t,t+1;V^*)} \leq \|Av\|_{L_2(t,t+1;V^*)} \leq \mu_0 \|v\|_{L_2(t,t+1;V)},$$

and using (37) we obtain

$$\|Av\|_{L_{4/3}(t,t+1;V^*)} \leq \mu_0 e^\gamma I_2^{1/2}. \tag{39}$$

By Lemma 2,

$$\|K_\varepsilon(v(s))\|_{V^*} \le c_0 \|v(s)\|_H^{1/2} \|v(s)\|_V^{3/2}, \quad \varepsilon \ge 0.$$

Hence, from the inclusion $v \in L_\infty(0, T; H) \cap L_2(0, T; V)$ it is easy to conclude that $\|K_\varepsilon(v)\|_{V^*}$ is integrable to the power 4/3 and to find an estimate for its norm:

$$\|K_\varepsilon(v)\|^{4/3}_{L_{4/3}(t,t+1;V^*)} = \int_t^{t+1} \|K_\varepsilon(v(s))\|^{4/3}_{V^*}\, ds \le c_0^{4/3} \int_t^{t+1} \|v(s)\|_H^{2/3} \|v(s)\|_V^2\, ds$$

$$\le c_0^{4/3} \|v\|^{2/3}_{L_\infty(t,t+1;H)} \|v\|^2_{L_2(0,T;V)} \le c_0^{4/3} e^{2\gamma} I_1^{2/3} I_2$$

(we have used (37) in the last inequality). As a result, we have

$$\|K_\varepsilon(v)\|_{L_{4/3}(t,t+1;V^*)} \le c_0 e^{3\gamma/2} I_1^{1/2} I_2^{3/4}, \quad \varepsilon \ge 0. \tag{40}$$

By Lemma 3

$$\|C(v, Z_\delta(v))(t)\|_{V^*} \le \mu_1\bigg(\int_0^t e^{-\frac{t-s}{\lambda}} \|v(s)\|_V\, ds\bigg)$$

$$\le \mu_1\bigg(\int_0^t e^{-(\lambda^{-1}-\gamma)(t-s)} e^{-\gamma(t-s)} \|v(s)\|_V\, ds\bigg)$$

$$\le \mu_1\bigg(\int_0^t e^{-2(\lambda^{-1}-\gamma)(t-s)}\, ds\bigg)^{\frac{1}{2}} \bigg(\int_0^t e^{-2\gamma(t-s)} \|v(s)\|_V^2\, ds\bigg)^{\frac{1}{2}}.$$

Since $\gamma < \lambda^{-1}$ (see Lemma 5), the first integral on the right-hand side is bounded:

$$\int_0^t e^{-2(\lambda^{-1}-\gamma)(t-s)}\, ds \le \frac{1}{2\left(\lambda^{-1}-\gamma\right)},$$

so that

$$\|C(v, Z_\delta(v))(t)\|_{V^*} \le c\bigg(\int_0^t e^{-2\gamma(t-s)} \|v(s)\|_V^2\, ds\bigg)^{\frac{1}{2}}.$$

We see that $\|C(v, Z_\delta(v))\|_{V^*}$ is bounded by a continuous function, so $C(v, Z_\delta(v))$ belongs to $L_\infty(0, T; V^*)$ and therefore also to $L_{4/3}(0, T; V^*)$; furthermore,

$$\|C(v, Z_\delta(v))\|_{L_{4/3}(t,t+1;V^*)} \le \|C(v, Z_\delta(v))\|_{L_\infty(t,t+1;V^*)} \le c I_2^{1/2}. \tag{41}$$

Finally,

$$\|f\|_{L_{4/3}(t,t+1;V^*)} \leq \|f\|_{L_2(t,t+1;V^*)}. \tag{42}$$

Thus the terms on the right-hand side of (38) belong to $L_{4/3}(0, T; V^*)$; hence v' also belongs to this space. Relations (39)–(42) yield

$$\|v'\|_{L_{4/3}(t,t+1;V^*)} \leq \mu_0 e^{\gamma} I_2^{1/2} + c_0 e^{3\gamma/2} I_1^{1/2} I_2^{3/4} + c I_2^{1/2} + \|f\|_{L_2(t,t+1;V^*)}.$$

Denoting the right-hand side by $\mathcal{M}(I_1, I_2)$ we arrive at (36). Obviously, $\mathcal{M}(I_1, I_2)$ is monotone in I_1 and I_2. □

Lemma 7 *Assume that* $\mu_0 - \mu_1\lambda > 0$. *Then, on any interval* $[0, T]$, $T > 1$ *for each* $v^0 \in V^*$ *problem (6), (7) has a weak solution satisfying (19) and (20).*

Proof Let $\varepsilon_m > 0$, $\varepsilon_m \to 0$, and let v_m be a solution of problem (13), (14) on the interval $[0, T]$, which exists by Theorem 4 in [7]. From Lemma 5 and the estimate

$$\|v_m\|^2_{L_2(0,T;V)} = \int_0^T \|v_m(s)\|_V^2\, ds \leq e^{2\gamma T} \int_0^T e^{-2\gamma(T-s)} \|v_m(s)\|_V\, ds \leq$$

$$C_2^2 e^{2\gamma T} \left(\|v^0\|_H^2 e^{-2\gamma T} + \sup_{\tau \in [0,T-1]} \|f\|^2_{L_2(\tau,\tau+1;V^*)} \right)$$

we see that the sequence $\{v_m\}$ is bounded in the norms of the spaces $L_\infty(0, T; H)$ and $L_2(0, T; V)$, and from Lemma 6 we obtain that the sequence of derivatives $\{v_m'\}$ is bounded in $L_{4/3}(0, T; V^*)$ (the sequences $\{v_m\}$ and $\{v_m'\}$ are bounded on any interval $[t, t+1]$ $\forall t \in [0, T-1]$ in Lemmas 5 and 6, thus we have that these sequences are bounded on interval $[0, T]$). As was shown in [7], this means that some subsequence $\{v_{m_k}\}$ converges to a limit function v^* weak-$*$ in $L_\infty(0, T; H)$ and weakly in $L_2(0, T; V)$, and the limit function v^* solves problem (6), (7). It remains to verify that the estimates (19) and (20) survive the passage to the limit. For each $t \in [0, T-1]$ the sequence $\{v_{m_k}\}$ converges to v^* weak-$*$ in $L_\infty(0, T; H)$ and by a property of the weak limit

$$\|v^*\|^2_{L_\infty(t,t+1;H)} \leq \varliminf_{k\to\infty} \|v_{m_k}\|^2_{L_\infty(t,t+1;H)} = \varliminf_{k\to\infty} \operatorname{vrai} \max_{s\in[t,t+1]} \|v_{m_k}(s)\|_H^2$$

$$\leq C_2^2 \left(\|v^0\|_H^2 e^{-2\gamma t} + \sup_{\tau\in[0,T-1]} \|f\|^2_{L_2(\tau,\tau+1;V^*)} \right),$$

which proves (19) for v^*. Now we choose some $t \in [0, T]$. Note that $v_{m_k} \rightharpoonup v^*$ weakly in $L_2(0, T; V)$, so that for fixed t the sequence $\{e^{-\gamma(t-s)} v_{m_k}(s)\}$ converges weakly to $e^{-\gamma(t-s)} v^*(s)$ in $L_2(0, T; V)$. n fact, multiplication by a bounded function does not take us outside $L_2(0, T; V)$ and for each function $\varphi \in L_2(0, T; V)$

$$\int_0^t \left(e^{-\gamma(t-s)} v_{m_k}(s)\right) \varphi(s)\, ds = \int_0^t v_{m_k}(s) \left(e^{-\gamma(t-s)} \varphi(s)\right) ds$$

$$\to \int_0^t v^*(s) \left(e^{-\gamma(t-s)} \varphi(s)\right) ds = \int_0^t \left(e^{-\gamma(t-s)} v^*(s)\right) \varphi(s)\, ds,$$

which proves the weak convergence. By a property of weak limits

$$\int_0^t e^{-2\gamma(t-s)} \|v^*(s)\|_V^2\, ds \leq \varliminf_{k\to\infty} \int_0^t e^{-2\gamma(t-s)} \|v_{m_k}(s)\|_V^2\, ds$$

$$\leq C_2 \left(\|v^0\|_H^2 e^{-2\gamma t} + \sup_{\tau\in[0,T-1]} \|f\|^2_{L_2(\tau,\tau+1;V^*)} \right),$$

which establishes (20) for the limit function v^*. □

4 The Trajectory Space of Problem (1)–(3)

We formulate the necessary further corollary of the theorem of Oben–Simon–Dubinsky (see [4, Corollary 4]):

Corollary 1 *Let X, B and Y be Banach spaces, $X \subset B \subset Y$ (the embedding $X \subset B$ is assumed to be compact). Let F be bounded in $L_\infty(0, T; X)$ and let $\frac{\partial F}{\partial t} = \{\frac{\partial f}{\partial t} : f \in F\}$ be bounded in $L_p(0, T; Y)$, where $p > 1$. Then F is relatively compact in $C(0, T; B)$.*

Throughout, we assume that $\mu_0 - \mu_1 \lambda > 0$.

We introduced the trajectory space in Definition 8. To show this definition is consistent we must demonstrate that for each $\sigma \in \Sigma$

$$\mathscr{H}_\sigma^+ \subset L_\infty(\mathbb{R}_+; H) \cap C(\mathbb{R}_+; V_\theta^*). \tag{43}$$

The inclusion $\mathscr{H}_\sigma^+ \subset L_\infty(\mathbb{R}_+; H)$ follows from the definition of the trajectory space. Each function $v \in \mathscr{H}^+$ satisfies (6) on every interval $[0, T]$, so its derivative belongs to $L_{4/3}(0, T; V^*)$ by Lemma 6. Applying Corollary 1, to the three spaces $H \subset V_\theta^* \subset V^*$ we see that v is a continuous function from $[0, T]$ into V_θ^*. Since the interval $[0, T]$ and the function $v \in \mathscr{H}^+$ are arbitrary, we have $\mathscr{H}_\sigma^+ \subset C(\mathbb{R}_+; V_\theta^*)$. This proves the inclusion (43).

We must also show that $\mathscr{H}^+$ is nonempty. We obtain this from the following result.

Theorem 5 *Let $\sigma \in \Sigma$ is the some symbol. Then for each $v_0 \in H$ there exists a trajectory $v \in \mathscr{H}_\sigma^+$ such that $v(0) = v_0$.*

Proof For a given $\sigma \in \Sigma$, consider the system

$$v' + A(v) - K(v) + C(v, Z_\delta(v)) = \sigma, \tag{44}$$

$$v(0) = v_0 \tag{45}$$

and let v_m be a solution of this system on the segment $[0, T_m]$, where $\{T_m\}$ is some increasing sequence that tends to ∞. By Lemma 7 the solutions v_m exist and satisfy the inequalities

$$\|v_m\|_{L_\infty(t,t+1;H)} \leq C_2\Big(\|v^0\|_H^2 e^{-2\gamma t} + \sup_{\tau\in[0,T-1]} \|\sigma\|^2_{L_2(\tau,\tau+1;V^*)}\Big)^{1/2}, \tag{46}$$

$$\int_0^t e^{-2\gamma(t-s)}\|v_m(s)\|_V^2\, ds \leq C_2^2\Big(\|v^0\|_H^2 e^{-2\gamma t} + \sup_{\tau\in[0,T-1]} \|\sigma\|^2_{L_2(\tau,\tau+1;V^*)}\Big)$$

for $t \in [0, T_m - 1]$ and $t \in [0, T_m]$ respectively.

Note that v_m is a continuous function on $[0, T_m]$ taking values in V_θ^*. In fact, it follows from (46) that v_m belongs to the space $L_\infty(0, T_m; H)$, then by Lemma 6 its derivative belongs to $L_{4/3}(0, T_m; V^*)$, so that applying Corollary 1 to the three spaces $H \subset V_\theta^* \subset V^*$ we obtain the necessary continuity. Now we extend the function v_m continuously to the half-axis $\mathbb{R}_+$ by setting

$$\tilde{v}_m(t) = \begin{cases} v_m(t), & 0 \leq t \leq T_m, \\ v_m(m), & t \geq T_m. \end{cases}$$

We claim that the sequence $\{\tilde{v}_m\}$ is relatively compact in $C(\mathbb{R}_+; V_\theta^*)$. To prove this we use the following fact: a subset $P \subset C(\mathbb{R}_+; E_0)$ is relatively compact in $C(\mathbb{R}_+; E_0)$ if and only if the set $\Pi_T P$ is relatively compact in $C([0, T]; E_0)$ for each $T > 0$, where $\Pi_T P$ is the restriction of the set of functions P to $[0, T]$. In view of this, it is sufficient to show that the sequence $\{\Pi_T \tilde{v}_m\}$ of the restrictions of these functions to an arbitrary interval $[0, T]$ is relatively compact in $C([0, T]; V_\theta^*)$. If m is sufficiently large, the restriction $\Pi_T \tilde{v}_m$ is a solution of problems (44), (45), so we can apply Lemmas 5 and 6 to these functions; these ensure that the sequence is bounded in the norm of $L_\infty(0, T; H)$ and the sequence of derivatives is bounded in the norm of $L_{4/3}(0, T; V^*)$. By Corollary 1 the sequence $\{\Pi_T \tilde{v}_m\}$ is relatively compact in $C([0, T]; V_\theta^*)$. Since T is arbitrary, the sequence $\{\tilde{v}_m\}$ is relatively compact in $C(\mathbb{R}_+; V_\theta^*)$. Hence it contains a subsequence $\tilde{v}_{m_k}$ converging to some function v in $C(\mathbb{R}_+; V_\theta^*)$.

Now we show that $v \in \mathscr{H}_\sigma^+$. Let $T > 0$. Since $\tilde{v}_{m_k} \to v$ in $C(\mathbb{R}_+; V_\theta^*)$, the restrictions $\Pi_T \tilde{v}_{m_k}$ converge to $\Pi_T v$ in $C([0, T]; V_\theta^*)$. For all sufficiently large k the functions $\Pi_T \tilde{v}_{m_k}$ are solutions of (44), (45), therefore for $t \in [0, T-1]$ and $t \in [0, T]$ respectively, we have

$$\|\Pi_T\tilde{v}_{m_k}\|_{L_\infty(t,t+1;H)} \le C_2\left(\|v^0\|_H^2 e^{-2\gamma t} + \sup_{\tau\in[0,T-1]}\|\sigma\|^2_{L_2(\tau,\tau+1;V^*)}\right)^{1/2}, \quad (47)$$

$$e^{-2\gamma t}\int_0^t \|\Pi_T\tilde{v}_{m_k}(s)\|_V^2\,ds \le \int_0^t e^{-2\gamma(t-s)}\|\Pi_T\tilde{v}_{m_k}(s)\|_V^2\,ds \le$$

$$\le C_2\left(\|v^0\|_H^2 e^{-2\gamma t} + \sup_{\tau\in[0,T-1]}\|\sigma\|^2_{L_2(\tau,\tau+1;V^*)}\right) \quad (48)$$

by Lemma 5.

In that way the sequence $\{\Pi_T\tilde{v}_{m_k}\}$ is bounded in the norms of the spaces $L_\infty(0, T; H)$ and $L_2(0, T; V)$, so it converges to the limit function $\Pi_T v$ weak-$*$ in $L_\infty(0, T; H)$ and weakly in $L_2(0, T; V)$. Moreover, by Lemma 6 the sequence of derivatives $\{\Pi_T\tilde{v}'_{m_k}\}$ is bounded in $L_{4/3}(0, T; V^*)$. As in the proof of the limit transition in the approximation equations, we deduce from this that the limit function is a solution of (44), (45). Passing to the limit in inequalities (47) and (48), for the limit function we obtain

$$\|\Pi_T v\|_{L_\infty(t,t+1;H)} \le C_2\left(\|v^0\|_H^2 e^{-2\gamma t} + \sup_{\tau\in[0,T-1]}\|\sigma\|^2_{L_2(\tau,\tau+1;V^*)}\right)^{1/2},$$

$$\int_0^t e^{-2\gamma(t-s)}\|\Pi_T v(s)\|_V^2\,ds \le C_2\left(\|v^0\|_H^2 e^{-2\gamma t} + \sup_{\tau\in[0,T-1]}\|\sigma\|^2_{L_2(\tau,\tau+1;V^*)}\right)$$

for $t \in [0, T-1]$ and $t \in [0, T]$ respectively.

Since T is arbitrary, these inequalities hold for all $t \in \mathbb{R}_+$, and therefore $v \in L_\infty(\mathbb{R}_+; H)$. Since $\sup_{\tau\in[0,T-1]}\|\sigma\|_{L_2(\tau,\tau+1;V^*)} \le \|\sigma\|_{\mathscr{X}} \le \|f\|_{\mathscr{X}}$ the last two estimates are even stronger than (11) and (12). Hence $v \in \mathscr{H}_\sigma^+$ and v satisfies initial condition $v(0) = v^0$.

We have proved that the trajectory space lies in $L_\infty(\mathbb{R}_+; H) \cap C(\mathbb{R}_+; V_\theta^*)$ and is nonempty, that is, Definition 8 is consistent. □

Below we shall need the constant

$$C_3 = \mathscr{M}(2C_1, 4C_1^2),$$

where $\mathscr{M}$ is the function in Lemma 6.

We look at the set

$$P = \Big\{v \in L_\infty(\mathbb{R}_+; H) \cap C(\mathbb{R}_+; V_\theta^*):$$

$$\forall t \ge 0\ \|v\|_{L_\infty(t,t+1;H)} + \|v'\|_{L_{4/3}(t,t+1;V^*)} \le 2C_1 + C_3\Big\}.$$

Lemma 8 *The set P is a uniformly trajectory semi-attractor for a family of trajectory spaces* $\{\mathscr{H}_\sigma^+ : \sigma \in \Sigma\}$ *of (1)–(3).*

Proof We shall verify that P satisfies the conditions in Definition 3.

(i) By the definition of P, for each function $v \in P$ and any $t \geq 0$,

$$\|v\|_{L_\infty(t,t+1;H)} \leq 2C_1 + C_3,$$

so that

$$\|v\|_{L_\infty(\mathbb{R}_+,H)} \leq 2C_1 + C_3,$$

which means that P is bounded in $L_\infty(\mathbb{R}_+, H)$.

Let P_t be the set of restrictions of functions in P to the interval $[t, t+1]$. It immediately follows from the definition of P that for each $v \in P_t$

$$\|v\|_{L_\infty(t,t+1;H)} \leq 2C_1 + C_3,$$
$$\|v'\|_{L_{4/3}(t,t+1;V^*)} \leq 2C_1 + C_3,$$

and the set P_t is relatively compact in $C([t, t+1]; V_\theta^*)$ by Corollary 1.

Since the restrictions of functions in P to an arbitrary interval $[t, t+1]$ form a relatively compact subset of $C([t, t+1]; V_\theta^*)$, the set of restrictions of functions in P to an interval of the form $[0, T]$ is a relatively compact subset of $C([0, T]; V_\theta^*)$, and as already mentioned, this ensures that P is relatively compact in $C(\mathbb{R}_+, V_\theta^*)$.

To prove that P is compact it remains to verify that it is closed in $C(\mathbb{R}_+, V_\theta^*)$. Assume that a sequence $\{v_m\} \subset P$ converges to a function v_* in $C(\mathbb{R}_+, V_\theta^*)$. We must show that $v_* \in P$. Note that the sequence $\{v_m\}$ is bounded in $L_\infty(\mathbb{R}_+, H)$, so it converges to v_* weak-$*$ in $L_\infty(\mathbb{R}_+, H)$. Hence $v_* \in L_\infty(\mathbb{R}_+, H)$.

Now we only need to prove that

$$\|v_*\|_{L_\infty(t,t+1;H)} + \|v_*'\|_{L_{4/3}(t,t+1;V^*)} \leq 2C_1 + C_3$$

for any $t \geq 0$.

Since $\{v_m'\}$ is a bounded sequence in the norm of $L_{4/3}(t, t+1; V^*)$, it converges weakly in this space. By the properties of weak convergence

$$\begin{aligned} &\|v_*\|_{L_\infty(t,t+1;H)} + \|v_*'\|_{L_{4/3}(t,t+1;V^*)} \\ &\leq \varliminf_{m\to\infty} \|v_m\|_{L_\infty(t,t+1;H)} + \varliminf_{m\to\infty} \|v_m'\|_{L_{4/3}(t,t+1;V^*)} \\ \leq &\varliminf_{m\to\infty} \left(\|v_m\|_{L_\infty(t,t+1;H)} + \|v_m'\|_{L_{4/3}(t,t+1;V^*)}\right) \leq 2C_1 + C_3, \end{aligned}$$

as required.

Thus $v^* \in P$, which demonstrates that P is closed and therefore compact in the space $C(\mathbb{R}_+, V_\theta^*)$.

(ii) If $v \in P$, $h \geq 0$, then $\mathrm{T}(h)v \in L_\infty(\mathbb{R}_+; H) \cap C(\mathbb{R}_+; V_\theta^*)$ and furthermore

$$\|\mathrm{T}(h)v\|_{L_\infty(t,t+1;H)} + \|\mathrm{T}(h)v'\|_{L_{4/3}(t,t+1;V^*)}$$
$$= \|v\|_{L_\infty(t+h,t+h+1;H)} + \|v'\|_{L_{4/3}(t+h,t+h+1;V^*)} \leq 2C_1 + C_3,$$

that is, $\mathrm{T}(h)v \in P$. Thus we have proved that P is translationally invariant.

(iii) We claim that P is a uniformly absorbing set. Let $B \subset \mathscr{H}_\Sigma^+$ be a set bounded in the norm of $L_\infty(\mathbb{R}_+, H)$; say, $\|v\|_{L_\infty(\mathbb{R}_+,H)} \leq R$ for $v \in B$. We take t_0 such that

$$1 + R^2 e^{-2\gamma t_0} \leq 4.$$

Let $h \geq t_0$. For $v \in B$, using (11) we obtain

$$\|\mathrm{T}(h)v\|_{L_\infty(t,t+1;H)} = \|v\|_{L_\infty(t+h,t+h+1;H)}$$
$$\leq C_1 \left(1 + \|v\|^2_{L_\infty(\mathbb{R}_+,H)} e^{-2\gamma(t+h)}\right)^{1/2} \leq C_1 \left(1 + R^2 e^{-2\gamma t_0}\right)^{1/2} \leq 2C_1. \quad (49)$$

We find an estimate for the derivative with the use of Lemma 6:

$$\|\mathrm{T}(h)v'\|_{L_{4/3}(t,t+1;V^*)} = \|v'\|_{L_{4/3}(t+h,t+h+1;V^*)} \leq \mathscr{M}(I_1(t+h), I_2(t+h)). \quad (50)$$

We estimate $I_1(t+h)$ and $I_1(t+h)$ by means of (11) and (12):

$$I_1(t+h) = \operatorname{vrai} \max_{s \in [t+h,t+h+1]} \|v(s)\|_H \leq C_1 \left(1 + R^2 e^{-2\gamma(t+h)}\right)^{1/2} \leq 2C_1;$$

$$I_2(t+h) = \max_{\tau \in [t+h,t+h+1]} \int_0^\tau e^{-2\gamma(\tau-s)} \|v(s)\|_V^2 \, ds$$
$$\leq C_1^2 \max_{\tau \in [t+h,t+h+1]} \left(1 + R^2 e^{-2\gamma\tau}\right) = C_1^2 \left(1 + R^2 e^{-2\gamma(t+h)}\right)$$
$$\leq C_1^2 \left(1 + R^2 e^{-2\gamma t_0}\right) = 4C_1^2.$$

Since $\mathscr{M}$ is monotone, it follows from (50) that

$$\|\mathrm{T}(h)v'\|_{L_{4/3}(t,t+1;V^*)} \leq \mathscr{M}(2C_1, 4C_1^2) = C_3.$$

By this inequality and (49),

$$\|\mathrm{T}(h)v\|_{L_\infty(t,t+1;H)} + \|\mathrm{T}(h)v'\|_{L_{4/3}(t,t+1;V^*)} \leq 2C_1 + C_3,$$

so that $\mathrm{T}(h)v \in P$ for $h \geq t_0$. We have proved that P is a uniformly absorbing and therefore it is a uniformly attracting set. Thus P is a uniform trajectory semi-attractor. □

Now we derive the proof of our main theorems.

Proof of Theorem 3. By Theorem 1 a family of trajectory spaces has a minimal uniform trajectory attractor if it has a trajectory semi-attractor. That a uniform semi-attractor exists for $\{\mathscr{H}_\sigma^+ : \sigma \in \Sigma\}$ was proved in Lemma 8, so this trajectory space also has a minimal uniform trajectory attractor. □

Proof of Theorem 4. By Theorem 2 a uniform global attractor of a family of trajectory spaces exists if this family of trajectory spaces has a minimal uniform trajectory attractor. Hence Theorem 4 is a consequence of Theorem 3. □

Acknowledgements This research was supported by Russian Foundation for Basic Research (project No. 20-01-00051).

References

1. Chepyzhov, V.V., Vishik, M.I.: Evolution equations and their trajectory attractors. J. Math. Pures Appl. **76**(10), 913–964 (1997)
2. Dmitrienko, V.T., Zvyagin, V.G.: Construction of a regularization operator in models of the motion of viscoelastic media. Vestn. Voronezh. Gos. Univ., Ser. Fiz. Mat., **2**, 148–153 (2004) (Russian)
3. Sell, G.R.: Nonautonomous differential equations and topological dynamics. I. The basic theory. Trans. Am. Math. Soc. **127**(2), 241–262 (1967)
4. Simon, J.: Compact sets in the space $L_p(0, T; B)$. Ann. Mat. Pura Appl. **146**, 65–96 (1987)
5. Temam, R.: Navier–Stokes Equations. Theory and Numerical Analysis. Studies in Mathematics and its Applications, vol. 2. North-Holland, Amsterdam, New York (1979)
6. Vorotnikov, D.A., Zvyagin, V.G.: Uniform attractors for non-autonomous motion equations of viscoelastic medium. J. Math. Anal. Appl. **325**(1), 438–458 (2007)
7. Zvyagin, V.G., Dmitrienko, V.T.: On weak solutions of a regularized model of a viscoelastic fluid. Differ. Equ. **38**(12), 1731–1744 (2002)
8. Zvyagin, V.G., Dmitrienko, V.T.: Approximating-Topological Approach to Investigation of Problems of Hydrodynamics. Navier–Stokes System, Editorial USSR, Moscow (2004)
9. Zvyagin, V.G., Kondrat'ev, S.K.: Attractors of weak solutions to the regularized system of equations of motion of fluid media with memory. Sbornik: Math. **203**(11), 1611–1630 (2012)
10. Zvyagin, V.G., Kondrat'ev, S.K.: Attractors of equations of non-Newtonian fluid dynamics. Russian Math. Surv. **69**(5), 845–913 (2014)
11. Zvyagin, V.G., Vorotnikov, D.A.: Topological Approximation Methods for Evolutionary Problems of Nonlinear Hydrodynamics. De Gruyter Series in Nonlinear Analysis and Applications, vol. 12, de Gruyter, Berlin (2008)

Investigation of Finite-Difference Analogue of the Integral Geometry Problem with a Weight Function

Galitdin B. Bakanov

Abstract Finite—difference analogue of the two-dimensional problem of integral geometry with a weight function are studied. The stability estimate for the considered problem are obtained.

Keywords Integral geometry · Finite-difference problem · Stability estimate · Uniqueness of the solution

1 Introduction

The problems of integral geometry are to find the functions, which are determined on certain variety, through its integrals on certain set of subvarieties with lower dimension.

Additionally, the problems of integral geometry are correlated with various solutions (data interpretation objectives of exploration seismology, electro-exploration, acoustics, and inverse problems of kinetic equations, widely used in plasma physics and astrophysics). In recent years, the studies on problems of integral geometry have critical significance for tomography, which is intensively developing scientific—technic pillar that has several applications in medicine and industry. Therefore, development of various solution methods for the integral geometry problems is actual issue.

One of the stimuli for studying such problems is their connection with multidimensional inverse problems for differential equations [1]. In some inverse problems for hyperbolic equations were shown to reduce to integral geometry problems and, in particular, a problem of integral geometry was considered in the case of shift-invariant curves. Mukhometov [2] showed the uniqueness and estimated the stability of the solution of a two-dimensional integral geometry on the whole. His results were

G. B. Bakanov (✉)
Khoja Akhmet Yassawi International Kazakh-Turkish University, 29 B. Sattarkhanov Avenue, Turkistan 161200, Kazakhstan
e-mail: galitdin.bakanov@ayu.edu.kz

A. Ashyralyev et al. (eds.), *Functional Analysis in Interdisciplinary Applications—II*, Springer Proceedings in Mathematics & Statistics 351,
https://doi.org/10.1007/978-3-030-69292-6_2

mainly based on the reduction of the two-dimensional integral geometry problem

$$V(\gamma, z) = \int\limits_{K(\gamma,z)} U(x, y)\,\rho(x, y, z)\,ds, \quad \gamma \in [0, l], \quad z \in [0, l], \tag{1}$$

where $U \in C^2\left(\overline{D}\right)$, $\rho(x, y, z)$ is a known function to the boundary value problem

$$\frac{\partial}{\partial z}\left(\frac{\partial W}{\partial x}\frac{\cos\theta}{\rho} + \frac{\partial W}{\partial y}\frac{\sin\theta}{\rho}\right) = 0, \ (x, y, z) \in \Omega_1, \tag{2}$$

$$W(\xi(\gamma), \eta(\gamma), z) = V(\gamma, z), \, V(z, z) = 0, \gamma, z \in [0, l]. \tag{3}$$

Here, D is a plane bounded by a simply connected domain with a smooth boundary Γ :

$$x = \xi(z), \ y = \eta(z), \ z \in [0, l], \ \xi(0) = \xi(l), \ \eta(0) = \eta(l),$$

where a parameter z is the length of the curve Γ:

$$\Omega_1 = \Omega / \{(\xi(\gamma), \ \eta(\gamma), z) : z \in [0, l]\}, \ \Omega = \bar{D} \times [0, l];$$

$K(x, y, z)$ is the part of the curve $K(\gamma, z)$ included between the points $(x, y) \in \overline{D}$ and $(\xi(z), \eta(z)), \ z \in [0, l]$;

$$W(x, y, z) = \int\limits_{K(x,y,z)} U(x, y)\,\rho(x, y, z)\,ds,$$

$\theta(x, y, z)$ is an angle between the tangent to $K(x, y, z)$ at the point (x, y) and the x—axis and the variable parameter s is the curve length.

2 Formulation of the Finite—Difference Problem

Suppose that the requirements on the family of curves $K(\gamma, z)$ and the domain D are necessary for the problem (1) to reduce to problem (2), (3) are met [2, 3]. Assume also that every line parallel to either the x– or the y—axis intersects boundary of D at no more than two points. Let

$$a_1 = \inf_{(x,y)\in D} x, \ b_1 = \sup_{(x,y)\in D} x, \ a_2 = \inf_{(x,y)\in D} y, \ b_2 = \sup_{(x,y)\in D} y,$$

$$h_j = \left(b_j - a_j\right)/N_j, \ j = 1, 2; \ h_3 = l/N_3,$$

and the N_j, $j = 1, 2, 3$, are natural number. Suppose that

$$0 < \varepsilon < \min\{(b_1 - a_1)/3, (b_2 - a_2)/3\},$$

$$D^{\varepsilon} = \left\{(x, y) \in D : \min_{(\alpha,\beta)\in\Gamma} \rho((x, y), (\alpha, \beta)) > \varepsilon\right\},$$

$$R_h = \left\{\left(x_i, y_j\right) : x_i = a_1 + ih_1, y_j = a_2 + jh_2, \ i = 1, \ldots, N_1, j = 1, \ldots, N_2\right\}.$$

A neighborhood $B(ih_1, jh_2)$ of the point $(a_1 + ih_1, a_2 + jh_2)$ is defined as the five-point set

$$\{(a_1 + ih_1, a_2 + jh_2), (a_1 + (i \pm 1)h_1, a_2 + (j \pm 1)h_2)\}.$$

A set D_h^{ε} consists of all points $(a_1 + ih_1, a_2 + jh_2)$ such that their neighborhoods $B(ih_1, jh_2)$ are contained in $D^{\varepsilon} \cap R_h$. A set Γ_h^{ε} is made of all points $(a_1 + ih_1, a_2 + jh_2) \in D_h^{\varepsilon}$ such that $B(ih_1, jh_2) \cap (D^{\varepsilon} \cap R_h) \setminus D_h^{\varepsilon} \neq 0$. Finally,

$$\Delta_h^{\varepsilon} = \bigcup_{\Gamma_h^{\varepsilon}} B(ih_1, jh_2), D_h = R_h \cap D,$$

$$\Omega_h^{\varepsilon} = \{(a_1 + ih_1, a_2 + jh_2, kh_3) : (a_1 + ih_1, a_2 + jh_2) \in D_h^{\varepsilon}, \ k = 0, 1, \ldots, N_3 - 1\}.$$

From here on we suppose that the coefficients and the solution of problem (2), (3) have the following properties:

$$W(x, y, z) \in C^3(\Omega^{\varepsilon}), \theta(x, y, z) \in C^2(\Omega^{\varepsilon}), \Omega^{\varepsilon} = \overline{D^{\varepsilon}} \times [0, l], \tag{4}$$

$$\rho(x, y, z) \in C^2(\Omega), \rho(x, y, z) > c_1 > 0, \frac{\partial\theta}{\partial z} > \left|\frac{\partial\rho}{\partial z}\frac{1}{\rho}\right|. \tag{5}$$

We consider the finite-difference problem of finding the functions $\Phi^k_{i,j}$, which satisfy the equation

$$\left[\Phi_{\overset{\circ}{x}}\frac{A}{C} + \Phi_{\overset{\circ}{y}}\frac{B}{C}\right]_z = 0, \quad (a_1 + ih_1, a_2 + jh_2, kh_3) \in \Omega_h^{\varepsilon}, \tag{6}$$

and the boundary condition

$$\Phi^k_{i,j} = F^k_{i,j}, \ (a_1 + ih_1, a_2 + jh_2) \in \Delta_h^{\varepsilon}, \ k = 1, \ldots, N_3 - 1, \tag{7}$$

$$\Phi^0_{i,j} = \Phi^{N_3}_{i,j}, \quad (a_1 + ih_1, a_2 + jh_2) \in D_h^{\varepsilon}, \tag{8}$$

where

$$\Phi^k_{i,j} = \Phi(x_i, y_j, z_k) = \Phi(a_1 + ih_1, a_2 + jh_2, kh_3),$$

$$\Phi_{\overset{\circ}{x}} = (\Phi_{i+1,j} - \Phi_{i-1,j})/2h_1,$$

$$\Phi_{\overset{\circ}{y}} = (\Phi_{i,j+1} - \Phi_{i,j-1})/2h_2,$$

$$f_z = \frac{\left(f_{i,j}^{k+1} - f_{i,j}^{k}\right)}{h_3}, \quad A = \cos\theta_{i,j}^{k}, \quad B = \sin\theta_{i,j}^{k},$$

$$\theta_{i,j}^{k} = \theta\left(a_1 + ih_1, a_2 + jh_2, kh_3\right), \quad C = \rho_{i,j}^{k} = \rho\left(a_1 + ih_1, a_2 + jh_2, kh_3\right).$$

We note that in the finite-difference formulation information on the solution is given not only on the boundary Γ but also in its ε—neighborhood, because the partial derivatives θ_z, W_{xz}, W_{yz}, W_{xy} have singularities of the type

$$\left[(x - \xi(z))^2 + (y - \eta(z))^2\right]^{-\frac{1}{2}}$$

in a neighborhood of an arbitrary point $(\xi(z), \eta(z), z)$ (see [2]).

3 Main Results

It is not difficult to verify the following assertion.

Lemma 1 *If u and v are mesh functions, then*

$$\left(\frac{u}{v}\right)_z = \frac{u_z v^k - u^k v_z}{v^k v^{k+1}}, \tag{9}$$

$$(uv)_z = u^k v_z + u_z v^k + h_3 u_z v_z, \tag{10}$$

$$(uv)_{\overset{\circ}{x}} = u_{\overset{\circ}{x}} v_i + u_i v_{\overset{\circ}{x}} + \frac{h_1^2}{2}[u_x v_x]_{\bar{x}}. \tag{11}$$

Theorem 1 *Suppose that the solution of problem (6)–(8) exists on Ω_h^{ε} and $\left|\Phi_{\overset{\circ}{x}z}\right| \le c_2$, $\left|\Phi_{\overset{\circ}{y}z}\right| \le c_2$, where c_2 is constant, and that*

$$(AB_z - A_z B) - \left|\frac{C_z}{C}\right| \ge \alpha > 0$$

for all N_j, $j = 1, 2, 3$. Then, there exists a positive constant N^{**} such that, for all $N_j > N^{**}$, $j = 1, 2, 3$,

$$\sum_{\Omega_h^\varepsilon}\left(\Phi_{\overset{\circ}{x}}^2+\Phi_{\overset{\circ}{y}}^2\right)h_1\,h_2\,h_3 \le c_3\sum_{\Delta_h^\varepsilon}\left(F_{\overset{\circ}{x}}^2h_1h_3+F_{\overset{\circ}{y}}^2h_2h_3+F_z^2(h_2+h_1)h_3\right)+c_2h_3^2. \tag{12}$$

Here, c_3 is a constant dependent on the function $\rho\,(x, y, z)$ and on the family of curves $K\,(\gamma, z)$.

Proof Multiplying both sides of (6) by $2C\left(-B\Phi_{\overset{\circ}{x}}+A\Phi_{\overset{\circ}{y}}\right)$ [4–7], we get $J_1+J_2=0$, where

$$J_1=J_2=C\left[-\Phi_{\overset{\circ}{x}}B+\Phi_{\overset{\circ}{y}}A\right]\left[\Phi_{\overset{\circ}{x}}\frac{A}{C}+\Phi_{\overset{\circ}{y}}\frac{B}{C}\right]_z.$$

Using the product differentiation formula (10), we obtain

$$J_1=\left[C\left(-\Phi_{\overset{\circ}{x}}B+\Phi_{\overset{\circ}{y}}A\right)\left(\Phi_{\overset{\circ}{x}}\frac{A}{C}+\Phi_{\overset{\circ}{y}}\frac{B}{C}\right)\right]_z-$$

$$-\left[C\left(-\Phi_{\overset{\circ}{x}}B+\Phi_{\overset{\circ}{y}}A\right)\right]_z\left(\Phi_{\overset{\circ}{x}}\frac{A}{C}+\Phi_{\overset{\circ}{y}}\frac{B}{C}\right)-$$

$$-h_3\left[C\left(-\Phi_{\overset{\circ}{x}}B+\Phi_{\overset{\circ}{y}}A\right)\right]_z\left[\Phi_{\overset{\circ}{x}}\frac{A}{C}+\Phi_{\overset{\circ}{y}}\frac{B}{C}\right]_z=0.$$

Exposing brackets in expression J_1 and J_2 taking into account formulas (9), (10) and equality (6), taking into account that

$$\left(1-\frac{C^k}{C^{k+1}}\right)\approx o\,(h_3)\,,\left(\frac{1}{C^{k+1}}-\frac{1}{C^k}\right)\approx o\,(h_3)\,,\ \left(\frac{1}{C^{k+1}}+\frac{1}{C^k}\right)\approx\frac{2}{C^k}+o\,(h_3)\,,$$

$$D=2AB=2\cos\theta\,\sin\theta=\sin 2\theta,\ \mathrm{E}=A^2-B^2=\cos^2\theta-\sin^2\theta=\cos 2\theta,$$

using the formulas

$$\Phi_{\overset{\circ}{x}}\Phi_{\overset{\circ}{x}z}=\frac{1}{2}\left(\Phi_{\overset{\circ}{x}}^2\right)_z-\frac{h_3}{2}\Phi_{\overset{\circ}{x}z}^2,\ \Phi_{\overset{\circ}{y}}\Phi_{\overset{\circ}{y}z}=\frac{1}{2}\left(\Phi_{\overset{\circ}{y}}^2\right)_z-\frac{h_3}{2}\Phi_{\overset{\circ}{y}z}^2,$$

we get

$$J_3+J_4+J_5+J_6+J_7+J_8=0, \tag{13}$$

where

$$J_3=\frac{1}{2}\left\{\Phi_{\overset{\circ}{x}}^2\left[(AB_z-A_zB)+D\frac{C_z}{C}\right]-2\Phi_{\overset{\circ}{y}}^{k+1}\mathrm{E}\frac{C_z}{C}+\right.$$

$$\left.+\left(\Phi_{\overset{\circ}{y}}^{k+1}\right)^2\left[(AB_z-A_zB)-D\frac{C_z}{C}\right]\right\},$$

$$J_4 = \frac{1}{2}\left\{\left(\Phi^{k+1}_{\overset{\circ}{y}}\right)^2\left[(AB_z - A_zB) + D\frac{C_z}{C}\right] - 2\Phi^{k+1}_{\overset{\circ}{x}}\Phi_{\overset{\circ}{y}}\mathrm{E}\frac{C_z}{C} + \right.$$

$$\left. + \Phi^2_{\overset{\circ}{y}}\left[(AB_z - A_zB) - D\frac{C_z}{C}\right]\right\},$$

$$J_5 = -\frac{h_3^2}{2}\Phi^2_{\overset{\circ}{xz}}\left[(\mathrm{AB}_z - \mathrm{A}_z\mathrm{B}) + D\frac{C_z}{C}\right] - \frac{h_3^2}{2}\Phi^2_{\overset{\circ}{yz}}\left[(\mathrm{AB}_z - \mathrm{A}_z\mathrm{B}) - D\frac{C_z}{C}\right] +$$

$$+\Phi^2_{\overset{\circ}{x}}(\mathrm{A}_z\mathrm{B} + \mathrm{AB}C_z)\,o(h_3) + \Phi^2_{\overset{\circ}{y}}(\mathrm{AB}_z - \mathrm{AB}C_z)\,o(h_3) +$$

$$+\Phi_{\overset{\circ}{x}}\Phi^{k+1}_{\overset{\circ}{y}}B^2C_z o(h_3) - \Phi_{\overset{\circ}{y}}\Phi^{k+1}_{\overset{\circ}{x}}A^2C_z o(h_3).$$

$$J_6 = \Phi_{\overset{\circ}{x}}\Phi_{\overset{\circ}{y}}BB_z\left(1 - \frac{C^k}{C^{k+1}}\right) + \Phi_{\overset{\circ}{x}}\Phi_{\overset{\circ}{y}}AA_z\left(\frac{C^k}{C^{k+1}} - 1\right) -$$

$$-h_3\Phi_{\overset{\circ}{x}}\Phi_{\overset{\circ}{yz}}\left(AA_z + BB_z\frac{C^k}{C^{k+1}}\right) + h_3\Phi_{\overset{\circ}{y}}\Phi_{\overset{\circ}{xz}}\left(AA_z\frac{C^k}{C^{k+1}} + BB_z\right),$$

$$J_7 = \left[\left(-\Phi_{\overset{\circ}{x}}B + \Phi_{\overset{\circ}{y}}A\right)\left(\Phi_{\overset{\circ}{x}}A + \Phi_{\overset{\circ}{y}}B\right)\right]_z - \Phi_{\overset{\circ}{x}}\Phi_{\overset{\circ}{yz}} + \Phi_{\overset{\circ}{y}}\Phi_{\overset{\circ}{xz}},$$

$$J_8 = h_3\Phi^2_{\overset{\circ}{x}}AB_z\frac{C_z}{C} + h_3^2\Phi_{\overset{\circ}{x}}\Phi_{\overset{\circ}{xz}}AB_z\frac{C_z}{C} + h_3\Phi_{\overset{\circ}{x}}\Phi_{\overset{\circ}{y}}AA_z\frac{C_z}{C} + h_3\Phi_{\overset{\circ}{x}}\Phi_{\overset{\circ}{y}}BB_z\frac{C_z}{C} -$$

$$-h_3^2\Phi_{\overset{\circ}{x}}\Phi_{\overset{\circ}{yz}}AA_z\frac{C_z}{C} + h_3^2\Phi_{\overset{\circ}{y}}\Phi_{\overset{\circ}{xz}}BB_z\frac{C_z}{C} - h_3\Phi^2_{\overset{\circ}{y}}A_zB\frac{C_z}{C} - h_3^2\Phi_{\overset{\circ}{y}}\Phi_{\overset{\circ}{yz}}A_zB\frac{C_z}{C}.$$

Now we will transform and we estimate each of these elements.

The expression for J_3 and J_4 is a quadratic form with respect to $\Phi_{\overset{\circ}{x}}$ and $\Phi^{k+1}_{\overset{\circ}{y}}$, to $\Phi^{k+1}_{\overset{\circ}{x}}$ and $\Phi_{\overset{\circ}{y}}$, whose determinant

$$\begin{vmatrix} (AB_z - A_zB) + D\frac{C_z}{C} & -\mathrm{E}\frac{C_z}{C} \\ -\mathrm{E}\frac{C_z}{C} & (AB_z - A_zB) - D\frac{C_z}{C} \end{vmatrix} = (AB_z - A_zB)^2 - D^2\left|\frac{C_z}{C}\right|^2 - \mathrm{E}^2\left|\frac{C_z}{C}\right|^2 =$$

$$(AB_z - A_zB)^2 - \left|\frac{C_z}{C}\right|^2,$$

because $\mathrm{E}^2 + D^2 = 1$, where $\mathrm{E} = \cos 2\theta$, $D = sin2\theta$. Then, from the condition $(AB_z - A_zB) - \left|\frac{C_z}{C}\right| \geq \alpha > 0$ the positive definiteness of the quadratic form J_3 and J_4 follows. Using the inequality

$$ax^2 + 2bxy + cy^2 \geq \frac{2\left(ac - b^2\right)}{a + c + \sqrt{(a-c)^2 + 4b^2}}\left(x^2 + y^2\right)$$

for a positively definite quadratic form $ax^2 + 2bxy + cy^2$, we obtain

$$J_3 \geq \frac{1}{2}\left[(AB_z - A_zB) - \left|\frac{C_z}{C}\right|\right]\left[\left(\Phi^k_{\overset{\circ}{x}}\right)^2 + \left(\Phi^{k+1}_{\overset{\circ}{y}}\right)^2\right], \tag{14}$$

$$J_4 \geq \frac{1}{2}\left[AB_z - A_zB - \left|\frac{C_z}{C}\right|\right]\left[\left(\Phi^{k+1}_{\overset{\circ}{x}}\right)^2 + \left(\Phi^k_{\overset{\circ}{y}}\right)^2\right]. \tag{15}$$

Taking into account that $\left(1 - \frac{C^k}{C^{k+1}}\right) \approx o\left(h_3\right)$, we have

$$J_6 = \Phi_{\overset{\circ}{x}}\Phi_{\overset{\circ}{y}}BB_zo\left(h_3\right) + \Phi_{\overset{\circ}{x}}\Phi_{\overset{\circ}{y}}AA_zo\left(h_3\right) - h_3\Phi_{\overset{\circ}{x}}\Phi_{\overset{\circ}{yz}}\left(AA_z + \frac{BB_zC^k}{C^{k+1}}\right) +$$

$$+h_3\Phi_{\overset{\circ}{y}}\Phi_{\overset{\circ}{xz}}\left(\frac{AA_zC^k}{C^{k+1}} + BB_z\right) = \Phi_{\overset{\circ}{x}}\Phi_{\overset{\circ}{y}}BB_zo\left(h_3\right) + \Phi_{\overset{\circ}{x}}\Phi_{\overset{\circ}{y}}AA_zo\left(h_3\right)\left(AA_z + BB_z\right) \times$$

$$\times\left(\Phi_{\overset{\circ}{y}}\Phi_{\overset{\circ}{xz}} - \Phi_{\overset{\circ}{x}}\Phi_{\overset{\circ}{yz}}\right) - h_3\Phi_{\overset{\circ}{x}}\Phi_{\overset{\circ}{yz}}BB_zo\left(h_3\right) + h_3\Phi_{\overset{\circ}{y}}\Phi_{\overset{\circ}{xz}}AA_zo\left(h_3\right) =$$

$$= \Phi_{\overset{\circ}{x}}\Phi_{\overset{\circ}{y}}\left(AA_z + BB_z\right)o\left(h_3\right) + \left(AA_z + BB_z\right)\left(\Phi^k_{\overset{\circ}{y}}\Phi^{k+1}_{\overset{\circ}{x}} - \Phi^k_{\overset{\circ}{x}}\Phi^{k+1}_{\overset{\circ}{y}}\right) -$$

$$-\left(\Phi^k_{\overset{\circ}{x}}\Phi^{k+1}_{\overset{\circ}{y}} + \Phi^k_{\overset{\circ}{x}}\Phi^k_{\overset{\circ}{y}}\right)BB_zo\left(h_3\right) + \left(\Phi^k_{\overset{\circ}{y}}\Phi^{k+1}_{\overset{\circ}{x}} - \Phi^k_{\overset{\circ}{y}}\Phi^k_{\overset{\circ}{x}}\right)AA_zo\left(h_3\right).$$

Using the inequality $|ab| \leq (a^2 + b^2)/2$, we get

$$\begin{aligned} J_6 \leq \tfrac{1}{2}\Bigg\{&\left[\left(\Phi^k_{\overset{\circ}{x}}\right)^2 + \left(\Phi^k_{\overset{\circ}{y}}\right)^2\right]\left(AA_z + BB_z\right)o\left(h_3\right) + \\ &+\left[\left(\Phi^k_{\overset{\circ}{x}}\right)^2 + \left(\Phi^{k+1}_{\overset{\circ}{y}}\right)^2\right]\left(AA_z + BB_z\right) + \\ &+\left[\left(\Phi^k_{\overset{\circ}{x}}\right)^2 + \left(\Phi^{k+1}_{\overset{\circ}{y}}\right)^2 + \left(\Phi^k_{\overset{\circ}{x}}\right)^2 + \left(\Phi^k_{\overset{\circ}{y}}\right)^2\right]BB_zo\left(h_3\right) + \\ &+\left[\left(\Phi^k_{\overset{\circ}{y}}\right)^2 + \left(\Phi^{k+1}_{\overset{\circ}{x}}\right)^2 + \left(\Phi^k_{\overset{\circ}{y}}\right)^2 + \left(\Phi^k_{\overset{\circ}{x}}\right)^2\right]AA_zo\left(h_3\right)\Bigg\}. \end{aligned} \tag{16}$$

Taking into account that $(AB_z - A_zB) - \left|\frac{C_z}{C}\right| \geq \alpha > 0$, conditions $\left|\Phi_{\overset{\circ}{x}z}\right| \leq c_2$, $\left|\Phi_{\overset{\circ}{y}z}\right| \leq c_2$ and the inequality $|ab| \leq (a^2 + b^2)/2$, we obtain

$$
\begin{gathered}
J_5 \leq \Phi^2_{\overset{\circ}{x}}\left(A_zB + ABC_z\right)o\left(h_3\right) + \Phi^2_{\overset{\circ}{y}}\left(AB_z - ABC_z\right)o\left(h_3\right) + \\
+\tfrac{1}{2}\left\{\left[\left(\Phi^k_{\overset{\circ}{x}}\right)^2 + \left(\Phi^{k+1}_{\overset{\circ}{x}}\right)^2\right]\left(A_zB + ABC_z\right)o\left(h_3\right) + \right. \\
+\left[\left(\Phi^k_{\overset{\circ}{y}}\right)^2 + \left(\Phi^{k+1}_{\overset{\circ}{y}}\right)^2\right] \times \left(AB_z + ABC_z\right)o\left(h_3\right) + \\
+\left[\left(\Phi^k_{\overset{\circ}{x}}\right)^2 + \left(\Phi^{k+1}_{\overset{\circ}{y}}\right)^2\right]B^2C_z o\left(h_3\right) + \\
\left. +\left[\left(\Phi^k_{\overset{\circ}{y}}\right)^2 + \left(\Phi^{k+1}_{\overset{\circ}{x}}\right)^2\right]A^2C_z o\left(h_3\right)\right\} + c_2h_3^2.
\end{gathered}
\tag{17}
$$

Due to formula (11), we have

$$
\begin{gathered}
\Phi_{\overset{\circ}{y}}\Phi_{\overset{\circ}{x}z} = \left[\Phi_{\overset{\circ}{y}}\Phi_z\right]_{\overset{\circ}{x}} - \Phi_{\overset{\circ}{x}\overset{\circ}{y}}\Phi_z - \tfrac{h_1^2}{2}\left[\Phi_{\overset{\circ}{y}x}\Phi_{zx}\right]_{\bar{x}}, \\
-\Phi_{\overset{\circ}{x}}\Phi_{\overset{\circ}{y}z} = -\left[\Phi_{\overset{\circ}{x}}\Phi_z\right]_{\overset{\circ}{y}} + \Phi_{\overset{\circ}{x}\overset{\circ}{y}}\Phi_z + \tfrac{h_2^2}{2}\left[\Phi_{\overset{\circ}{x}y}\Phi_{zy}\right]_{\bar{y}}.
\end{gathered}
$$

Consequently,

$$
\begin{gathered}
J_7 = \left[(-\Phi_{\overset{\circ}{x}}B + \Phi_{\overset{\circ}{y}}A)(\Phi_{\overset{\circ}{x}}A + \Phi_{\overset{\circ}{y}}B)\right]_z - \Phi_{\overset{\circ}{x}}\Phi_{\overset{\circ}{y}z} + \Phi_{\overset{\circ}{y}}\Phi_{\overset{\circ}{x}z} = \\
= \left[\left(-\Phi_{\overset{\circ}{x}}B + \Phi_{\overset{\circ}{y}}A\right)\left(\Phi_{\overset{\circ}{x}}A + \Phi_{\overset{\circ}{y}}B\right)\right]_z + \left[\Phi_{\overset{\circ}{y}}\Phi_z\right]_{\overset{\circ}{x}} - \\
-\left[\Phi_{\overset{\circ}{x}}\Phi_z\right]_{\overset{\circ}{y}} - \tfrac{h_1^2}{2}\left[\Phi_{\overset{\circ}{y}x}\Phi_{zx}\right]_{\bar{x}} + \tfrac{h_2^2}{2}\left[\Phi_{\overset{\circ}{x}y}\Phi_{zy}\right]_{\bar{y}}.
\end{gathered}
\tag{18}
$$

Now we will transform and estimate J_8:

$$
\begin{gathered}
J_8 = h_3\Phi^2_{\overset{\circ}{x}}AB_z\frac{C_z}{C} + h_3\Phi^k_{\overset{\circ}{x}}\Phi^{k+1}_{\overset{\circ}{x}}AB_z\frac{C_z}{C} - h_3\Phi^2_{\overset{\circ}{x}}AB_z\frac{C_z}{C} - h_3\Phi_{\overset{\circ}{x}}\Phi_{\overset{\circ}{y}}AA_z\frac{C_z}{C} + \\
+h_3\Phi_{\overset{\circ}{x}}\Phi_{\overset{\circ}{y}}BB_z\frac{C_z}{C} - h_3\Phi^k_{\overset{\circ}{x}}\Phi^{k+1}_{\overset{\circ}{y}}AA_z\frac{C_z}{C} + h_3\Phi_{\overset{\circ}{x}}\Phi_{\overset{\circ}{y}}AA_z\frac{C_z}{C} + h_3\Phi^k_{\overset{\circ}{y}}\Phi^{k+1}_{\overset{\circ}{x}}BB_z\frac{C_z}{C} - \\
-h_3\Phi_{\overset{\circ}{y}}\Phi_{\overset{\circ}{x}}BB_z\frac{C_z}{C} - h_3\Phi^2_{\overset{\circ}{y}}A_zB\frac{C_z}{C} - h_3\Phi^k_{\overset{\circ}{y}}\Phi^{k+1}_{\overset{\circ}{y}}A_zB\frac{C_z}{C} + h_3\Phi^2_{\overset{\circ}{y}}A_zB\frac{C_z}{C}.
\end{gathered}
$$

Using the inequality $|ab| \leq (a^2 + b^2)/2$, we obtain

$$
\begin{gathered}
J_8 \leq \tfrac{h_3}{2}\left\{\left[\left(\Phi^k_{\overset{\circ}{x}}\right)^2 + \left(\Phi^{k+1}_{\overset{\circ}{x}}\right)^2\right]AB_z\tfrac{C_z}{C} + \left[\left(\Phi^k_{\overset{\circ}{x}}\right)^2 + \left(\Phi^{k+1}_{\overset{\circ}{y}}\right)^2\right]AA_z\tfrac{C_z}{C} + \right. \\
\left. +\left[\left(\Phi^k_{\overset{\circ}{y}}\right)^2 + \left(\Phi^{k+1}_{\overset{\circ}{x}}\right)^2\right]BB_z\tfrac{C_z}{C} + \left[\left(\Phi^k_{\overset{\circ}{y}}\right)^2 + \left(\Phi^{k+1}_{\overset{\circ}{y}}\right)^2\right]A_zB\tfrac{C_z}{C}\right\}.
\end{gathered}
\tag{19}
$$

Supposing that A, B, C enough smooth limit functions and taking into account the expressions (14)–(19), from (13) we get

$$\begin{aligned}\tfrac{1}{2}\Big[(AB_z - A_zB) - \Big|\tfrac{C_z}{C}\Big|\Big] \times \Big[\Big(\Phi^k_{\overset{\circ}{x}}\Big)^2 + \Big(\Phi^{k+1}_{\overset{\circ}{x}}\Big)^2 + \Big(\Phi^k_{\overset{\circ}{y}}\Big)^2 + \Big(\Phi^{k+1}_{\overset{\circ}{y}}\Big)^2\Big] \le \\ \le \tfrac{h_3}{2}K\Big[\Big(\Phi^k_{\overset{\circ}{x}}\Big)^2 + \Big(\Phi^{k+1}_{\overset{\circ}{x}}\Big)^2 + \Big(\Phi^k_{\overset{\circ}{y}}\Big)^2 + \Big(\Phi^{k+1}_{\overset{\circ}{y}}\Big)^2\Big] + R^k_{i,j} + c_2h_3^2,\end{aligned} \tag{20}$$

where

$$R^k_{i,j} = \Big[\Phi_{\overset{\circ}{x}}\Phi_z\Big]_{\overset{\circ}{y}} - \Big[\Phi_{\overset{\circ}{y}}\Phi_z\Big]_{\overset{\circ}{x}} - \Big[\Big(\Phi_{\overset{\circ}{x}}A + \Phi_{\overset{\circ}{y}}B\Big)\Big(-\Phi_{\overset{\circ}{x}}B + \Phi_{\overset{\circ}{y}}A\Big)\Big]_z +$$

$$+\frac{h_1^2}{2}\Big[\Phi_{zx}\Phi_{\overset{\circ}{yx}}\Big]_{\bar{x}} - \frac{h_2^2}{2}\Big[\Phi_{zy}\Phi_{\overset{\circ}{xy}}\Big]_{\bar{y}}.$$

Let

$$\Big[(AB_z - A_zB) - \Big|\frac{C_z}{C}\Big|\Big] \ge \alpha > 0, \;\; N_j > 9, \;\; j = 1, 2, \;\; \mathrm{K}h_3 < \frac{\alpha}{2},$$

i.e. $N_3 > \frac{2\mathrm{K}l}{\alpha}$, because $h_3 = \frac{l}{N_3}$, where α and K is a constant.

Then, from (20) we have

$$\sum_{\Omega_h^\varepsilon}\Big(\Phi^2_{\overset{\circ}{x}} + \Phi^2_{\overset{\circ}{y}}\Big)h_1\,h_2\,h_3 \le \frac{2}{\alpha}\sum_{\Omega_h^\varepsilon} R^k_{i,j} + c_2h_3^2. \tag{21}$$

Using conditions (7), (8) and inequality $|ab| \le (a^2 + b^2)/2$, we can transform (21) so as to obtain the desired estimate (12). Thus, the proof of Theorem is completed.

References

1. Lavrent'ev, M.M., Romanov, V.G.: Three linearized inverse problems for hyperbolic equations. Sov. Math. Dokl. **7**, 1650–1652 (1966)
2. Mukhometov, R.G.: On a problem of integral geometry. Math. Probl. Geophys. **6**, 212–242 (1975). (in Russian)
3. Romanov, V.: Integral Geometry and Inverse Problems for Hyperbolic Equations. Springer, Berlin (1974)
4. Kabanikhin, S.I., Bakanov, G.B.: On the stability of a finite-difference analogue of a two dimensional problem of integral geometry. Sov. Math. Dokl. **35**, 16–19 (1987)
5. Kabanikhin, S.I., Bakanov, G.B.: Discrete analogy of Gel'fand-Levitan method. Dokl. Akad. Nauk. **356**, 157–160 (1997)
6. Bakanov, G.B.: On the stability estimation of differential- difference analogue of the integral geometry problem with a weight function. In: AIP Conference Proceedings, vol. 1759, p. 020067 (2016)

7. Bakanov, G.: On the stability of a differential- difference analogue of a two-dimensional problem of integral geometry. Filomat **32**(3), 933–938 (2018). https://doi.org/10.2298/FIL1803933B

Six Point Implicit Methods for the Approximation of the Derivatives of the Solution of First Type Boundary Value Problem for Heat Equation

Suzan C. Buranay and Lawrence A. Farinola

Abstract We construct six point symmetric implicit difference boundary value problems for the first derivative of the solution $u(x, t)$ of the first type boundary value problem for one dimensional heat equation with respect to the time variable t and for the pure second derivative of the solution with respect to the spatial variable x. Furthermore, for the approximation of the pure second derivative of the solution with respect to time t, and for the approximation of the mixed derivative of the solution special six point implicit difference methods are proposed. It is proved that the constructed implicit schemes converge with the order $O\left(h^2 + \tau^2\right)$ where, h is the step size in x and τ is the step size in time to the corresponding exact derivatives under the assumption that the given boundary value problem of one dimensional heat equation has the initial function belonging to the Hölder space $C^{10+\alpha}$, $0 < \alpha < 1$, the heat source function from the Hölder space $C_{x,t}^{8+\alpha,4+\frac{\alpha}{2}}$, the boundary functions from $C^{5+\frac{\alpha}{2}}$, and between the initial and the boundary functions the conjugation conditions of orders $q = 0, 1, 2, 3, 4, 5$ are satisfied. Theoretical results are justified by numerical examples.

Keywords 65M06 · 65M12 · 65M22

1 Introduction

In mathematical physics calculating the derivatives of a solution of a differential equation is as important as finding the solution, because they provide information about some physical phenomena [3]. Such as the first derivatives of the potential function defines the electrostatic field [13]. As another example, in [19] a theory of

S. C. Buranay · L. A. Farinola (✉)
Department of Mathematics, Faculty of Arts and Sciences, Eastern Mediterranean University, Via Mersin 10, Famagusta, North Cyprus, Turkey
e-mail: lawrence.farinola@emu.edu.tr

S. C. Buranay
e-mail: suzan.buranay@emu.edu.tr

A. Ashyralyev et al. (eds.), *Functional Analysis in Interdisciplinary Applications—II*, Springer Proceedings in Mathematics & Statistics 351,
https://doi.org/10.1007/978-3-030-69292-6_3

the drying wood adopts the fundamental hypothesis that the rate of which transfusion takes place transversely with respect to the wood fibres ($\frac{\partial \theta}{\partial t}$) is proportional to the slope of the moisture gradient ($\frac{\partial^2 \theta}{\partial x^2}$), where θ is the moisture content expressed as a percentage of the oven-dry weight of the wood.

To find highly accurate computations of the derivatives of an unknown solution of a differential equation is problematic because the differentiation operation is ill-conditioned. Also, it is well known that accuracy of the approximate derivatives depends on the accuracy of the approximate solution. The study of approximate derivatives using finite differences is first investigated in [14] where, it was proved that the high order difference derivatives uniformly converge to the corresponding derivatives of the exact solution for the 2D Laplace equation in any strictly interior subdomain with the same order $O(h)$ (h is the grid step) of which the difference solution converges on the given domain. For the Dirichlet problem of the Laplace equation on a rectangle in [20], $O\left(h^2\right)$ order uniform convergence of the solution of the difference equation and its first and pure second difference derivatives to the solution and corresponding derivatives of the solution respectively, were proved over the whole grid domain. Later for this problem in [7] special difference schemes were constructed to approximate the first and pure second order derivatives of the solution. Therein, it was proved that the solutions of these problems converge to the respective exact derivatives with order $O\left(h^4\right)$.

For the three dimensional Laplace equation difference schemes for obtaining the solution of the Dirichlet problem, its first derivatives and pure second derivatives on a cubic grid with uniform accuracy $O\left(h^2\right)$ are constructed in [21] under the agreement that the boundary functions belong to $C^{4,\lambda}, 0 < \lambda < 1$ on the faces, are continuous on the edges, and their second order derivatives satisfy the compatibility condition. In [8] difference schemes for the approximation of the first and pure second derivatives of the solution of the Dirichlet problem in a rectangular parallelepiped which converge uniformly on the cubic grid of order $O\left(h^4\right)$ are proposed when the boundary functions belong to $C^{6,\lambda}, 0 < \lambda < 1$ on the faces, are continuous on the edges, and have second and fourth order derivatives satisfying the compatibility conditions. Most recently, in [9] difference schemes on a cubic grid for obtaining the solution of the same problem, its first and pure second derivatives difference schemes are constructed of which the approximate values of the first and pure second derivatives converge with orders $O\left(h^6 \left|\ln h\right|\right)$ and $O\left(h^{5+\lambda}\right), 0 < \lambda < 1$, respectively. Meanwhile in [10], $O\left(h^{p-1}\right), p \in [4, 5]$ order of approximation for the first order derivatives of the solution of the 3D Laplace equation is proven.

The construction and justification of highly accurate approximate methods for the solution and the derivatives of the solution of the parabolic equation are important to develop the theory of these methods and to improve the use of these methods in applied problems such as in fluid mechanics or in computational biological problems. Some examples are the numerical solution of heat conduction problems in melting solids [11, 16, 17], computational fluid flow solution over endothelial cells inside the capillary vascular system [1], modified Crank–Nicolson difference schemes for parabolic equations with non smooth data arising in biomechanics [2]. Nowadays, special difference problems of four point and six point symmetric implicit difference schemes for the derivative of the solution $u(x,t)$ of the first type boundary value problem for one dimensional heat equation with respect to the spatial variable x are proposed in [4]. It is required that the given first type boundary value problem for one dimensional heat equation has the initial function belonging to the Hölder space C^{s+2}, the heat source function from the Hölder space $C_{x,t}^{s,\frac{s}{2}}$, the boundary functions from $C^{1+\frac{s}{2}}$, where $s=3+\alpha$, $0<\alpha<1$ for the construction of the four point implicit scheme and $s=5+\alpha, 0<\alpha<1$ for the construction of the six point symmetric scheme. Also between the initial and boundary functions the conjugation conditions upto second order ($q=0,1,2$) and upto third order ($q=0,1,2,3$) are required to be satisfied for the construction of the four point and six point symmetric implicit schemes respectively. It is proved that the solution of the constructed four point and six point difference schemes converge uniformly to the exact value of $\frac{\partial u}{\partial x}$ on the grids of order $O\left(h^2+\tau\right)$ and $O\left(h^2+\tau^2\right)$ respectively, where, h is the step size in spatial variable x and τ is the step size in time.

In this study we have extended the approach given in [4] to find the first difference derivative of $u(x,t)$ with respect to t and its second order difference derivatives with $O\left(h^2+\tau^2\right)$ order of convergence over the whole grid domain to the corresponding respective exact derivatives where, $u(x,t)$ is the solution of the first type boundary value problem for one dimensional heat equation. The general idea of this research is presented as an extended abstract in [5] and in the doctoral thesis [12]. The work in this study is organized as follows: In Sect. 2, basic notations and the first type boundary value problem of one dimensional heat equation is given under the agreement that the initial function belongs to $C^{10+\alpha}$, $0<\alpha<1$ the heat source function is from $C_{x,t}^{8+\alpha,4+\frac{\alpha}{2}}$, the boundary functions are from $C^{5+\frac{\alpha}{2}}$, and between the initial and the boundary functions of the conjugation conditions of orders $q=0,1,2,3,4,5$ are satisfied. In Sect. 3 we construct boundary value problems for $\frac{\partial u}{\partial t}$ and $\frac{\partial^2 u}{\partial x^2}$ of which $u(x,t)$ is the solution of the given first type boundary value problem for one dimensional heat equation based on the assumption in Sect. 2. Difference problems of six point symmetric implicit schemes (Crank and Nicolson) approximating $\frac{\partial u}{\partial t}$, $\frac{\partial^2 u}{\partial x^2}$ are constructed, and we give a pointwise estimation for the error function depending on $\rho(x,t)$ which is the distance from the current grid point in the domain to the boundary. In Sects. 4, and 5, special six point implicit difference problems for the approximation of $\frac{\partial^2 u}{\partial t^2}$, $\frac{\partial^2 u}{\partial x \partial t}$ respectively are proposed and it is proved that the solutions of the constructed difference schemes converge uniformly to the exact values of the respective derivatives on the grids of order $O\left(h^2+\tau^2\right)$. In Sect. 6, we construct a numerical example to justify the theoretical results and the obtained results are presented through tables and figures. Concluding remarks are given in Sect. 7.

2 Basic Notations and One Dimensional Heat Equation

Let $\Omega = (0, b)$, $\sigma_T = (0, T)$, and $\overline{\Omega}, \overline{\sigma}_T$ be the closure of these sets respectively. Also denote by $Q_T = \{(x, t) : 0 < x < b, 0 < t \leq T\}$, $\gamma_1 = \{(0, t) : t \in \overline{\sigma}_T\}$, $\gamma_2 = \left\{(x, 0) : x \in \overline{\Omega}\right\}$, $\gamma_3 = \{(b, t) : t \in \overline{\sigma}_T\}$. Let $\gamma = \bigcup_{i=1}^{3} \gamma_i$, represent the boundary of Q_T, and $\overline{Q}_T = Q_T \cup \gamma$. We use the notations $\partial_t^k = \frac{\partial^k}{\partial t^k}$, $\partial_x^k = \frac{\partial^k}{\partial x^k}$ and $D_t^k = \frac{d^k}{dt^k}$, $D_x^k = \frac{d^k}{dx^k}$ to present $k - th$ partial and ordinary derivatives respectively with respect to the time variable t, spatial variable x. Let s be a noninteger positive number, $C_{x,t}^{s,s/2}\left(\overline{Q}_T\right)$ denotes the Hölder space of functions $u(x, t)$ that are continuous in $\overline{Q}_T$ together with all derivatives of the form $\partial_t^{j_0} \partial_x^j$ for $2j_0 + j < s$ and have a finite norm (see [15]). $C^s\left(\overline{\Omega}\right)$ is the space whose elements are continuous functions g in $\overline{\Omega}$ having in $\overline{\Omega}$ continuous derivatives up to order $[s]$ inclusively, and have finite norm as defined in [15]. We consider the first type boundary value problem for one dimensional heat equation:

$$Lu = f(x, t) \text{ on } Q_T, \tag{1}$$
$$u(x, 0) = u_0(x) \text{ on } \gamma_2, \tag{2}$$
$$u(0, t) = u_1(t) \text{ on } \gamma_1, \ u(b, t) = u_2(t) \text{ on } \gamma_3, \tag{3}$$

where, $L \equiv \frac{\partial}{\partial t} - a\frac{\partial^2}{\partial x^2}$ and a is positive constant. Let q be nonnegative integer. We introduce the notations

$$u^{(0)}(x) = u_0(x), \ u^{(q)}(x) = \partial_t^q u(x, t)\big|_{t=0}, \ q = 1, 2, \ldots, \tag{4}$$
$$u_1^{(0)}(0) = u_1(t)|_{t=0}, \ u_2^{(0)}(0) = u_2(t)|_{t=0}, \tag{5}$$
$$u_1^{(q)}(0) = D_t^q u_1(t)\big|_{t=0}, \ u_2^{(q)}(0) = D_t^q u_2(t)\big|_{t=0}, q = 1, 2, \ldots \tag{6}$$

and the functions $u^{(q)}(x)$, $q = 1, 2, \ldots$ in (4) are defined as

$$u^{(1)}(x) = aD_x^2 u_0(x) + f(x, 0), \tag{7}$$
$$u^{(q)}(x) = a\partial_x^2 u^{(q-1)}(x) + f^{(q-1)}(x), \tag{8}$$

where, $f^{(0)}(x) = f(x, 0)$ and $f^{(q)}(x) = \partial_t^q f(x, t)\big|_{t=0}$. The conjugation conditions of order q between the initial and boundary conditions are defined as follows (see [15]),

$$u^{(q)}(0) = u_1^{(q)}(0), \tag{9}$$
$$u^{(q)}(b) = u_2^{(q)}(0) \tag{10}$$

Theorem 1 (see [12]) *Let s be a positive noninteger. Suppose that the initial function* $u_0(x) \in C^{2+s}(\overline{\Omega})$, *the heat source function* $f(x,t) \in C_{x,t}^{s,\frac{s}{2}}(\overline{Q}_T)$ *and the boundary functions* $u_1(t), u_2(t) \in C^{1+\frac{s}{2}}(\overline{\sigma}_T)$ *also, between the initial and boundary functions the conjugation conditions (9), (10) up to order* $1+\left[\frac{s}{2}\right]$ *are satisfied. Then, the first type boundary value problem (1)–(3) has a unique solution* $u(x,t)$ *belonging to the space* $C_{x,t}^{2+s,1+\frac{s}{2}}(\overline{Q}_T)$.

Proof The proof of Theorem 1 follows from Theorem 5.2, Sect. 5, Chap. IV in [15].

Problem 1 The boundary value problem (1)–(3) with the assumption

$$u_0(x) \in C^{10+\alpha}(\overline{\Omega}), \ f(x,t) \in C_{x,t}^{8+\alpha,4+\frac{\alpha}{2}}(\overline{Q}_T) \text{ and } u_i(t) \in C^{5+\frac{\alpha}{2}}(\overline{\sigma}_T), i=1,2, \tag{11}$$

satisfying the conjugation conditions (9), (10) of order 0, 1, 2, 3, 4, 5 and $\alpha \in (0,1)$.

We remark that the following results in this study were obtained and presented in the doctoral thesis [12].

3 Crank and Nicolson Schemes for the Approximation of $\frac{\partial u}{\partial t}$ and $\frac{\partial^2 u}{\partial x^2}$

Given the Problem 1, let $w = \partial_t u$ and $\phi = \partial_x^2 u$ then w and ϕ satisfy the following boundary value problems respectively, see also [5].

Problem 2

$$Lw = \partial_t f(x,t) = F(x,t) \text{ on } Q_T, \tag{12}$$
$$w(x,0) = u^{(1)}(x) = w_0(x) \text{ on } \gamma_2, \tag{13}$$
$$w(0,t) = D_t u_1(t) = w_1(t) \text{ on } \gamma_1, \tag{14}$$
$$w(b,t) = D_t u_2(t) = w_2(t) \text{ on } \gamma_3. \tag{15}$$

Problem 3

$$L\phi = \partial_x^2 f(x,y) = G(x,t) \text{ on } Q_T, \tag{16}$$
$$\phi(x,0) = \partial_x^2 u_0(x) = \phi_0(x) \text{ on } \gamma_2, \tag{17}$$
$$\phi(0,t) = \frac{1}{a}(D_t u_1(t) - f(0,t)) = \phi_1(t) \text{ on } \gamma_1, \tag{18}$$
$$\phi(b,t) = \frac{1}{a}(D_t u_2(t) - f(b,t)) = \phi_2(t) \text{ on } \gamma_3, \tag{19}$$

where, $f(x,t)$ is the heat source function given in (1), $u_0(x)$ and $u_1(t), u_2(t)$ are the initial and boundary functions given in (2), (3) respectively, also $u^{(1)}(x)$ is as defined in (7). Also,

$$F(x,t) \in C_{x,t}^{6+\alpha,3+\frac{\alpha}{2}}\left(\overline{Q}_T\right),\ w_0(x) \in C^{8+\alpha}\left(\overline{\Omega}\right),\ w_i(t) \in C^{4+\frac{\alpha}{2}}\left(\overline{\sigma}_T\right),\ i = 1, 2. \tag{20}$$

$$G(x,t) \in C_{x,t}^{6+\alpha,3+\frac{\alpha}{2}}\left(\overline{Q}_T\right),\phi_0(x) \in C^{8+\alpha}\left(\overline{\Omega}\right),\ \phi_i(t) \in C^{4+\frac{\alpha}{2}}\left(\overline{\sigma}_T\right),\ i = 1, 2, \tag{21}$$

and between the initial and the boundary functions of Problems 2 and 3 the conjugation conditions (9), (10) of orders 0, 1, 2, 3, 4 are satisfied.

Lemma 1 *The Problem 1 has unique solution* $u \in C_{x,t}^{10+\alpha,5+\frac{\alpha}{2}}\left(\overline{Q}_T\right)$ *and the constructed boundary value Problems 2 and 3 have unique solution* w *and* ϕ *respectively, belonging to the space* $C_{x,t}^{8+\alpha,4+\frac{\alpha}{2}}\left(\overline{Q}_T\right)$.

Proof From Theorem 1, Problem has unique solution $u \in C_{x,t}^{10+\alpha,5+\frac{\alpha}{2}}\left(\overline{Q}_T\right)$. Taking into account that Problems 2 and 3 are also first type boundary value problems analogous to the problem (1)–(3) on the basis of Theorem 1 the proof follows.

Assume that $c, c_1, c_2, \ldots$ are positive constants independent from h and τ, in each section, those constants are enumerated anew. For the numerical solution of the Problems 2 and 3 we use the implicit six point symmetric difference scheme (Crank and Nicolson Scheme) [18] on the grid $\overline{\omega}_{h,\tau} = \overline{\omega}_h \times \omega_\tau$ where,

$$\overline{\omega}_h = \left\{x_m = mh, h = \frac{b}{N}, m = 0, \ldots, N\right\},$$

$$\omega_\tau = \left\{t_j = j\tau,\ \tau = \frac{T}{M},\ j = 0, \ldots, M\right\},$$

and the set of internal nodes is denoted by

$$\omega_{h,\tau} = \omega_h \times \omega_\tau = \left\{\left(x_m, t_j\right) : m = 1, \ldots, N-1,\ j = 1, \ldots, M\right\}.$$

We present the set of nodes on γ_i $i = 1, 2, 3$ by

$$\omega_{0,\tau} = \left\{\left(0, t_j\right) : t_j = j\tau, \tau = \frac{T}{M}, j = 0, \ldots, M\right\},$$

$$\overline{\omega}_{h,0} = \left\{(x_m, 0) : x_m = mh, h = \frac{b}{N}, m = 0, \ldots, N\right\},$$

$$\omega_{b,\tau} = \left\{\left(b, t_j\right) : t_j = j\tau, \tau = \frac{T}{M}, j = 0, \ldots, M\right\},$$

respectively. To realize the numerical solution of the Problems 2, and 3 we propose the following implicit six point difference problems, of which the solution of (22)–(24) is $\widetilde{w}$, and the solution of (28)–(30) is $\widetilde{\phi}$ see also [5].

$$\widetilde{w}_{\bar{t},m}^{h,\tau} = a\Lambda \widetilde{w}_m^{h,\tau} + \varphi_{\widetilde{w},m} \text{ on } \omega_{h,\tau}, \tag{22}$$

$$\widetilde{w}_m^0 = w_0(x_m) \text{ on } \overline{\omega}_{h,0}, \tag{23}$$

$$\widetilde{w}_0^j = w_1(t_j) \text{ on } \omega_{0,\tau},\ \widetilde{w}_N^j = w_2(t_j), \text{ on } \omega_{b,\tau}, \tag{24}$$

where, $\widetilde{w}_m^0 = \widetilde{w}(x_m, 0)$, $\widetilde{w}_0^j = \widetilde{w}(0, t_j)$, $\widetilde{w}_N^j = \widetilde{w}(b, t_j)$ and

$$\widetilde{w}_{\bar{t},m}^{h,\tau} = \frac{1}{\tau}\left(\widetilde{w}_m^{j+1} - \widetilde{w}_m^j\right) \tag{25}$$

$$\Lambda \widetilde{w}_m^{h,\tau} = \frac{1}{2h^2}\left(\widetilde{w}_{m-1}^{j+1} - 2\widetilde{w}_m^{j+1} + \widetilde{w}_{m+1}^{j+1}\right) + \frac{1}{2h^2}\left(\widetilde{w}_{m-1}^j - 2\widetilde{w}_m^j + \widetilde{w}_{m+1}^j\right) \tag{26}$$

$$\varphi_{\widetilde{w},m} = F\left(x_m, t_{j+\frac{1}{2}}\right) = \partial_t f\left(x_m, t_{j+\frac{1}{2}}\right). \tag{27}$$

$$\widetilde{\phi}_{\bar{t},m}^{h,\tau} = a\Lambda \widetilde{\phi}_m^{h,\tau} + \varphi_{\widetilde{\phi},m} \text{ on } \omega_{h,\tau}, \tag{28}$$

$$\widetilde{\phi}_m^0 = \phi_0(x_m) \text{ on } \overline{\omega}_{h,0}, \tag{29}$$

$$\widetilde{\phi}_0^j = \phi_1(t_j) \text{ on } \omega_{0,\tau},\ \widetilde{\phi}_N^j = \phi_2(t_j) \text{ on } \omega_{b,\tau}, \tag{30}$$

where $\widetilde{\phi}_{\bar{t},m}^{h,\tau}$, $\Lambda \widetilde{\phi}_m^{h,\tau}$ are formulae analogous to (25) and (26) respectively, and $\widetilde{\phi}_m^0 = \widetilde{\phi}(x_m, 0)$, $\widetilde{\phi}_0^j = \widetilde{\phi}(0, t_j)$, $\widetilde{\phi}_N^j = \widetilde{\phi}(b, t_j)$ also $\varphi_{\widetilde{\phi},m} = G\left(x_m, t_{j+\frac{1}{2}}\right) = \partial_x^2 f\big|_{\left(x_m, t_{j+\frac{1}{2}}\right)}$. Here $t_{j+\frac{1}{2}} = t_j + 0.5\tau$, $f(x,t)$ is the given function in (1) and $u_0(x)$ given in (2), $u_1(t)$, $u_2(t)$ given in (3) are the initial and boundary functions respectively. Using maximum principle the difference problems (22)–(24) and (28)–(30) have unique solution. Consider the following systems:

$$\widehat{q}_{\bar{t},m}^{h,\tau} = a\Lambda \widehat{q}_m^{h,\tau} + \widehat{g}^{h,\tau} \text{ on } \omega_{h,\tau}, \tag{31}$$

$$\widehat{q}_m^0 = 0 \text{ on } \overline{\omega}_{h,0}, \tag{32}$$

$$\widehat{q}_0^j = 0 \text{ on } \omega_{0,\tau},\ \widehat{q}_N^j = 0 \text{ on } \omega_{b,\tau}, \tag{33}$$

$$\overline{q}_{\bar{t},m}^{h,\tau} = a\Lambda \overline{q}_m^{h,\tau} + \overline{g}^{h,\tau} \text{ on } \omega_{h,\tau}, \tag{34}$$

$$\overline{q}_m^0 \geq 0 \text{ on } \overline{\omega}_{h,0}, \tag{35}$$

$$\overline{q}_0^j \geq 0 \text{ on } \omega_{0,\tau},\ \overline{q}_N^j \geq 0 \text{ on } \omega_{b,\tau}, \tag{36}$$

where $\widehat{g}^{h,\tau}$, $\overline{g}^{h,\tau}$ are given functions and $\left|\widehat{g}^{h,\tau}\right| \leq \overline{g}^{h,\tau}$ on $\omega_{h,\tau}$ also $\widehat{q}_{\bar{t},m}^{h,\tau}$, $\overline{q}_{\bar{t},m}^{h,\tau}$ are difference formulae analogous of (25) and $\Lambda \widehat{q}_m^{h,\tau}$, $\Lambda \overline{q}_m^{h,\tau}$ are difference formulae analogous of (26). Let $r = \frac{a\tau}{h^2}$.

Lemma 2 ([4]) *The solution $\widehat{q}$ of the system (31)–(33) and the solution $\overline{q}$ of the system (34)–(36) satisfy the inequality*

$$|\widehat{q}| \leq \overline{q} \text{ on } \overline{\omega}_{h,\tau} \tag{37}$$

for $r \leq 1$.

Lemma 3 ([4]) *For the solution of the problem*

$$\widehat{q}^{h,\tau}_{\bar{t},m} = a\Lambda \widehat{q}^{h,\tau}_{m} + h^2 + \tau^2 \text{ on } \omega_{h,\tau}, \tag{38}$$

$$\widehat{q}^{0}_{m} = 0 \text{ on } \overline{\omega}_{h,0}, \tag{39}$$

$$\widehat{q}^{j}_{0} = 0 \text{ on } \omega_{0,\tau},\ \widehat{q}^{j}_{N} = 0 \text{ on } \omega_{b,\tau}, \tag{40}$$

the following inequality holds

$$\widehat{q} \leq \rho d \left(h^2 + \tau^2\right) \text{ on } \overline{\omega}_{h,\tau} \tag{41}$$

for $r \leq 1$, where $d = \max\left[\frac{b}{2a}, 1\right]$, $\rho = \rho(x,t)$ is the distance from the current point $(x,t) \in \overline{\omega}_{h,\tau}$ to the boundary γ of Q_T.

Theorem 2 *Let w be the solution of the differential Problem 2 and $\widetilde{w}$ be the solution of the difference problem (22)–(24). The following pointwise estimation holds true*

$$|\widetilde{w} - w| \leq c_1 \rho \left(h^2 + \tau^2\right) \tag{42}$$

for $r \leq 1$.

Proof On the basis of Theorem 1 the exact solution $w \in C^{8+\alpha, 4+\frac{\alpha}{2}}_{x,t}\left(\overline{Q}_T\right)$. Let $\varepsilon^{h,\tau}_{w} = \widetilde{w} - w$ then the error function $\varepsilon^{h,\tau}_{w}$ satisfies the following difference problem

$$\varepsilon^{h,\tau}_{w,\bar{t},m} = a\Lambda \varepsilon^{h,\tau}_{w,m} + \Psi_w \text{ on } \omega_{h,\tau}, \tag{43}$$

$$\varepsilon^{0}_{w,m} = 0 \text{ on } \overline{\omega}_{h,0}, \tag{44}$$

$$\varepsilon^{j}_{w,0} = 0 \text{ on } \omega_{0,\tau},\ \varepsilon^{j}_{w,N} = 0, \text{ on } \omega_{b,\tau}, \tag{45}$$

where $\Psi_w = a\Lambda w - w_{\bar{t},m} + \varphi_{\widetilde{w},m}$ and $\varphi_{\widetilde{w},m}$ is as given in (27). Using Taylor's formula for the function $w(x,t)$ about the node $\left(x_m, t_{j+\frac{1}{2}}\right)$ gives that $\Psi_w = O\left(h^2 + \tau^2\right)$. Applying Lemma 2 to the problem (38)–(40) and (43)–(45) and on the basis of Lemma 3 we obtain $\left|\varepsilon^{h,\tau}_{w}\right| \leq c_1 \rho \left(h^2 + \tau^2\right)$.

Theorem 3 *The following inequality holds*

$$|\widetilde{\phi} - \phi| \leq c_2 \rho \left(h^2 + \tau^2\right) \tag{46}$$

for $r \leq 1$ *where,* ϕ *is the solution of the differential Problem 3 and* $\widetilde{\phi}$ *is the solution of the difference problem (28)–(30).*

Proof The proof is analogous to the proof of Theorem 2.

4 Implicit Scheme for the Approximation of $\partial_t^2 u$

When Problem 1 is given we setup the boundary value Problem 2 for $w = \partial_t u$ and use the difference system (22)–(24) for obtaining the approximate solution $\widetilde{w}$. We denote $q_i = \partial_t^2 u$ on γ_i, $i = 1, 2, 3$ and construct the following boundary value problem for $z = \partial_t^2 u$, see also [5].

Problem 4

$$Lz = \partial_t^2 f(x,t) \text{ on } Q_T, \tag{47}$$

$$z(x,0) = q_2 \text{ on } \gamma_2, \tag{48}$$

$$z(0,t) = q_1 \text{ on } \gamma_1,\ z(b,t) = q_3 \text{ on } \gamma_3, \tag{49}$$

where, $f(x,t)$ is the heat source function in (1).

We take

$$q_{1\tau} = D_t^2 u_1(t) = D_t w_1(t) \text{ on } \omega_{0,\tau}, \tag{50}$$

$$q_{2\tau}(\widetilde{w}) = \frac{1}{2\tau}\left(-3w_0(x) + 4\widetilde{w}(x,\tau) - \widetilde{w}(x,2\tau)\right) \text{ on } \omega_{h,0}, \tag{51}$$

$$q_{3\tau} = D_t^2 u_2(t) = D_t w_2(t) \text{ on } \omega_{b,\tau}, \tag{52}$$

where, $w_0(x) = u^{(1)}(x)$ is as defined in (7) and $u_1(t), u_2(t)$ are given boundary functions in (3).

Lemma 4 *The following inequality holds*

$$|q_{2\tau}(\widetilde{w}) - q_{2\tau}(w)| \leq c_1\left(h^2 + \tau^2\right) \tag{53}$$

for $r \leq 1$ *where,* w *is the solution of the differential Problem 2 and* $\widetilde{w}$ *is the solution of the difference problem (22)–(24).*

Proof From Theorem 2, we have

$$|q_{2\tau}(\widetilde{w}) - q_{2\tau}(w)| \leq \frac{1}{2\tau}\left(4(c\tau)\left(h^2 + \tau^2\right) + (c2\tau)\left(h^2 + \tau^2\right)\right) \leq c_1\left(h^2 + \tau^2\right) \tag{54}$$

Lemma 5 *The following inequality is true*

$$\max_{\overline{\omega}_{h,0}} |q_{2\tau}(\widetilde{w}) - q_2| \le c_2 \left(h^2 + \tau^2\right) \tag{55}$$

for $r \le 1$ where, $\widetilde{w}$ is the solution of the difference problem (22)–(24).

Proof On the basis of Lemma 1, $w \in C_{x,t}^{8+\alpha,4+\frac{\alpha}{2}}\left(\overline{Q}_T\right)$ and at the end points $(mh, 0) \in \overline{\omega}_{h,0}$ of each line segment $[(x,t) : 0 \le x \le b,\ 0 \le t \le \tau]$ the Eq. (51) gives the second order approximation of $\partial_t w$. From the truncation error formula (see [6]) it follows that

$$\max_{\overline{\omega}_{h,0}} |q_{2\tau}(w) - q_2| \le \frac{\tau^2}{3} \max_{\overline{Q}_T} \left|\partial_t^3 w\right| \le c_2 \tau^2. \tag{56}$$

Using Lemma 4 and the estimations (53), (56) follows (55).

We construct the following difference problem for the numerical solution of Problem 4

$$\widetilde{z}_{\bar{t},m}^{h,\tau} = a \Lambda \widetilde{z}_m^{h,\tau} + \varphi_{\widetilde{z},m} \text{ on } \omega_{h,\tau}, \tag{57}$$

$$\widetilde{z}_m^0 = q_{2\tau}(\widetilde{w}) \text{ on } \overline{\omega}_{h,0}, \tag{58}$$

$$\widetilde{z}_0^j = q_{1\tau} \text{ on } \omega_{0,\tau},\ \widetilde{z}_N^j = q_{3\tau} \text{ on } \omega_{b,\tau}, \tag{59}$$

where, $q_{i\tau}$ $i = 1, 2, 3$ are defined by (50)–(52) and $\varphi_{\widetilde{z},m} = \partial_t^2 f\Big|_{\left(x_m, t_{j+\frac{1}{2}}\right)}$.

Theorem 4 ([5]) *The solution $\widetilde{z}$ of the finite difference problem (57)–(59) satisfies*

$$\max_{\overline{\omega}_{h,\tau}} |\widetilde{z} - z| \le c_3 \left(h^2 + \tau^2\right), \tag{60}$$

for $r \le 1$, where $z = \partial_t^2 u$ is the exact solution of Problem 4.

Proof Let

$$\varepsilon_z^{h,\tau} = \widetilde{z} - z \text{ on } \overline{\omega}_{h,\tau} \tag{61}$$

where $z = \partial_t^2 u$. Denote by $\left\|\varepsilon_z^{h,\tau}\right\| = \max_{\overline{\omega}_{h,\tau}} |\widetilde{z} - z|$. From (57)–(59) and (61) we have

$$\varepsilon_{z,\bar{t},m}^{h,\tau} = a \Lambda \varepsilon_{z,m}^{h,\tau} + \Psi_z \text{ on } \omega_{h,\tau}, \tag{62}$$

$$\varepsilon_{z,m}^0 = q_{2\tau}(\widetilde{w}) - z \text{ on } \overline{\omega}_{h,0}, \tag{63}$$

$$\varepsilon_{z,0}^j = 0 \text{ on } \omega_{0,\tau},\ \varepsilon_{z,N}^j = 0 \text{ on } \omega_{b,\tau}, \tag{64}$$

where $\Psi_z = a\Lambda z - z_{\bar{t},m} + \varphi_{\widetilde{z},m}$. We take

$$\varepsilon_z^{h,\tau} = \varepsilon_z^{1,h,\tau} + \varepsilon_z^{2,h,\tau} \tag{65}$$

where,

$$\varepsilon_{z,\bar{t},m}^{1,h,\tau} = a\Lambda\varepsilon_{z,m}^{1,h,\tau} \text{ on } \omega_{h,\tau}, \tag{66}$$

$$\varepsilon_{z,m}^{1,0} = q_{2\tau}(\widetilde{w}) - z \text{ on } \overline{\omega}_{h,0}, \tag{67}$$

$$\varepsilon_{z,0}^{1,j} = 0 \text{ on } \omega_{0,\tau},\ \varepsilon_{z,N}^{1,j} = 0 \text{ on } \omega_{b,\tau} \tag{68}$$

$$\varepsilon_{z,\bar{t},m}^{2,h,\tau} = a\Lambda\varepsilon_{z,m}^{2,h,\tau} + \Psi_z \text{ on } \omega_{h,\tau}, \tag{69}$$

$$\varepsilon_{z,m}^{2,0} = 0 \text{ on } \overline{\omega}_{h,0}, \tag{70}$$

$$\varepsilon_{z,0}^{2,j} = 0 \text{ on } \omega_{0,\tau},\ \varepsilon_{z,N}^{2,j} = 0 \text{ on } \omega_{b,\tau}. \tag{71}$$

From Lemma 5 and by maximum principle for the solution of the system (66)–(68) we have

$$\max_{\overline{\omega}_{h,\tau}} \left|\varepsilon_z^{1,h,\tau}\right| \le \max_{\overline{\omega}_{h,\tau}} \left|q_{2\tau}(\widetilde{w}) - z\right| \le c_4\left(h^2 + \tau^2\right). \tag{72}$$

The solution $\varepsilon_z^{2,h,\tau}$ of system (69)–(71) is the error of the approximate solution obtained by the finite difference method for the boundary value Problem 4 when

$$q_2 \in C^{6+\alpha}\left(\overline{\Omega}\right), \partial_t^2 f(x,t) \in C_{x,t}^{4+\alpha,2+\frac{\alpha}{2}}\left(\overline{Q}_T\right), q_i \in C^{3+\frac{\alpha}{2}}\left(\overline{\sigma}_T\right), i = 1, 3. \tag{73}$$

and

$$\begin{cases} q_1^{(q)}(0) = z^{(q)}(0) \\ q_3^{(q)}(0) = z^{(q)}(b) \end{cases}, q = 0, 1, 2, 3 \tag{74}$$

Since the function $z = \partial_t^2 u$ satisfies the Eq. (47) with the initial function q_2 on γ_2 and boundary functions q_1, q_3 on γ_1 and γ_3 respectively, using (73), (74) and on the basis of Theorem 1, and maximum principle in Chap. 4 of [18] we obtain

$$\max_{\overline{\omega}_{h,\tau}} \left|\varepsilon_z^{2,h,\tau}\right| \le c_5\left(h^2 + \tau^2\right). \tag{75}$$

Using (65), (72) and (75) we obtain (60).

5 Implicit Scheme for the Approximation of $\partial_x\partial_t u$

Given the Problem 1, we setup the boundary value Problem 2 for $w = \partial_t u$ and use the difference system (22)–(24) for obtaining the approximate solution $\widetilde{w}$. We denote $p_i = \partial_x\partial_t u$ on γ_i, $i = 1, 2, 3$ respectively and give the following boundary value problem for $y = \partial_x\partial_t u$.

Problem 5

$$Ly = \partial_x \partial_t f(x,t) \text{ on } Q_T, \tag{76}$$

$$y(x,0) = p_2 \text{ on } \gamma_2, \tag{77}$$

$$y(0,t) = p_1 \text{ on } \gamma_1,\ y(b,t) = p_3 \text{ on } \gamma_3, \tag{78}$$

where, $f(x,t)$ is the given function in (1). We take

$$p_{1h}(\widetilde{w}) = \frac{1}{2h}\left(-3w_1(t) + 4\widetilde{w}(h,t) - \widetilde{w}(2h,t)\right) \text{ on } \omega_{0,\tau}, \tag{79}$$

$$p_{2h} = D_x w_0(x) \text{ on } \overline{\omega}_{h,0}, \tag{80}$$

$$p_{3h}(\widetilde{w}) = \frac{1}{2h}\left(3w_2(t) - 4\widetilde{w}(b-h,t) + \widetilde{w}(b-2h,t)\right) \text{ on } \omega_{b,\tau}, \tag{81}$$

where, $w_0(x) = u^{(1)}(x)$ and $w_1 = D_t u_1(t)$, $w_2 = D_t u_2(t)$.

We construct the following difference problem for the numerical solution of Problem 5 and denote this solution by $\widetilde{y}$

$$\widetilde{y}^{h,\tau}_{\bar{t},m} = a\Lambda \widetilde{y}^{h,\tau}_m + \varphi_{\widetilde{y},m} \text{ on } \omega_{h,\tau}, \tag{82}$$

$$\widetilde{y}^0_m = p_{2h} \text{ on } \overline{\omega}_{h,0}, \tag{83}$$

$$\widetilde{y}^j_0 = p_{1h}(\widetilde{w}) \text{ on } \omega_{0,\tau},\ \widetilde{y}^j_N = p_{3h}(\widetilde{w}) \text{ on } \omega_{b,\tau}. \tag{84}$$

Here, p_{ih} are defined by (79)–(81) $\widetilde{y}^{h,\tau}_{\bar{t},m}$, $\Lambda\widetilde{y}^{h,\tau}_m$ are formulae analogous to (25) and (26) respectively, and $\widetilde{y}^0_m = \widetilde{y}(x_m,0)$, $\widetilde{y}^j_0 = \widetilde{y}(0,t_j)$, $\widetilde{y}^j_N = \widetilde{y}(b,t_j)$ and $\varphi_{\widetilde{y},m} = \partial_x\partial_t f|_{\left(x_m,t_{j+\frac{1}{2}}\right)}$.

Lemma 6 *The following inequality holds*

$$|p_{ih}(\widetilde{w}) - p_{ih}(w)| \le c_1\left(h^2+\tau^2\right),\ i = 1,3. \tag{85}$$

for $r \le 1$ where w is the solution of the differential Problem 2 and $\widetilde{w}$ is the solution of the difference problem (22)–(24).

Proof From (79), (81) and Theorem 2, we have

$$|p_{ih}(\widetilde{w}) - p_{ih}(w)| \le \frac{1}{2h}\left(4(ch)\left(h^2+\tau^2\right) + (c2h)\left(h^2+\tau^2\right)\right) \le c_1\left(h^2+\tau^2\right),\ i = 1,3 \tag{86}$$

Lemma 7 *The following inequality is true*

$$\max_{\omega_{0,\tau}\cup\omega_{b,\tau}} |p_{ih}(\widetilde{w}) - p_i| \le c_2\left(h^2+\tau^2\right),\ i = 1,3. \tag{87}$$

for $r \le 1$ where, $\widetilde{w}$ is the solution of the difference problem (22)–(24).

Proof On the basis of Lemma 1, the exact solution w of Problem 2 belongs to $C_{x,t}^{8+\alpha,4+\frac{\alpha}{2}}\left(\overline{Q}_T\right)$. Then at the end points $(0,\sigma\tau)\in\omega_{0,\tau}$ and $(b,\sigma\tau)\in\omega_{b,\tau}$ of each line segment $[(x,t):0\le x\le b,\ 0\le t\le T]$ the Eqs. (79) and (81) give the second order approximation of $\partial_x\partial_t u$ respectively. From the truncation error formula (see [6]) it follows that

$$\max_{\omega_{0,\tau}\cup\omega_{b,\tau}}|p_{ih}(w)-p_i|\le\frac{h^2}{3}\max_{\overline{Q}_T}\left|\partial_x^3 w\right|\le c_3h^2,\ i=1,3. \tag{88}$$

On the basis of Lemma 6 using the estimation (85) and (88) follows (87).

Theorem 5 *The solution $\widetilde{y}$ of the finite difference problem (82)–(84) satisfies*

$$\max_{\overline{\omega}_{h,\tau}}|\widetilde{y}-y|\le c_4\left(h^2+\tau^2\right), \tag{89}$$

for $r\le 1$ where, $y=\partial_x\partial_t u$ is the exact solution of Problem 5.

Proof Let

$$\varepsilon_y^{h,\tau}=\widetilde{y}-y \text{ on } \overline{\omega}_{h,\tau}, \tag{90}$$

where $y=\partial_x\partial_t u$. Denote by $\left\|\varepsilon_y^{h,\tau}\right\|=\max_{\overline{\omega}_{h,\tau}}|\widetilde{y}-y|$. From (82)–(84) and (90) we have

$$\varepsilon_{y,\bar{t},m}^{h,\tau}=a\Lambda\varepsilon_{y,m}^{h,\tau}+\psi_y \text{ on } \omega_{h,\tau}, \tag{91}$$

$$\varepsilon_{y,m}^{0}=0 \text{ on } \overline{\omega}_{h,0}, \tag{92}$$

$$\varepsilon_{y,0}^{j}=p_{1h}(\widetilde{w})-y \text{ on } \omega_{0,\tau},\ \varepsilon_{y,N}^{j}=p_{3h}(\widetilde{w})-y \text{ on } \omega_{b,\tau}, \tag{93}$$

where $\psi_y=a\Lambda y-y_{\bar{t},m}+\varphi_{\widetilde{y},m}$. We take

$$\varepsilon_y^{h,\tau}=\varepsilon_y^{1,h,\tau}+\varepsilon_y^{2,h,\tau}, \tag{94}$$

where,

$$\varepsilon_{y,\bar{t},m}^{1,h,\tau}=a\Lambda\varepsilon_{y,m}^{1,h,\tau} \text{ on } \omega_{h,\tau}, \tag{95}$$

$$\varepsilon_{y,m}^{1,0}=0 \text{ on } \overline{\omega}_{h,0}, \tag{96}$$

$$\varepsilon_{y,0}^{1,j}=p_{1h}(\widetilde{w})-y \text{ on } \omega_{0,\tau},\ \varepsilon_{y,N}^{1,j}=p_{3h}(\widetilde{w})-y \text{ on } \omega_{b,\tau}, \tag{97}$$

$$\varepsilon_{y,\bar{t},m}^{2,h,\tau}=a\Lambda\varepsilon_{y,m}^{2,h,\tau}+\psi_y \text{ on } \omega_{h,\tau}, \tag{98}$$

$$\varepsilon_{y,m}^{2,0}=0 \text{ on } \overline{\omega}_{h,0}, \tag{99}$$

$$\varepsilon_{y,0}^{2,j}=0 \text{ on } \omega_{0,\tau},\ \varepsilon_{y,N}^{2,j}=0 \text{ on } \omega_{b,\tau}. \tag{100}$$

From Lemma 7 and by maximum principle for the solution of the system (95)–(97) we have

$$\max_{\overline{\omega}_{h,\tau}} \left|\varepsilon_y^{1,h,\tau}\right| \leq \max_{i=1,3} \max_{\overline{\omega}_{h,\tau}} |p_{ih}(\widetilde{w}) - y| \leq c_4\left(h^2+\tau^2\right). \tag{101}$$

The solution $\varepsilon_y^{2,h,\tau}$ of system (98)–(100) is the error of the approximate solution obtained by the finite difference method for the Problem 5 when

$$p_2 \in C^{7+\alpha}\left(\overline{\Omega}\right), \partial_x\partial_t f(x,t) \in C_{x,t}^{5+\alpha,\frac{5+\alpha}{2}}\left(\overline{Q}_T\right), p_i \in C^{\frac{7+\alpha}{2}}(\overline{\sigma}_T), i = 1, 3. \tag{102}$$

$$\begin{cases} p_1^{(q)}(0) = y^{(q)}(0) \\ p_3^{(q)}(0) = y^{(q)}(b) \end{cases}, q = 0, 1, 2, 3. \tag{103}$$

Since the function $y = \partial_x\partial_t u$ satisfies the Eq. (76) with the initial function p_2 on γ_2 and boundary functions p_1, p_3 on γ_1 and γ_3 respectively using (102) and (103) and on the basis of Theorem 1 and maximum principle in Chap. 4 of [18] we obtain

$$\max_{\overline{\omega}_{h,\tau}} \left|\varepsilon_y^{2,h,\tau}\right| \leq c_5\left(h^2+\tau^2\right) \tag{104}$$

using (94), (101) and (104) we obtain (89).

6 Numerical Results

We take $Q_T = \{(x,t) : 0 < x < 1, 0 < t \leq 1\}$, $\gamma_1 = \{(0,t) : 0 \leq t \leq 1\}$, $\gamma_2 = \{(x,0) : 0 \leq x \leq 1\}$, and $\gamma_3 = \{(1,t) : 0 \leq t \leq 1\}$ and consider the following boundary value problem

$$\begin{aligned} Lu &= f(x,t) \text{ on } Q_T, \\ u(x,0) &= 0.005 + \sin(2\pi x) \text{ on } \gamma_2, \\ u(0,t) &= 0.005\cos\left(t^{\frac{51}{10}}\right) \text{ on } \gamma_1, \\ u(1,t) &= 0.0005\sin\left(t^{\frac{51}{10}}\right) + 0.005\cos\left(t^{\frac{51}{10}}\right) + \sin(2\pi) \text{ on } \gamma_3, \end{aligned} \tag{105}$$

where,

$$\begin{aligned} f(x,t) = \sin\left(t^{\frac{51}{10}}\right)\left(-0.0005\frac{51}{5}\frac{46}{5}x^{\frac{41}{5}} - 0.005\frac{51}{10}t^{\frac{41}{10}}\right) + \\ 0.0005\frac{51}{10}x^{\frac{51}{5}}t^{\frac{41}{10}}\cos\left(t^{\frac{41}{10}}\right) + 4\pi^2\sin(2\pi x). \end{aligned} \tag{106}$$

The boundary value problem for $w = \frac{\partial u}{\partial t}$ is constructed using the proposed Problem 2. Further, the approximate solution $\widetilde{w}$ of the difference problem (22)–(24) is obtained at the same grid points. Using the exact solution

$$w(x,t) = 0.0005\frac{51}{10}x^{\frac{51}{5}}t^{\frac{41}{10}}\cos\left(t^{\frac{51}{10}}\right) - 0.005\frac{51}{10}t^{\frac{41}{10}}\sin\left(t^{\frac{51}{10}}\right) \tag{107}$$

we denote the maximum errors by $\left\|\varepsilon_w^{h,\tau}\right\| = \max_{\overline{\omega}_{h,\tau}} |\widetilde{w} - w|$. Table 1 presents the maximum errors and the order of convergence

$$\mathfrak{R}_{\widetilde{w}}^{h} = \frac{\left\|\varepsilon_w^{2^{-\mu},2^{-\lambda}}\right\|}{\left\|\varepsilon_w^{2^{-(\mu+1)},2^{-\lambda}}\right\|}. \tag{108}$$

of the approximate solution $\widetilde{w}$ to the exact solution $w = \frac{\partial u}{\partial t}$ in spatial variable x (with respect to h) when $h = 2^{-\mu}$, $\mu = 4, 5, 6, 7, 8$ and $\tau = 2^{-16}$. Table 2, shows the maximum errors and the order of convergence

$$\mathfrak{R}_{\widetilde{w}}^{h,\tau} = \frac{\left\|\varepsilon_w^{2^{-\mu},2^{-\lambda}}\right\|}{\left\|\varepsilon_w^{2^{-(\mu+1)},2^{-(\lambda+1)}}\right\|} \tag{109}$$

of $\widetilde{w}$ to the exact solution w in time variable t (with respect to τ) when $h = 2^{-\mu}$, $\mu = 4, 5, 6, 7, 8$ and $\tau = 2^{-\lambda}$, $\lambda = 13, 14, 15, 16, 17$ respectively. Note that $O(h^2 + \tau^2)$ order of convergence corresponds to $\asymp 2^2$ of the quantities defined by (108), and (109). The exact solution $w = \partial_t u$ and the grid function $w^{2^{-7},2^{-16}}$ presenting

Table 1 Maximum errors and the order of convergence of approximate solution $\tilde{w}$ to $w = \frac{\partial u}{\partial t}$, with respect to h

$(h = 2^{-\mu}, \tau = 2^{-\lambda})$ $(h = 2^{-(\mu+1)}, \tau = 2^{-\lambda})$	$\left\|\varepsilon_w^{2^{-\mu},2^{-\lambda}}\right\|$ $\left\|\varepsilon_w^{2^{-(\mu+1)},2^{-\lambda}}\right\|$	$\mathfrak{R}_{\widetilde{w}}^{h}$
$(2^{-4}, 2^{-16})$ $(2^{-5}, 2^{-16})$	$2.6585E-5$ $6.6997E-6$	3.9681
$(2^{-5}, 2^{-16})$ $(2^{-6}, 2^{-16})$	$6.6997E-6$ $1.6794E-6$	3.9893
$(2^{-6}, 2^{-16})$ $(2^{-7}, 2^{-16})$	$1.6794E-6$ $4.2013E-7$	3.9973
$(2^{-7}, 2^{-16})$ $(2^{-8}, 2^{-16})$	$4.2013E-7$ $1.0505E-7$	3.9993

Table 2 Maximum errors and the order of convergence of approximate solution $\tilde{w}$ with respect to h and τ.

$(h = 2^{-\mu}, \tau = 2^{-\lambda})$ $(h = 2^{-(\mu+1)}, \tau = 2^{-(\lambda+1)})$	$\left\Vert \varepsilon_w^{2^{-\mu},2^{-\lambda}} \right\Vert$ $\left\Vert \varepsilon_w^{2^{-(\mu+1)},2^{-(\lambda+1)}} \right\Vert$	$\Re^{h,\tau}_{\tilde{w}}$
$(2^{-4}, 2^{-13})$ $(2^{-5}, 2^{-14})$	$2.6585E-5$ $6.6996E-6$	3.9682
$(2^{-5}, 2^{-14})$ $(2^{-6}, 2^{-15})$	$6.6996E-6$ $1.6794E-6$	3.9893
$(2^{-6}, 2^{-15})$ $(2^{-7}, 2^{-16})$	$1.6794E-6$ $4.2013E-7$	3.9973
$(2^{-7}, 2^{-16})$ $(2^{-8}, 2^{-17})$	$4.2013E-7$ $1.0505E-7$	3.9993

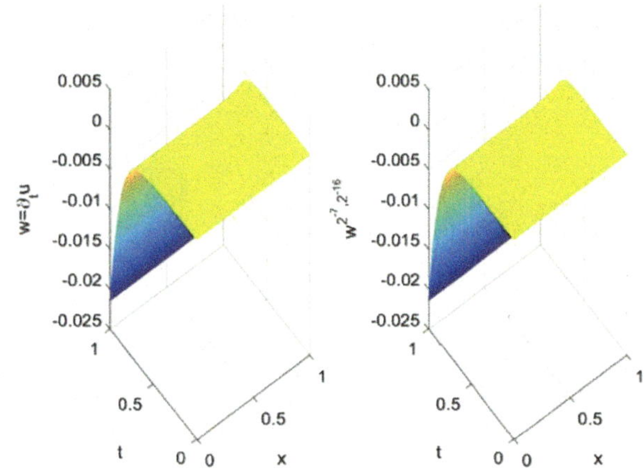

Fig. 1 The exact solution $w = \partial_t u$ and the grid function $w^{2^{-7},2^{-16}}$ presenting the approximate solution $\widetilde{w}$ when $h = 2^{-7}, \tau = 2^{-16}$

the approximate solution $\widetilde{w}$ when $h = 2^{-7}, \tau = 2^{-16}$ are shown in Fig. 1. The error function $\left| \varepsilon_w^{2^{-7},2^{-16}} \right| = |\widetilde{w} - w|$ for $h = 2^{-7}$, and $\tau = 2^{-16}$ is given in Fig. 2.

Next we give the numerical results for the approximate solution of pure second derivative of the solution $u(x, t)$ with respect to x. The boundary value problem for $\phi = \frac{\partial^2 u}{\partial x^2}$ is constructed using the proposed Problem 3. Then, the approximate solution $\widetilde{\phi}$ of the difference problem (28)–(30) is obtained at the same grid points. Using the exact solution

$$\phi(x, t) = 0.0005 \frac{51}{5} \frac{46}{5} x^{\frac{41}{5}} \sin\left(t^{\frac{51}{10}}\right) - 4\pi^2 \sin(2\pi x), \tag{110}$$

we denote the maximum error by $\left\Vert \varepsilon_\phi^{h,\tau} \right\Vert = \max_{\overline{\omega}_{h,\tau}} \left|\widetilde{\phi} - \phi\right|$. Table 3 shows the maximum errors and the order of convergence $\Re^{h}_{\widetilde{\phi}}$ of the approximate solution $\widetilde{\phi}$ to the exact

Fig. 2 The error function $\left|\varepsilon_{w}^{2^{-7},2^{-16}}\right| = |\widetilde{w} - w|$ for $h = 2^{-7}$, and $\tau = 2^{-16}$

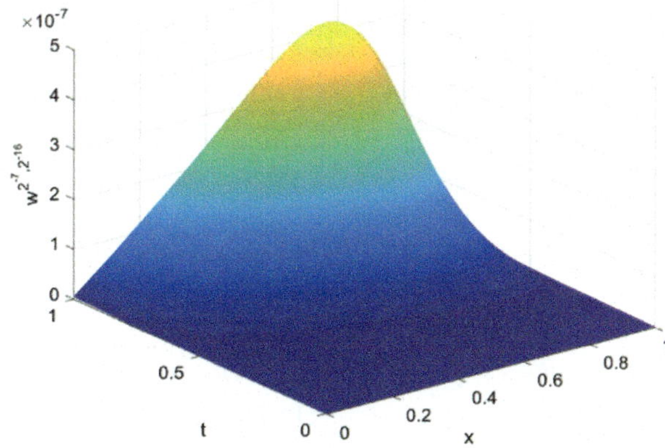

Table 3 Maximum errors and the order of convergence of approximate solution $\tilde{\phi}$ to $\phi = \frac{\partial^2 u}{\partial x^2}$, with respect to h

$(h = 2^{-\mu}, \tau = 2^{-\lambda})$ $(h = 2^{-(\mu+1)}, \tau = 2^{-\lambda})$	$\left\Vert\varepsilon_{\tilde{\phi}}^{2^{-\mu},2^{-\lambda}}\right\Vert$ $\left\Vert\varepsilon_{\tilde{\phi}}^{2^{-(\mu+1)},2^{-\lambda}}\right\Vert$	$\Re_{\tilde{\phi}}^{h}$
$(2^{-4}, 2^{-16})$ $(2^{-5}, 2^{-16})$	0.5116 0.1272	4.0220
$(2^{-5}, 2^{-16})$ $(2^{-6}, 2^{-16})$	0.1272 $3.1745E-2$	4.0069
$(2^{-6}, 2^{-16})$ $(2^{-7}, 2^{-16})$	$3.1745E-2$ $7.9334E-3$	4.0014
$(2^{-7}, 2^{-16})$ $(2^{-8}, 2^{-16})$	$7.9334E-3$ $1.9832E-3$	4.0003

solution $\phi = \frac{\partial^2 u}{\partial x^2}$ analogous to the formula (108) in spatial variable x (with respect to h) when $h = 2^{-\mu}, \mu = 4, 5, 6, 7, 8$ and $\tau = 2^{-16}$. Table 4, shows the maximum errors and the order of convergence $\Re_{\tilde{\phi}}^{h,\tau}$ of $\tilde{\phi}$ to the exact solution ϕ analogous to (109) when $h = 2^{-\mu}$, $\mu = 4, 5, 6, 7, 8$ and $\tau = 2^{-\lambda}$, $\lambda = 13, 14, 15, 16, 17$ respectively. Figure 3 demonstrates the exact solution $\phi = \partial_x^2 u$ and the grid function $\phi^{2^{-7},2^{-16}}$ presenting the approximate solution $\tilde{\phi}$ when $h = 2^{-7}, \tau = 2^{-16}$. The error function $\left|\varepsilon_{\phi}^{2^{-7},2^{-16}}\right| = \left|\tilde{\phi} - \phi\right|$ for $h = 2^{-7}$, and $\tau = 2^{-16}$ is given in Fig. 4.

Now, we present the numerical results for the approximate solution of pure second derivative of the solution $u(x, t)$ with respect to t. First we construct the boundary value Problem 2 and the approximate solution $\widetilde{w}$ is obtained by solving the system (22)–(24). Then the boundary value problem for $z = \frac{\partial^2 u}{\partial t^2}$ is constructed using the proposed Problem 4. Further, the approximate solution $\tilde{z}$ of the difference problems (57)–(59) is obtained at the same grid points. Using the exact solution

Table 4 Maximum errors and the order of convergence of approximate solution $\tilde{\phi}$ with respect to h and τ

$(h = 2^{-\mu}, \tau = 2^{-\lambda})$ $(h = 2^{-(\mu+1)}, \tau = 2^{-(\lambda+1)})$	$\left\|\varepsilon_{\phi}^{2^{-\mu},2^{-\lambda}}\right\|$ $\left\|\varepsilon_{\phi}^{2^{-(\mu+1)},2^{-(\lambda+1)}}\right\|$	$\Re_{\tilde{\phi}}^{h,\tau}$
$(2^{-4}, 2^{-13})$ $(2^{-5}, 2^{-14})$	0.5116 0.1272	4.0220
$(2^{-5}, 2^{-14})$ $(2^{-6}, 2^{-15})$	0.1272 $3.1745E-2$	4.0069
$(2^{-6}, 2^{-15})$ $(2^{-7}, 2^{-16})$	$3.1745E-2$ $7.9334E-3$	4.0014
$(2^{-7}, 2^{-16})$ $(2^{-8}, 2^{-17})$	$7.9334E-3$ $1.9832E-3$	4.0003

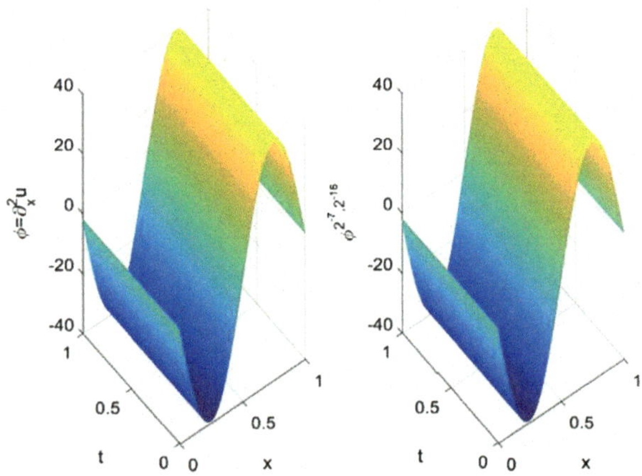

Fig. 3 The exact solution $\phi = \partial_x^2 u$ and the grid function $\phi^{2^{-7},2^{-16}}$ presenting the approximate solution $\widetilde{\phi}$ when $h = 2^{-7}$, $\tau = 2^{-16}$

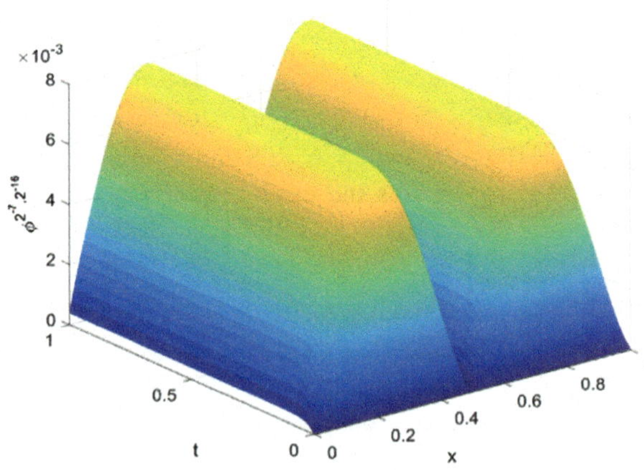

Fig. 4 The error function $\left|\varepsilon_{\phi}^{2^{-7},2^{-16}}\right| = |\widetilde{\phi} - \phi|$ for $h = 2^{-7}$, and $\tau = 2^{-16}$

Table 5 Maximum errors and the order of convergence of approximate solution $\tilde{z}$ to $z = \frac{\partial^2 u}{\partial t^2}$, with respect to h

$\left(h = 2^{-\mu}, \tau = 2^{-\lambda}\right)$ $\left(h = 2^{-(\mu+1)}, \tau = 2^{-\lambda}\right)$	$\left\Vert \varepsilon_z^{2^{-\mu},2^{-\lambda}} \right\Vert$ $\left\Vert \varepsilon_z^{2^{-(\mu+1)},2^{-\lambda}} \right\Vert$	$\Re_{\tilde{z}}^h$
$\left(2^{-4}, 2^{-16}\right)$ $\left(2^{-5}, 2^{-16}\right)$	$7.8539E-5$ $1.9792E-5$	3.9682
$\left(2^{-5}, 2^{-16}\right)$ $\left(2^{-6}, 2^{-16}\right)$	$1.9792E-5$ $4.9611E-6$	3.9894
$\left(2^{-6}, 2^{-16}\right)$ $\left(2^{-7}, 2^{-16}\right)$	$4.9611E-6$ $1.2411E-6$	3.9973
$\left(2^{-7}, 2^{-16}\right)$ $\left(2^{-8}, 2^{-16}\right)$	$1.2411E-6$ $3.1033E-7$	3.9993

$$z\left(x,t\right) = \sin\left(t^{\frac{51}{10}}\right)\left(-0.0005\left(\frac{51}{10}\right)^2 x^{\frac{51}{5}} t^{\frac{41}{5}} - 0.005\frac{51}{10}\frac{41}{10} t^{\frac{31}{10}}\right) +$$
$$\cos\left(t^{\frac{51}{10}}\right)\left(-0.005\left(\frac{51}{10}\right)^2 t^{\frac{41}{5}} + 0.0005\frac{51}{10}\frac{41}{10} x^{\frac{51}{5}} t^{\frac{31}{10}}\right), \quad (111)$$

we denote the maximum error by $\left\Vert \varepsilon_z^{h,\tau} \right\Vert = \max_{\overline{\omega}_{h,\tau}} \left|\tilde{z} - z\right|$. Table 5 presents the maximum errors and the order of convergence $\Re_{\tilde{z}}^h$ of the approximate solution $\tilde{z}$ to the exact solution $z = \frac{\partial^2 u}{\partial t^2}$ in spatial variable x when $h = 2^{-\mu}$, $\mu = 4, 5, 6, 7, 8$ and $\tau = 2^{-16}$ analogous to the formula (108). Table 6, shows the maximum errors and the order of convergence $\Re_{\tilde{z}}^{h,\tau}$ of $\tilde{z}$ to the exact solution z with respect to h and τ when $h = 2^{-\mu}$, $\mu = 4, 5, 6, 7, 8$ and $\tau = 2^{-\lambda}$, $\lambda = 13, 14, 15, 16, 17$ respectively, analogous to the formula (109). The exact solution $z = \partial_t^2 u$ and the grid function $z^{2^{-7},2^{-16}}$ presenting the approximate solution $\tilde{z}$ when $h = 2^{-7}, \tau = 2^{-16}$ are shown in Fig. 5 while the error function $\left|\varepsilon_z^{2^{-7},2^{-16}}\right| = \left|\tilde{z} - z\right|$ for $h = 2^{-7}$, and $\tau = 2^{-16}$ is presented in Fig. 6.

Next we give the numerical results for the approximate solution of mixed derivative of the solution $u\left(x,t\right)$. First we construct the boundary value Problem 2 and the approximate solution $\tilde{w}$ is obtained by solving the system (22)–(24). Then the boundary value problem for $y = \frac{\partial^2 u}{\partial x \partial t}$ is constructed using the proposed Problem 5 and the approximate solution $\tilde{w}$. Further, the approximate solution $\tilde{y}$ of the difference problems (82)–(84) is obtained at the same grid points. Using the exact solution

$$y\left(x,t\right) = 0.0005\frac{51}{5}\frac{51}{10} x^{\frac{46}{5}} t^{\frac{41}{10}} \cos\left(t^{\frac{51}{10}}\right) \quad (112)$$

we denote the maximum error by $\left\Vert \varepsilon_y^{h,\tau} \right\Vert = \max_{\overline{\omega}_{h,\tau}} \left|\tilde{y} - y\right|$. Table 7 presents the maximum errors and the order of convergence $\Re_{\tilde{y}}^h$ of the approximate solution $\tilde{y}$ to the exact

Table 6 Maximum errors and the order of convergence of approximate solution $\tilde{z}$ with respect to h and τ

$\left(h = 2^{-\mu}, \tau = 2^{-\lambda}\right)$ $\left(h = 2^{-(\mu+1)}, \tau = 2^{-(\lambda+1)}\right)$	$\left\Vert \varepsilon_z^{2^{-\mu},2^{-\lambda}} \right\Vert$ $\left\Vert \varepsilon_z^{2^{-(\mu+1)},2^{-(\lambda+1)}} \right\Vert$	$\Re_{\tilde{z}}^{h,\tau}$
$\left(2^{-4}, 2^{-13}\right)$ $\left(2^{-5}, 2^{-14}\right)$	$7.8539E-5$ $1.9792E-5$	3.9682
$\left(2^{-5}, 2^{-14}\right)$ $\left(2^{-6}, 2^{-15}\right)$	$1.9792E-5$ $4.9611E-6$	3.9894
$\left(2^{-6}, 2^{-15}\right)$ $\left(2^{-7}, 2^{-16}\right)$	$4.9611E-6$ $1.2411E-6$	3.9973
$\left(2^{-7}, 2^{-16}\right)$ $\left(2^{-8}, 2^{-17}\right)$	$1.2411E-6$ $3.1034E-7$	3.9992

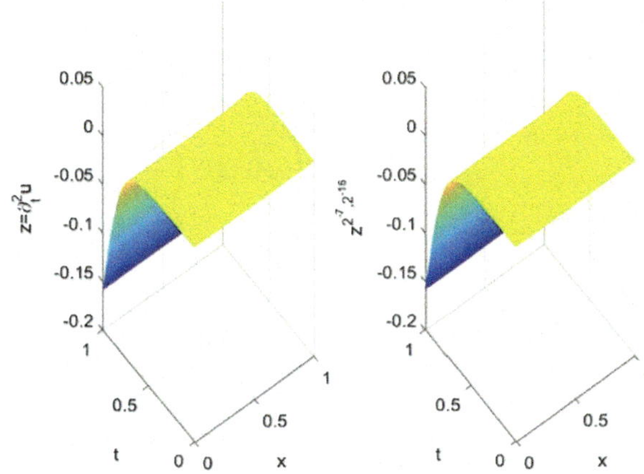

Fig. 5 The exact solution $z = \partial_t^2 u$ and the grid function $z^{2^{-7},2^{-16}}$ presenting the approximate solution $\tilde{z}$ when $h = 2^{-7}, \tau = 2^{-16}$

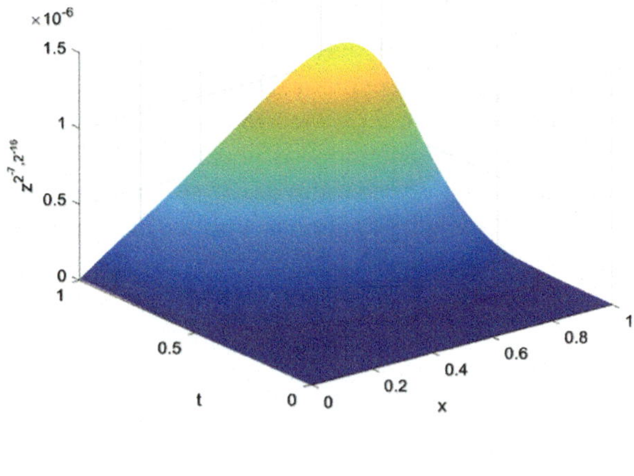

Fig. 6 The error function $\left|\varepsilon_z^{2^{-7},2^{-16}}\right| = |\tilde{z} - z|$ for $h = 2^{-7}$, and $\tau = 2^{-16}$

Table 7 Maximum errors and the order of convergence of approximate solution $\tilde{y}$ to $y = \frac{\partial^2 u}{\partial x \partial t}$, with respect to h

$(h = 2^{-\mu}, \tau = 2^{-\lambda})$ $(h = 2^{-(\mu+1)}, \tau = 2^{-\lambda})$	$\left\Vert \varepsilon_y^{2^{-\mu},2^{-\lambda}} \right\Vert$ $\left\Vert \varepsilon_y^{2^{-(\mu+1)},2^{-\lambda}} \right\Vert$	$\Re_{\tilde{y}}^{h}$
$(2^{-4}, 2^{-16})$ $(2^{-5}, 2^{-16})$	$7.8539E-5$ $1.9792E-5$	3.9682
$(2^{-5}, 2^{-16})$ $(2^{-6}, 2^{-16})$	$1.9792E-5$ $4.9611E-6$	3.9894
$(2^{-6}, 2^{-16})$ $(2^{-7}, 2^{-16})$	$4.9611E-6$ $1.2411E-6$	3.9973
$(2^{-7}, 2^{-16})$ $(2^{-8}, 2^{-16})$	$1.2411E-6$ $3.1033E-7$	3.9993

Table 8 Maximum errors and the order of convergence of approximate solution $\tilde{y}$ with respect to h and τ

$(h = 2^{-\mu}, \tau = 2^{-\lambda})$ $(h = 2^{-(\mu+1)}, \tau = 2^{-(\lambda+1)})$	$\left\Vert \varepsilon_y^{2^{-\mu},2^{-\lambda}} \right\Vert$ $\left\Vert \varepsilon_y^{2^{-(\mu+1)},2^{-(\lambda+1)}} \right\Vert$	$\Re_{\tilde{y}}^{h,\tau}$
$(2^{-4}, 2^{-13})$ $(2^{-5}, 2^{-14})$	$7.8539E-5$ $1.9792E-5$	3.9682
$(2^{-5}, 2^{-14})$ $(2^{-6}, 2^{-15})$	$1.9792E-5$ $4.9611E-6$	3.9894
$(2^{-6}, 2^{-15})$ $(2^{-7}, 2^{-16})$	$4.9611E-6$ $1.2411E-6$	3.9973
$(2^{-7}, 2^{-16})$ $(2^{-8}, 2^{-17})$	$1.2411E-6$ $3.1034E-7$	3.9992

solution $y = \frac{\partial^2 u}{\partial x \partial t}$ in spatial variable x when $h = 2^{-\mu}, \mu = 4, 5, 6, 7, 8$ and $\tau = 2^{-16}$. Table 8, shows the maximum errors and the order of convergence $\Re_{\widetilde{y}}^{h,\tau}$ of $\widetilde{y}$ to the exact solution y with respect to h and τ when $h = 2^{-\mu}$, $\mu = 4, 5, 6, 7, 8$ and $\tau = 2^{-\lambda}$, $\lambda = 13, 14, 15, 16, 17$ respectively.The exact solution $y = \partial_x \partial_t u$ and the grid function $y^{2^{-7},2^{-16}}$ illustrating the approximate solution $\widetilde{y}$ when $h = 2^{-7}, \tau = 2^{-16}$ are given in Fig. 7 and the error function $\left| \varepsilon_y^{2^{-7},2^{-16}} \right| = |\widetilde{y} - y|$ for $h = 2^{-7}$, and $\tau = 2^{-16}$ is given in Fig. 8. Figure 9 shows the maximum errors $\left\Vert \varepsilon_y^{h,2^{-16}} \right\Vert$ when $\tau = 2^{-16}$, with respect to h.

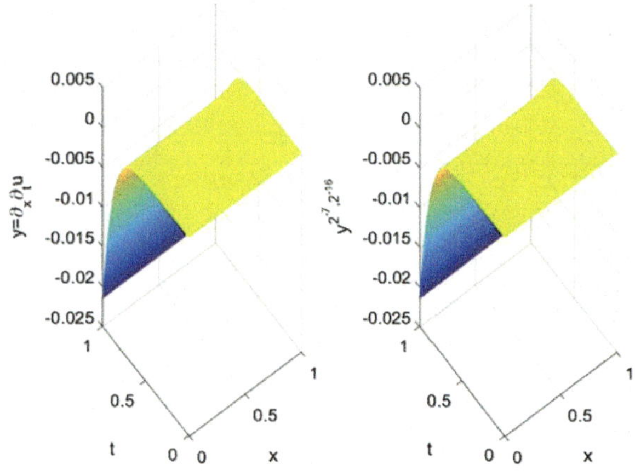

Fig. 7 The exact solution $y = \partial_x \partial_t u$ and the grid function $y^{2^{-7},2^{-16}}$ presenting the approximate solution $\widetilde{y}$ when $h = 2^{-7}$, $\tau = 2^{-16}$

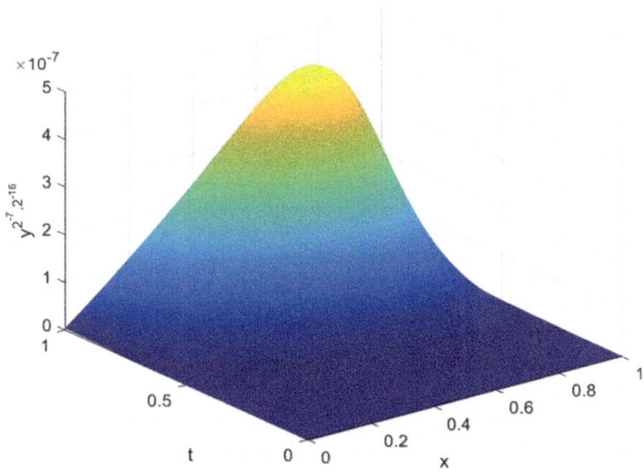

Fig. 8 The error function $\left|\varepsilon_y^{2^{-7},2^{-16}}\right| = |\widetilde{y} - y|$ for $h = 2^{-7}$, and $\tau = 2^{-16}$

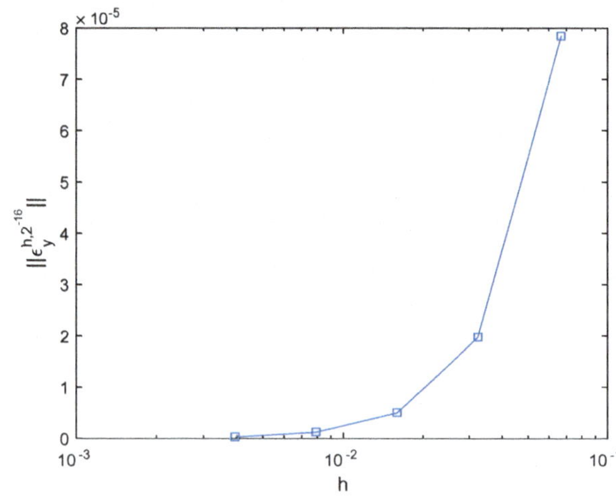

Fig. 9 The maximum errors $\left\|\varepsilon_y^{h,2^{-16}}\right\|$ for $\tau = 2^{-16}$, with respect to h

7 Conclusion

It is assumed that the given first type boundary value problem of one dimensional heat equation has the initial function belonging to the Hölder space $C^{10+\alpha}$, $0 < \alpha < 1$, the heat source function from the Hölder space $C_{x,t}^{8+\alpha,4+\frac{\alpha}{2}}$, the boundary functions from $C^{5+\frac{\alpha}{2}}$, satisfying the conjugation conditions of orders $q = 0, 1, 2, 3, 4, 5$. Difference problems of six point symmetric implicit schemes (Crank and Nicolson) approximating $\frac{\partial u}{\partial t}, \frac{\partial^2 u}{\partial x^2}$ are constructed, and a pointwise estimation for the error function depending on $\rho(x, t)$ which is the distance from the current grid point in the domain to the boundary is given. Furthermore, special six point implicit difference problems for the approximation of $\frac{\partial^2 u}{\partial t^2}, \frac{\partial^2 u}{\partial x \partial t}$ are proposed and it is proved that the solutions of the constructed difference schemes converge uniformly to the exact values of the respective derivatives on the grids of order $O\left(h^2 + \tau^2\right)$.

References

1. Arslan, N., Erdogan, A.S., Ashyralyev, A.: Computational fluid flow solution over endothelial cells inside the capillary vascular system. Int. J. Numer. Methods Eng. **74**(11), 1679–1689 (2008)
2. Ashyralyev, A., Erdogan, A.S., Arslan, N.: On the modified Crank Nicholson difference schemes for parabolic equation with non-smooth data arising in biomechanics. Int. J. Numer. Methods Biomed. Eng. **26**, 501–510 (2010)
3. Bateman, H.: Partial Differential Equations of Mathematical Physics. Dover Publications, New York (1944)
4. Buranay, S.C., Farinola, L.A.: Implicit methods for the first derivative of the solution to heat equation. Adv. Differ. Equ. **2018**, 430 (2018). https://doi.org/10.1186/s13662-018-1887-1
5. Buranay, S.C., Farinola, L.A.: Six point implicit methods for the pure second derivatives of the solution of first type boundary value problem for one dimensional heat equation. In: International Conference on Analysis and Applied Mathematics (ICAAM 2018). AIP Conference Proceedings, vol. 1997, pp. 020077-1–020077-6. https://doi.org/10.1063/1.5049071
6. Burden, R.L., Faires, J.D.: Numerical Analysis. Brooks/Cole, Cengage Learning, Boston (2011)
7. Dosiyev, A.A., Sadeghi, M.M.H.: A fourth order accurate approximation of the first and pure second derivatives of the Laplace equation on a rectangle. Adv. Differ. Equ. **2015**(67), 1–11 (2015)
8. Dosiyev, A.A., Sadeghi, M.M.H.: On a highly accurate approximation of the first and pure second derivatives of the Laplace equation in a rectangular parellelpiped. Adv. Differ. Equ. **2016**(145), 1–13 (2016)
9. Dosiyev, A.A., Abdussalam, A.: On the high order convergence of the difference solution of Laplace's equation in a rectangular parallelepiped. Filomat **32**(3), 893–901 (2018)
10. Dosiyev, A.A., Sarikaya, H.: 14-Point difference operator for the approximation of the first derivatives of a solution of Laplace's equation in a rectangular parallelepiped. Filomat **32**(3), 791–800 (2018)
11. Douglas, J., Rachford, H.H.: On the numerical solution of heat conduction problems in two and three space variables. Trans. Am. Math. Soc. **82**(2), 421–439 (1956)
12. Farinola, L.A.: Grid approximation of derivatives of the solution of heat conduction equation. Doctoral thesis, Eastern Mediterranean University (2019)
13. Il'in, V.P.: Numerical methods solving the electrophysics problems. Nauka Navosibirsk (1974) (in Russian)

14. Lebedev, V.I.: Evaluation of the error involved in the grid method for Newmann's two dimensional problem. Sov. Math. Dokl. **1**, 703–705 (1960)
15. Ladyženskaja, O.A., Solonnikov, V.A., Ural'ceva, N.N.: Linear and Quasi-linear Equations of Parabolic Type. Translation of Mathematical Monographs. American Mathematical Society, Providence (1967)
16. Landau, H.G.: Heat conduction in a melting solid. Q. Appl. Math. **8**, 81–94 (1950)
17. Lotkin, M.: The calculation of heat flow in melting solids. Q. Appl. Math. **18**(1), 79–85 (1960)
18. Samarskii, A.A.: The Theory of Difference Schemes. Marcel Dekker Inc., New York (2001)
19. Tuttle, F.: A mathematical theory of the drying of wood. J. Frankl. Inst. **200**, 609–614 (1925)
20. Volkov, E.A.: On convergence in C_2 of a difference solution of the Laplace equation on a rectangle. Russ. J. Numer. Anal. Math. Model. **14**(3), 291–298 (1999)
21. Volkov, E.A.: On the grid method by approximating the derivatives of the solution of the Dirichlet problem for the Laplace equation on the rectangular parallelpiped. Russ. J. Numer. Anal. Math. Model. **19**(3), 209–278 (2004)

Identification Elliptic Problem with Dirichlet and Integral Conditions

Charyyar Ashyralyyev

Abstract In the present paper, identification elliptic problem in a Hilbert space with Dirichlet and integral boundary conditions is discussed. Identification problem is reduced to auxiliary nonlocal boundary value problem with nonlocal integral condition. Operator approach is used to prove stability and coercive stability inequalities for solution of source identification elliptic problem in the case of a self-adjoint positive definite operator in a differential equation. As applications, four mixed boundary value problems for strongly elliptic multidimensional partial differential equation are investigated. Theorems on stability of solutions of these boundary value problems are established.

1 Identification Problem

In last decades, the subject of the identification problems for partial differential equations (PDEs) is quickly growing interest for many scientists. In particular, the general theory and methods of solutions problems for determining unknown parameters in PDEs have been extensively studied by several researchers (see [1–3, 5, 7–15, 17, 18], and references therein).

Let H be any Hilbert space, A and I be the self-adjoint positive definite (SAPD) and identity operators in H , $F(H)$ be the some space of $H-$valued smooth functions on the segment $[0, 1]$, δ be positive real number such that $A > \delta I$. Let elements $\vartheta, \eta, \xi \in H$, and $f \in F(H)$ be given. Definition of the SAPD operator and its properties can be found in [16].

C. Ashyralyyev (✉)
Department of Mathematical Engineering, Gumushane University, Gumushane 29100, Turkey
e-mail: charyyar@gumushane.edu.tr; charyar@gmail.com

TAU, Ashgabat, Turkmenistan

A. Ashyralyev et al. (eds.), *Functional Analysis in Interdisciplinary Applications—II*,
Springer Proceedings in Mathematics & Statistics 351,
https://doi.org/10.1007/978-3-030-69292-6_4

Assume that $\rho : [0, 1] \to R$ is given function such that

$$\int_0^1 |\rho(\tau)|\, d\tau \le \frac{1}{2}, \tag{1}$$

and $\gamma \in (0, 1)$ is known number. Consider a source identification problem to find an element $p \in H$ and a function u for the next elliptic differential equation

$$-u_{tt}(t) + A\, u(t) = p + f(t),\ 0 < t < 1 \tag{2}$$

with Dirichet and integral conditions

$$u(0) = \vartheta,\ u(1) = \int_0^1 \rho(\tau) u(\tau) d\tau + \eta, u(\gamma) = \xi. \tag{3}$$

We say that (u, p) is the solution of the problem (2), (3) in $F(H) \times H_1$ if the next conditions are valid:

a. $u''(t),\ Au(t) \in F(H),\ p \in H_1 \subset H,$
b. (u, p) satisfies the relationship (2) and three conditions (3).

In future, we need notations $C(H)$ and $C_{01}^{\alpha,\alpha}(H)$ $(0 < \alpha < 1)$, for the Banach spaces of H-valued smooth functions q on $[0, 1]$ with suitable norms

$$\begin{aligned} \|q\|_{C(H)} &= \max_{0\le t\le 1} \|q(t)\|_H, \\ \|q\|_{C_{01}^{\alpha,\alpha}(H)} &= \|q\|_{C(H)} + \sup_{0\le t<t+s\le 1} (t+s)^\alpha (1-t)^\alpha s^{-\alpha} \|q(t+s) - q(t)\|_H. \end{aligned} \tag{4}$$

Suppose that the function $v(t)$ satisfies the following equation

$$-v_{tt}(t) + Av(t) = f(t),\ 0 < t < 1 \tag{5}$$

for known $f(t)$. If $v(0)$ and $v(1)$ are given, then the solution of (5) is derived by formula ([4], p. 205, 5.11)

$$v(t) = P^{-1}\left[e^{-tB}(I - e^{-(1-t)B})\, v(0) + e^{-(1-t)B}\, (I - e^{-2tB})\, v(1)\right] + F(B, f|\, t), \tag{6}$$

where

$$B = A^{\frac{1}{2}}, P = I - e^{-2B},$$

$$F(B, f|t) = -P^{-1}e^{-(1-t)B}\left(I - e^{-2tB}\right)(2B)^{-1}\int_0^1 e^{-(1-s)B}(I - e^{-2sB})f(s)ds$$

$$+(2B)^{-1}\int_0^t e^{-(t-s)B}f(s)ds + (2B)^{-1}\int_t^1 e^{-(s-t)B}ds - (2B)^{-1}\int_0^1 e^{-(t+s)B}f(s)ds.$$

In [1], overdetermined problem for Eq. (2) with Dirichlet conditions in three points such as $u(0) = \vartheta$, $u(1) = \eta$, $u(\gamma) = \xi$ was studied. The exact stability estimates for its solution in some Hölder norms were established. In the papers [5–7], identification problems in case of Neumann boundary conditions were studied on well-posedness. In the papers [8, 10], the authors discussed inverse problems for elliptic equation with multi-point nonlocal boundary conditions. Recently, in [9], the authors presented stability estimates for solution of Neumann-type elliptic overdetermined problem with integral condition.

This paper is devoted to the elliptic identification problem for (2) with corresponding Dirichlet and integral conditions (3). In this work, we obtain stability and coercive stability estimates for its solution. Moreover, in application, four mixed boundary value problems (BVPs) for strongly elliptic multidimensional PDE are studied.

2 Stability Estimates

Existence and boundedness of A^{-1} follows from self-adjoint property and positivity of A. So, by using

$$u(t) = A^{-1}p + v(t), \tag{7}$$

it can be obtained nonlocal BVP for suitable function $v(t)$:

$$\begin{cases} -v_{tt}(t) + Av(t) = f(t),\ 0 < t < 1, \\ v(0) - v(\gamma) = \vartheta - \xi, \\ v(1) - v(0)\mu(\rho) - \int\limits_0^1 \rho(\tau)v(\tau)d\tau = -\vartheta\,\mu(\rho) + \eta, \end{cases} \tag{8}$$

where $\mu(\rho) = 1 - \int\limits_0^1 \rho(\tau)d\tau$.

Now, we prove that solution of nonlocal BVP (8) exists. Applying (6), we get linear system equation with unknown elements $v(0)$ and $v(1)$:

$$\left[I - e^{-2B} - e^{-\gamma B} + e^{-(2-\gamma)B}\right]v(0) + e^{-(1-\gamma)B}\left[-I + e^{-2\gamma B}\right]v(1)$$

$$= P\vartheta - P\xi - PF(B, f|\gamma),$$

$$\left[-\mu(\rho) I - \int_0^1 \rho(\tau)e^{-\tau B}(I - e^{-2(1-\tau)B})d\tau\right] v(0)$$

$$+\left[I - e^{-2B} - \int_0^1 \rho(\tau)e^{-(1-\tau)B}(I - e^{-2\tau B})d\tau\right] v(1) \tag{9}$$

$$= -P\vartheta\, \mu(\rho) + P\eta + \int_0^1 \rho(\tau)F(B, f|\tau)d\tau.$$

Determinant operator of linear system Eq. (9) is derived by

$$D = \left(I - e^{-\gamma B}\right)\left(I + e^{-(2-\gamma)B}\right) P$$

$$\times \left[I - P^{-1} \int_0^1 \rho(\tau)e^{-(1-\tau)B}(I - e^{-2\tau B})d\tau\right] \tag{10}$$

$$+e^{-(1-\gamma)B}\left(I - e^{-2\gamma B}\right)\left[I + \int_0^1 \rho(\tau)\left(e^{-(1-\tau)B}(I - e^{-2\tau B}) - I\right)d\tau\right].$$

Since A is SAPD operator and (1) is valid, then

$$B > 0, I - e^{-2B} > 0, I + \int_0^1 \rho(\tau)\left(e^{-(1-\tau)B}(I - e^{-2\tau B}) - I\right)d\tau > 0,$$

$$I - P^{-1} \int_0^1 \rho(\tau)e^{-(1-\tau)B}(I - e^{-2\tau B})d\tau > 0,$$

$$\forall\gamma \in (0, 1), I - e^{-\gamma B} > 0, I + e^{-(2-\gamma)B} > 0, e^{-(1-\gamma)B} > 0, I - e^{-2\gamma B} > 0. \tag{11}$$

From (11) it follows that $D > 0$. So, we can conclude that the operator D has a bounded inverse D^{-1} :

$$\|D^{-1}\|_{H\to H} \leq M(\delta). \tag{12}$$

Finally, we have solution of (9) by

$$v(0) = D^{-1} \left\{ \left[I - e^{-2B} - \int_0^1 \rho(\tau)(e^{-(1-\tau)B} - e^{-(1+\tau)B})d\tau \right] \right.$$

$$\times \left[P\vartheta - P\xi - PF(B, f|\gamma) \right] - e^{-(1-\gamma)B} \left(I - e^{-2\gamma B} \right)$$

$$\left. \times \left[I + \int_0^1 \rho(\tau) \left(e^{-(1-\tau)B}(I - e^{-2\tau B}) - I \right) d\tau \right] \right\},$$

$$v(1) = D^{-1} \left\{ e^{-(1-\gamma)B} \left(I - e^{-2\gamma B} \right) \right.$$

$$\times \left[I + \int_0^1 \rho(\tau) \left(e^{-(1-\tau)B}(I - e^{-2\tau B}) - I \right) d\tau \right]$$

$$\times \left[P\vartheta - P\xi - PF(B, f|\gamma) \right] + \left(I - e^{-\gamma B} \right) \left(I + e^{-(2-\gamma)B} \right) \left(I - e^{-2B} \right)$$

$$\left. \times \left[-P\vartheta\ \mu(\rho) + P\eta + \int_0^1 \rho(\tau)F(B, f|\tau)d\tau \right] \right\}.$$

Therefore, solution of nonlocal BVP (8) is derived by:

$$v(t) = P^{-1}(e^{-tB} - e^{-(2-t)B})\, D^{-1} \left\{ \left[I - e^{-2B} \right. \right.$$

$$\left. - \int_0^1 \rho(\tau)(e^{-(1-\tau)B} - e^{-(1+\tau)B})d\tau \right] \left[P\vartheta - P\xi - PF(B, f|\gamma) \right]$$

$$\left. -e^{-(1-\gamma)B} \left(I - e^{-2\gamma B} \right) \left[I + \int_0^1 \rho(\tau) \left(e^{-(1-\tau)B}(I - e^{-2\tau B}) - I \right) d\tau \right] \right\},$$

$$+P^{-1}(e^{-(1-t)B} - e^{-(1+t)B})D^{-1} \left\{ e^{-(1-\gamma)B} \left(I - e^{-2\gamma B} \right) \right. \tag{13}$$

$$\times \left[I + \int_0^1 \rho(\tau) \left(e^{-(1-\tau)B}(I - e^{-2\tau B}) - I \right) d\tau \right]$$

$$\times \left[P\vartheta - P\xi - PF(B, f|\gamma) \right] + \left(I - e^{-\gamma B} \right) \left(I + e^{-(2-\gamma)B} \right)$$

$$\left. \times \left(I - e^{-2B} \right) \left[-P\vartheta\ \mu(\rho) + P\eta + \int_0^1 \rho(\tau)F(B, f|\tau)d\tau \right] \right\} + F(B, f|t).$$

So, a unique solution of (8), $v(t)$, exists and and it is defined by formula (13). After solving nonlocal problem (8), an unknown element p can be computed by virtue

$$p = A\vartheta - AD^{-1}\left\{\left[I - e^{-2B} - \int_0^1 \rho(\tau)(e^{-(1-\tau)B} - e^{-(1+\tau)B})d\tau\right]\right.$$
$$\times\left[P\vartheta - P\xi - PF(B, f|\gamma)\right] - e^{-(1-\gamma)B}\left(I - e^{-2\gamma B}\right) \tag{14}$$
$$\left.\times\left[I + \int_0^1 \rho(\tau)\left(e^{-(1-\tau)B}(I - e^{-2\tau B}) - I\right)d\tau\right]\right\}.$$

By using method of [4] to (13), we can obtain estimates for solution of nonlocal problem (8):

$$\|v\|_{C(H)} \le M(\delta)\left[\|\vartheta\|_H + \|\xi\|_H + \|\eta\|_H + \|f\|_{C(H)}\right], \tag{15}$$

$$\|v''\|_{C_{01}^{\alpha,\alpha}(H)} + \|Av\|_{C_{01}^{\alpha,\alpha}(H)}$$
$$\le M(\delta)\left[\|A\vartheta\|_H + \|A\xi\|_H + \|A\eta\|_H + [\alpha(1-\alpha)]^{-1}\|f\|_{C_{01}^{\alpha,\alpha}(H)}\right]. \tag{16}$$

With the help of operational calculus, triangle inequality, formula (13), one can show that

$$\|p\|_H \le M(\delta)\left[\|A\vartheta\|_H + \|A\xi\|_H + \|A\eta\|_H + [\alpha(1-\alpha)]^{-1}\|f\|_{C_{01}^{\alpha,\alpha}(H)}\right]. \tag{17}$$

Then, by using formulas (7), (13),(14), (15), (16), (17), and operational calculus, we can conclude the following results.

Theorem 1 *Let $\xi, \vartheta, \eta \in D(A)$, $\alpha \in (0, 1)$, $f(t) \in C_{01}^{\alpha,\alpha}(H)$ and inequality (1) be valid. Accordingly, for the solution (u, p) of identification elliptic problem (2), (3) the stability estimates*

$$\|u\|_{C(H)} \le M(\delta)\left[\|\vartheta\|_H + \|\eta\|_H + \|\xi\|_H + \|f\|_{C(H)}\right], \tag{18}$$

$$\|p\|_H \le M(\delta)\left[\|A\vartheta\|_H + \|A\eta\|_H + \|A\xi\|_H + [\alpha(1-\alpha)]^{-1}\|f\|_{C_{01}^{\alpha,\alpha}(H)}\right] \tag{19}$$

hold, where $M(\delta)$ does not depend on $\xi, \vartheta, \eta, \alpha$, and $f(t)$.

Here and in future throughout paper, M denotes positive constants, which is not a subject of precision. We will use the notation $M(\delta)$ to stress the fact that the constant depends only on δ.

Theorem 2 *Assume that $\xi, \vartheta, \eta \in D(A)$, $\alpha \in (0, 1)$ $f(t) \in C_{01}^{\alpha,\alpha}(H)$ are given and inequality (1) is satisfied. Then, the solution (u, p) of identification elliptic problem (2), (3) obeys coercive stability estimate:*

$$\begin{aligned}&\left\|u''\right\|_{C_{01}^{\alpha,\alpha}(H)}+\|Au\|_{C_{01}^{\alpha,\alpha}(H)}+\|p\|_{H}\\ &\leq M\left(\delta\right)\left[[\alpha(1-\alpha)]^{-1}\|f\|_{C_{01}^{\alpha,\alpha}(H)}+\|A\vartheta\|_{H}\ +\ \|A\eta\|_{H}\ +\ \|A\xi\|_{H}\right].\end{aligned} \tag{20}$$

3 Applications to BVPs in Case of Strongly Elliptic PDE

Denote by $\Omega=(0,1)\times(0,1)...\times(0,1)\in R^{n}$ open unit cube with boundary $S=S_{1}\cup S_{2},\ \ \overline{\Omega}=S\cup\Omega$, where

$$\begin{aligned}S_{1}&=\left\{x\mid x_{i}=0,\ x_{j}\in[0,1]\,,\,j\neq i,\ 1\leq i\leq n\right\},\\ S_{2}&=\left\{x\mid x_{i}=1,\quad x_{j}\in(0,1]\,,\,j\neq i,\ 1\leq i\leq n\right\}.\end{aligned}$$

Suppose that a_{r},ξ,ϑ,η be given function from Ω to R, f be known function from $(0,1)\ \times\Omega$ to R. Moreover, $a_{r}(x)\geq a_{0}>0$ for any $x\in\Omega$ and $\kappa>0,\gamma\in(0,1)$ are given numbers.

Let $C_{01}^{\alpha,\alpha}(L_{2}(\overline{\Omega}))$ be the Banach space $L_{2}(\overline{\Omega})$-valued smooth functions υ on the segment $[0,1]$ with the norm defined by (4) for $H=L_{2}(\overline{\Omega})$. Now, we apply abstract results on well-posedness of identification elliptic problem to several overdetermined BVPs for multidimensional strongly elliptic PDE with Dirichlet and integral conditions.

Firstly, in $[0,1]\times\Omega$, we will consider the following identification problem with Dirichlet and integral conditions for multidimensional elliptic PDE:

$$\begin{cases}-u_{tt}(t,x)\ -\ \sum\limits_{r=1}^{n}(a_{r}(x)u_{x_{r}}(t,x))_{x_{r}}+\kappa u(t,x)\ =\ p(x)\ +\ f(t,x),\\ (t,x)\in(0,1)\times\Omega,\\ u\left(0,x\right)=\vartheta\left(x\right),\ u\left(1,x\right)=\int\limits_{0}^{1}\rho\left(\tau\right)u\left(\tau,x\right)d\tau+\eta(x),\\ u\left(\gamma,x\right)=\xi(x),\ \ x\in\overline{\Omega},\\ u(t,x)\ =\ 0,\ \ (t,x)\ \in[0,1]\times S.\end{cases} \tag{21}$$

Assume that all compatibility conditions are satisfied. The expression

$$A^{x}u(x)=-\sum_{r=1}^{n}(a_{r}(x)u_{x_{r}}(x))_{x_{r}}+\kappa u(x) \tag{22}$$

defines A^{x} which is SAPD operator acting on the space $L_{2}\left(\overline{\Omega}\right)$ with the appropriate domain

$$D(A^x) = \left\{ u \in W_2^2\left(\overline{\Omega}\right),\ u = 0 \ on\ S \right\}. \tag{23}$$

Thus, identification problem (21) can be reduced to abstract form (2), (3) in the Hilbert space $H = L_2\left(\overline{\Omega}\right)$. Applying Theorems 1 and 2, one can establish the following statements on stability of solution identification problem (21).

Theorem 3 *Let $D(A^x)$ be defined by (23), $\xi, , \vartheta, \eta \in D(A^x)$, $f \in C(L_2(\overline{\Omega}))$, ρ be given scalar function under condition (1). Then, the solution of identification problem (21) obeys the following stability estimate*

$$\|u\|_{C(L_2(\overline{\Omega}))} \leq M\left(\delta\right)\ [\|\vartheta\|_{L_2(\overline{\Omega})} + \|\xi\|_{L_2(\overline{\Omega})} + \|\eta\|_{L_2(\overline{\Omega})} + \|f\|_{C(L_2(\overline{\Omega}))}]. \tag{24}$$

Theorem 4 *Let $D(A^x)$ be defined by (23), $\xi, \vartheta, \eta \in D(A^x)$, $f \in C_{01}^{\alpha,\alpha}(L_2(\overline{\Omega}))$, ρ be given scalar function under condition (1). Then, for the solution of identification problem (21) the coercive stability estimate*

$$\left\|u''\right\|_{C_{01}^{\alpha,\alpha}(L_2(\overline{\Omega}))} + \|u\|_{C_{01}^{\alpha,\alpha}(W_2^2(\overline{\Omega}))} + \|p\|_{L_2(\overline{\Omega})}$$

$$\leq M\left(\delta\right)\ [\alpha(1-\alpha)]^{-1}\ \|f\|_{C_{01}^{\alpha,\alpha}(L_2(\overline{\Omega}))} + \|\vartheta\|_{W_2^2(\overline{\Omega})} + \|\xi\|_{W_2^2(\overline{\Omega})} + \|\eta\|_{W_2^2(\overline{\Omega})}] \tag{25}$$

is satisfied.

Secondly, in $[0, 1] \times \Omega$, we will discuss the following overdetermined mixed BVP for strongly elliptic multidimensional PDE:

$$\begin{cases} -u_{tt}(t,x) - \sum\limits_{r=1}^{n}(a_r(x)u_{x_r}(t,x))_{x_r} + \kappa u(t,x) = p(x) + f(t,x), \\ (t,x) \in (0,1) \times \Omega, \\ u\left(0,x\right) = \vartheta\left(x\right),\ u\left(1,x\right) = \int\limits_0^1 \rho\left(\tau\right) u\left(\tau,x\right) d\tau + \eta(x), \\ u\left(\gamma,x\right) = \xi(x), x \in \overline{\Omega}, \\ \frac{\partial u(t,x)}{\partial \overrightarrow{n}} = 0,\ (t,x)\ \in [0,1] \times S. \end{cases} \tag{26}$$

Let all compatability conditions be fulfilled.

The expression (22) defines SAPD operator A^x acting on space $L_2\left(\overline{\Omega}\right)$ with the corresponding domain

$$D(A^x) = \left\{ u \in W_2^2\left(\overline{\Omega}\right),\ \frac{\partial u}{\partial \overrightarrow{n}} = 0 \ on\ S \right\}. \tag{27}$$

So, by reducing problem (26) to problem (2), (3) and applying Theorems 1 and 2, it can be proved the following statements on stability.

Theorem 5 *Suppose that* $D(A^x)$ *is defined by (27),* $\xi, \vartheta, \eta \in D(A^x)$, $f \in C(L_2(\overline{\Omega}))$, *and function* ρ *is under condition (1. Then, for the solution of problem (26) the stability inequality (24) is valid.*

Theorem 6 *Assume that* $D(A^x)$ *is defined by (27),* $\xi, \vartheta, \eta \in D(A^x)$, $f \in C_{01}^{\alpha,\alpha}(L_2(\overline{\Omega}))$, *and function* ρ *is under condition (1). Then, for the solution of problem (26) coercive stability inequality (25) is satisfied.*

Thirdly, in $[0, 1] \times \Omega$, we will study the following overdetermined mixed BVP for multidimensional elliptic PDE:

$$\begin{cases} -u_{tt}(t,x) - \sum\limits_{r=1}^{n} (a_r(x)u_{x_r}(t,x))_{x_r} + \kappa u(t,x) = p(x) + f(t,x), \\ (t,x) \in (0,1) \times \Omega, \\ u(0,x) = \vartheta(x),\ u(1,x) = \int\limits_0^1 \rho(\tau) u(\tau,x)\,d\tau + \eta(x), \\ u(\gamma,x) = \xi(x),\ x \in \overline{\Omega}, \\ u(t,x) = 0,\ (t,x) \in [0,1] \times S_1,\ \frac{\partial u(t,x)}{\partial \vec{n}} = 0,\ (t,x) \in [0,1] \times S_2. \end{cases} \tag{28}$$

Suppose that all compatibility conditions are satisfied. The differential expression (22) defines A^x which is a SAPD operator acting on $L_2(\overline{\Omega})$ with the suitable domain

$$D(A^x) = \left\{ u \in W_2^2(\overline{\Omega}),\ u = 0 \ on\ S_1,\ \frac{\partial u}{\partial \vec{n}} = 0 \ on\ S_2 \right\}. \tag{29}$$

By reducing problem (28) to problem (2), (3) and applying Theorems 1 and 2, we can establish the following statements.

Theorem 7 *Suppose that* $D(A^x)$ *is defined by (29),* $\xi, \vartheta, \eta \in D(A^x)$, $f \in C_{01}^{\alpha,\alpha}(L_2(\overline{\Omega}))$, *and function* ρ *is under condition (1). Then, for the solution of identification problem (28) the stability inequality (24) is satisfied.*

Theorem 8 *Assume that* $D(A^x)$ *is defined by (29),* $\xi, \vartheta, \eta \in D(A^x)$, $f \in C_{01}^{\alpha,\alpha}(L_2(\overline{\Omega}))$, *and scalar function* ρ *is under restriction (1). Then, the solution of problem (28) obeys coercive stability inequality (25).*

Finally, in $[0, 1] \times \Omega$, we will examine the next identification problem

$$\begin{cases} -u_{tt}(t,x) - \sum\limits_{r=1}^{n}(a_r(x)u_{x_r}(t,x))_{x_r} + \kappa u(t,x) = p(x) + f(t,x), \\ (t,x) \in (0,1) \times \Omega, \\ \\ u(0,x) = \vartheta(x),\ u(1,x) = \int\limits_0^1 \rho(\tau)u(\tau,x)d\tau + \eta(x), \\ u(\gamma,x) = \xi(x), x \in \overline{\Omega}, \\ \frac{\partial u(t,x)}{\partial \overrightarrow{n}} = 0,\ (t,x) \in [0,1] \times S_1,\ u(t,x) = 0,\ (t,x) \in [0,1] \times S_2. \end{cases} \tag{30}$$

Assume that all compatibility conditions are satisfied. Then, in the similar manner, the expression (22) defines a SAPD operator A^x acting on $L_2\left(\overline{\Omega}\right)$ with the coincident domain

$$D(A^x) = \left\{ u \in W_2^2\left(\overline{\Omega}\right),\ \frac{\partial u}{\partial \overrightarrow{n}} = 0 \ on\ S_1, u = 0\ on\ S_2 \right\}. \tag{31}$$

Therefore, problem (30) can be reduced to problem (2), (3). Applying Theorems 1 and 2, we have the following Theorems.

Theorem 9 *Assume that $D(A^x)$ is defined by (31), $\xi, \vartheta, \eta \in D(A^x)$, $f \in C(L_2(\overline{\Omega}))$, and function ρ is under restriction (1). Then, for the solution of problem (30) the stability inequality (24) is valid.*

Theorem 10 *Suppose that $D(A^x)$ is defined by (31), $\xi, \vartheta, \eta \in D(A^x)$, $f \in C_{01}^{\alpha,\alpha}(L_2(\overline{\Omega}))$, and scalar function ρ is under restriction (1). Then, the solution of problem (30) obeys coercive stability inequality (25).*

References

1. Ashyralyev, A., Ashyralyyev, C.: On the problem of determining the parameter of an elliptic equation in a Banach space. Nonlinear Anal. Model. Control **19**, 350–366 (2014)
2. Ashyralyev, A., Emharab, F.: Identification hyperbolic problems with the Neumann boundary condition. Bull. Karaganda Univ.-Math. **91**, 89–98 (2018)
3. Ashyralyev, A., Sazaklioglu, A.U.: Investigation of a time-dependent source identification inverse problem with integral overdetermination. Numer. Funct. Anal. Optim. **38**, 1276–1294 (2017)
4. Ashyralyev, A., Sobolevskii, P.E.: New Difference Schemes for Partial Differential Equations. Operator Theory Advances and Applications. Birkhäuser Verlag, Basel, Boston, Berlin (2004)
5. Ashyralyyev, C.: Stability estimates for solution of Neumann type overdetermined elliptic problem. Numer. Funct. Anal. Optim. **38**, 1226–1243 (2017)
6. Ashyralyyev, C.: Numerical solution to Bitsadze–Samarskii type elliptic overdetermined multipoint NBVP. Bound. Value Probl. **2017**, 74 (2017)
7. Ashyralyyev, C.: A fourth order approximation of the Neumann type overdetermined elliptic problem. Filomat **31**, 967–980 (2017)
8. Ashyralyyev, C., Akyuz, G.: Finite difference method for Bitsadze–Samarskii type overdetermined elliptic problem with Dirichlet conditions. Filomat **32**, 859–872 (2018)

9. Ashyralyyev, C., Cay, A.: Well-posedness of Neumann-type elliptic overdetermined problem with integral condition. AIP Conf. Proc. **1997**, 020026 (2018)
10. Ashyralyyev, C., Akyuz, G., Dedeturk, M.: Approximate solution for an inverse problem of multidimensional elliptic equation with multipoint nonlocal and Neumann boundary conditions. Electron. J. Differ. Equ. **2017**, 197 (2017)
11. Ashyralyyeva, M.A., Ashyraliyev, M.: On the numerical solution of identification hyperbolic-parabolic problems with the Neumann boundary condition. Bull. Karaganda Univ.-Math. **91**, 69–74 (2018)
12. Erdogan, A.S., Kusmangazinova, D., Orazov, I., Sadybekov, M.A.: On one problem for restoring the density of sources of the fractional heat conductivity process with respect to initial and final temperatures. Bull. Karaganda Univ.-Math. **91**, 31–44 (2018)
13. Kabanikhin, S.I.: Inverse and Ill-Posed Problems: Theory and Applications. Walter de Gruyter, Berlin (2011)
14. Kirane, M., Malik, S.A., Al-Gwaiz, M.A.: An inverse source problem for a two dimensional time fractional diffusion equation with nonlocal boundary conditions. Math. Methods Appl. Sci. **36**, 056–069 (2013)
15. Klibanov, M.V., Romanov, V.G.: Two reconstruction procedures for a 3D phaseless inverse scattering problem for the generalized Helmholtz equation. Inverse Probl. **32**, 1 (2016)
16. Krein, S.G.: Linear Differential Equations in Banach Space. Nauka, Moscow, Russia (1966)
17. Orazov, I., Sadybekov, M.A.: On a class of problems of determining the temperature and density of heat sources given initial and final temperature. Sib. Math. J. **53**, 146–151 (2012)
18. Sazaklioglu, A.U., Ashyralyev, A., Erdogan, A.S.: Existence and uniqueness results for an iinverse problem for semilinear parabolic equations. Filomat **31**, 1057–1064 (2017)

On Solvability of Some Boundary Value Problems with Involution for the Biharmonic Equation

Valery V. Karachik and Batirkhan Kh. Turmetov

Abstract In this paper we study new classes of well-posed boundary-value problems for the biharmonic equation. The considered problems are Bitsadze–Samarskii type nonlocal boundary value problems. The investigated problems are solved by reducing them to the Neumann and Dirichlet type problems. In this paper, theorems on existence and uniqueness of the solution are proved, and exact conditions for solvability of the problems are found. In addition, integral representations of the solution are obtained.

Keywords Biharmonic equation · Nonlocal problem · Involution · Neumann type problem

1 Introduction

Significant number of mathematical models in physics and engineering lead to partial differential equations. The steady processes of various physical nature are described by the partial differential equations of elliptic type. One of the important special cases of fourth order elliptic equations is the biharmonic equation $\Delta^2 u(x) = f(x)$. Investigation of mathematical models of the plane deformation of the elasticity theory in many cases is reduced to integration of the biharmonic equation with the appropriate boundary conditions and under some uniqueness conditions for the unknown function.

Moreover investigation of many mathematical models of continuum mechanics are reduced to solving the harmonic and biharmonic equations. Application of biharmonic problems in mathematical models of mechanics and physics can be found in

V. V. Karachik
South Ural State University, Prosp. Lenina 76, Chelyabinsk 454080, Russian Federation
e-mail: karachik@susu.ru

B. Kh. Turmetov (✉)
Akhmet Yassawi University, 29 B.Sattarkhanov avenue, Turkistan 161200, Kazakhstan
e-mail: turmetovbh@mail.ru

A. Ashyralyev et al. (eds.), *Functional Analysis in Interdisciplinary Applications—II*,
Springer Proceedings in Mathematics & Statistics 351,
https://doi.org/10.1007/978-3-030-69292-6_5

the numerous scientific investigations (see, for example, [2, 9, 23]). Multiple applications of boundary value problems for the biharmonic equation in mathematical models of mechanics and physics encourages investigation of various formulations of boundary value problems for the biharmonic equation. There is a big interest in studying of boundary value problems (see [7, 24, 33]) for the biharmonic equation.

The Dirichlet problem (see, for example, [3]) is the well known boundary value problem for biharmonic equation. In recent years other types of boundary value problems for the biharmonic equation, such as the problems by Riquier (see [11]), by Neumann (see [12]), by Robin (see [17]) and etc. [1] are actively studied.

Nonlocal boundary value problems for elliptic equations in which boundary conditions are given in the form of a connection between the values of the unknown function and its derivatives at various points of the boundary are called the problems of Bitsadze–Samarskii type [6]. Numerous applications of nonlocal boundary value problems for elliptic equations in problems of physics, engineering, and other branches of science are described in detail in [28, 29]. Solvability of nonlocal boundary value problems for elliptic equations is discussed in [4, 8, 10, 20, 21, 25]. Boundary problems with involution for elliptic equations of the second and fourth order, as a special case of nonlocal problems, are considered in [18, 22, 26, 27, 32].

Let $\Omega = \{x \in R^n : |x| < 1\}$ be the unit ball, $n \geq 2$, and $\partial\Omega$ be the unit sphere. For any point $x = (x_1, x_2, \ldots, x_n) \in \Omega$ we consider the point $x^* = Sx$, where S is a real orthogonal matrix $S \cdot S^T = E$. Suppose also that $S^2 = E$.

In this paper we study the following nonlocal boundary value problem

$$\Delta^2 u(x) = f(x), \ x \in \Omega, \tag{1}$$

$$D_\nu^m u(x) + \alpha D_\nu^m u(x^*) = g_1(x), \ x \in \partial\Omega, \tag{2}$$

$$D_\nu^{m+1} u(x) + \beta D_\nu^{m+1} u(x^*) = g_2(x), \ x \in \partial\Omega, \tag{3}$$

where $0 \leq m \leq 2$, a, b are real numbers, $D_\nu^m = \frac{\partial^m}{\partial \nu^m}$, ν is a unit vector of the outward normal to $\partial\Omega$, $D_\nu^0 = I$ is a unit operator.

By a solution of the problem (1)–(3) we mean a function $u(x) \in C^4(\Omega) \cap C^{m+1}(\bar{\Omega})$ satisfying conditions (1)–(3) in the classical sense. In the case $\alpha = \beta = 0$ when $m = 0$ we obtain the well-known Dirichlet problem and when a Neumann type problem [1, 12, 30, 31].

Content is constructed as follows. In Sect. 1, in Lemmas 1 and 2, we give some auxiliary propositions. In Sect. 2, in Theorems 1 and 2, we give results on solvability of Neumann type problems. In Sect. 3, we prove Theorem 3 on uniqueness of the solution of problem (1)–(3). In Sect. 4 we prove Theorem 4 on existence of the solution of problem (1)–(3), and in Theorem 5 we give a method for constructing a solution of problem (1)–(3) with homogeneous boundary conditions.

2 Auxiliary Statements

First we note that if $x \in \Omega$, or $x \in \partial\Omega$, then $x^* = Sx \in \Omega$, or $x^* = Sx \in \partial\Omega$, respectively, since the transformation of the space R^n by the matrix S preserves the norm $|x^*|^2 = |Sx|^2 = (Sx, Sx) = (S^T Sx, x) = |x|^2$.

The case $x^* = -x$ investigated in [18, 22, 26, 27, 32] is a particular case of the situation considered here since for $S = -E$ we have $S \cdot S^T = -E(-E) = E$.

Consider the operator

$$I_S u(x) = u(Sx) = u(x^*).$$

In view of what has been said above, this operator is defined on functions $u(x), x \in \Omega$. We also consider the operator $\Lambda u = \sum_{i=1}^{n} x_i u_{x_i}(x)$ that is homogeneous, preserves the biharmonicity of function $u(x)$, and has the property $D_\nu^m u|_{\partial\Omega} = \Lambda^{[m]} u|_{\partial\Omega}$, where $\Lambda^{[m]} = \Lambda(\Lambda - 1)\dots(\Lambda - m + 1)$ [14]. Let S_{col}^i and S_{row}^i be the ith column and ith row of the matrix S, respectively.

We prove two simple lemmas. Let $u(x)$ be a twice continuously differentiable function in Ω.

Lemma 1 *Operators Λ and I_S are commutative $\Lambda I_S u(x) = I_S \Lambda u(x)$, and also the equality $\nabla I_S = I_S S^T \nabla$ holds, and operators Δ and I_S are also commutative.*

Proof We can write the operator Λ in the form $\Lambda u = (x, \nabla)u$. Since

$$\frac{\partial}{\partial x_i} I_S u(x) = \frac{\partial}{\partial x_i} u(x) = \frac{\partial}{\partial x_i} u((S_{row}^1, x), \dots, (S_{row}^n, x)) = \sum_{j=1}^{n} s_{ji} I_S u_{x_j}(x)$$

$$= (S_{col}^i, I_S \nabla u(x)) = I_S (S_{col}^i, \nabla) u(x). \tag{4}$$

then

$$\Lambda I_S u(x) = \Lambda u(Sx) = \sum_{i=1}^{n} x_i \frac{\partial}{\partial x_i} u(Sx) = \sum_{i=1}^{n} x_i \left(S_{col}^i, I_S \nabla u(x)\right)$$

$$= \left(\sum_{i=1}^{n} x_i S_{col}^i, I_S \nabla u(x)\right) = (Sx, I_S \nabla u(x)) = I_S(x, \nabla u(x)) = I_S \Lambda u(x).$$

Further, due to the formula (4), we find

$$\frac{\partial^2}{\partial x_i^2} I_S u(x) = \frac{\partial}{\partial x_i} I_S (C_{col}^i, \nabla) u(x) = I_S (C_{col}^i, \nabla)^2 u(x)$$

and therefore

$$\Delta I_S u(x) = \sum_{i=1}^{n} I_S (S^i_{col}, \nabla)^2 u(x) = I_S \big|((S^1_{col}, \nabla), \ldots, (S^n_{col}, \nabla))\big|^2 u(x)$$

$$= I_S \big|S^T \nabla\big|^2 u(x) = I_S (S^T \nabla, S^T \nabla) u(x) = I_S (S S^T \nabla, \nabla) u(x) = I_S \Delta u(x).$$

At last,

$$\nabla I_S u(x) = I_S ((S^1_{col}, \nabla), \ldots, (S^n_{col}, \nabla)) u(x) = I_S (S^T \nabla) u(x).$$

Lemma is proved.

Corollary 1 *If the function $u(x)$ is biharmonic in Ω, then the function $u(x^*) = I_S u(x)$ is also biharmonic in Ω.*

Indeed, due to Lemma 1, $\Delta^2 u(x) = 0 \Rightarrow \Delta^2 I_S u(x) = I_S \Delta^2 u(x) = 0$.

Lemma 2 *The operator $1 + \alpha I_S$, when $\alpha \neq \pm 1$ is invertible and the operator*

$$J_\alpha = \frac{1}{1 - \alpha^2} (1 - \alpha I_S) \tag{5}$$

is inverse to $1 + \alpha I_S$.

Proof It is easy to see that

$$J_\alpha (1 + \alpha I_S) u(x) = \left(\frac{1 - \alpha I_S}{1 - \alpha^2} \right) (1 + \alpha I_S)\, u(x) = \frac{1}{1 - \alpha^2} (u(x) - \alpha^2 I_S{}^2 u(x))$$

$$= \frac{1}{1 - \alpha^2} (1 - \alpha^2) u(x).$$

Thus, if $\alpha^2 \neq 1$ then we can divide both sides of the equality by $1 - \alpha^2$ and hence the operator J_α is inverse to $1 + \alpha I_S$. Lemma is proved.

3 Neumann Type Problems

In this section we study the following problem

$$\Delta^2 v(x) = \varphi(x), \quad x \in \Omega, \tag{6}$$

$$D^m_\nu v(x)|_{\partial\Omega} = \psi_1(x), \quad x \in \partial\Omega, \tag{7}$$

$$D^{m+1}_\nu v(x)|_{\partial\Omega} = \psi_2(x), \quad x \in \partial\Omega, \tag{8}$$

where $0 \leq m \leq 2$.

Problem (6)–(8) in the case $m = 0$ is the Dirichlet problem, and in the cases $m = 1$ or $m = 2$ is a problem of Neumann type. It is known (see, e.g. [1]) that the Dirichlet problem is unconditionally solvable, but for the solvability of the Neumann type problems it is necessary to fulfill so called "orthogonality conditions" for the given functions [5].

Bitsadze A. V. in his work [5] established that for the solvability of the problem (6)–(8) in the case $\varphi(x) = 0$ and $m = 1$ the following condition is necessary and sufficient to be fulfilled

$$\int\limits_{\partial\Omega} [\psi_2(x) - \psi_1(x)]\, dS_x = 0.$$

Furthermore, in [19] this result was improved for the inhomogeneous equation. The following proposition is true.

Theorem 1 *Let $m = 1$, $\varphi(x) \in C(\bar{\Omega})$, $\psi_1(x) \in C^1(\partial\Omega)$, $\psi_2(x) \in C(\partial\Omega)$. Then for the solvability of the problem (6)–(8) the following condition is necessary and sufficient*

$$\int\limits_{\partial\Omega} [\psi_2(x) - \psi_1(x)]\, dS_x = \frac{1}{2} \int\limits_{\partial\Omega} \left(1 - |x|^2\right) \psi(x)\, dx. \tag{9}$$

If solution of the problem exists, then it is unique up to a constant.

The case $m = 2$ is considered in [30] and the following proposition is established.

Theorem 2 *Let $m = 2$, $0 < \lambda < 1$, $n \geq 3$, $\varphi(x) \in C^{\lambda+1}(\bar{\Omega})$, $\psi_1(x) \in C^{\lambda+1}(\partial\Omega)$, $\psi_2(x) \in C^{\lambda}(\partial\Omega)$. Then for the solvability of the problem (6)–(8) the following conditions are necessary and sufficient*

$$\int\limits_{\partial\Omega} \psi_2(x)\, dS_x = \frac{n-1}{2} \int\limits_{\Omega} |x|^2 \varphi(x)\, dx - \frac{n-3}{2} \int\limits_{\Omega} \varphi(x)\, dx, \tag{10}$$

$$\int\limits_{\partial\Omega} x_j [\psi_2(x) - \psi_1(x)]\, dS_x = \frac{n-1}{2} \int\limits_{\Omega} x_j |x|^2 \varphi(x) dx - \frac{n-3}{2} \int\limits_{\Omega} x_j \varphi(x)\, dx. \tag{11}$$

If solution of the problem exists, then it is unique up to a first order polynomial.

Note that the case when the boundary conditions of the problem have the form $D_\nu v(x)|_{\partial\Omega} = \psi_1(x)$ and $D_\nu^3 v(x)|_{\partial\Omega} = \psi_2(x)$, where $x \in \partial\Omega$ is considered in [15].

4 Uniqueness

In this section we investigate the uniqueness of the solution of the problem (1)–(3). The following proposition is true.

Theorem 3 *Let $\alpha^2 \neq 1$, $\beta^2 \neq 1$, and solution of the problem (1)–(3) exists. Then*
(1) if $m = 0$, then solution of the problem is unique;
(2) if $m = 1$, then solution of the problem is unique up to a constant;
(3) if $m = 2$, then solution of the homogeneous problem (1)–(3) is a function of the form

$$u(x) = c_0 + \sum_{j=1}^{n} c_j x_j,$$

where c_j for $j = 0, \dots, n$ are arbitrary constants.

Proof To prove the uniqueness of the solution of problem (1)–(3), consider a function $u(x)$—a solution of the homogeneous problem (1)–(3) (all right-hand sides in the problem are zero). If the problem (1)–(3) has at least two solutions such a function exists. It is clear that $u(x)$ is a biharmonic function, satisfying the following homogeneous conditions

$$D_\nu^m u(x) + \alpha D_\nu^m u(x^*)\Big|_{\partial\Omega} = \Lambda^{[m]}(1 + \alpha I_S)u(x)\Big|_{\partial\Omega} = 0,\ D_\nu^{m+1} u(x) + \beta D_\nu^{m+1} u(x^*)\Big|_{\partial\Omega}$$
$$= \Lambda^{[m+1]}(1 + \beta I_S)u(x)\Big|_{\partial\Omega} = 0. \tag{12}$$

Since $\alpha^2 \neq 1$ and $\beta^2 \neq 1$, then applying to the equality (12) the operators J_α and J_β from (5) and using Lemma 1, we get

$$0 = J_\alpha \Lambda^{[m]}(1 + \alpha I_S)u(x) = \Lambda^{[m]} J_\alpha(1 + \alpha I_S)u(x) = \Lambda^{[m]}u(x) = D_\nu^m u(x),$$

$$0 = J_\beta \Lambda^{[m+1]}(1 + \alpha I_S)u(x) = \Lambda^{[m+1]} J_\beta(1 + \beta I_S)u(x) = \Lambda^{[m+1]}u(x) = D_\nu^{m+1} u(x),$$

where $x \in \partial\Omega$, or

$$D_\nu^m u(x)\big|_{\partial\Omega} = D_\nu^{m+1} u(x)\big|_{\partial\Omega} = 0.$$

Therefore, if $u(x)$ is a solution of the homogenous problem (1)–(3), then it is also a solution of the homogeneous problem (6)–(8). Then, due to uniqueness of the solution of the Dirichlet problem (the case $m = 0$), we obtain the uniqueness of the solution of the problem (1)–(3). Similarly, by the statements of Theorems 1 and 2, we obtain the remaining statements of this theorem. Theorem is proved.

Remark 1 If $\alpha^2 = 1$ and $\beta^2 = 1$, then the homogeneous problem (1)–(3) can have an infinite number of linearly independent solutions. For example, suppose that $m =$

$0, \alpha = \beta = 1$, $S = -E$. Consider the function $u_k(x) = |x|^2 H_{2k-1}(x)$, where $k \in N$ and $H_{2k-1}(x)$ is a homogeneous harmonic polynomial of degree $2k - 1$. It is obvious that $\Delta^2 u_k(x) = 0$, $x \in \Omega$. Further, since $u_k(x^*) = (-1)^{2k-1} u_k(x) = -u_k(x)$ and $\Lambda u_k(x) = (2k+1) u_k(x)$, then the function $u_k(x)$ for all $k \in N$ satisfies the homogeneous conditions of the problem (1)–(3)

$$u_k(x) + u_k(x^*) = u_k(x) - u_k(x) = 0,$$

$$\Lambda u_k(x) + \Lambda u_k(x^*) = \Lambda u_k(x) + (\Lambda u_k)(x^*) = (2k+1)(u_k(x) + u_k(x^*)) = 0.$$

5 Existence

In this section we present a statement on the existence of a solution of the problem (1)–(3).

Theorem 4 *Let $\alpha \neq \pm 1$, $\beta \neq \pm 1$ and $f(x)$, $g_1(x)$, $g_2(x)$ be smooth enough functions. Then*

(1) if $m = 0$, then solution of the problem (1)–(3) exists and is unique;

(2) if $m = 1$, and $\alpha \neq -1$, $\beta \neq -1$, then the necessary and sufficient condition for solvability of the problem (1)–(3) has the form

$$\frac{1}{2} \int\limits_{\Omega} (1 - |x|^2) f(x)\, dx = \int\limits_{\partial\Omega} \left[\frac{g_2(x)}{1+b} - \frac{g_1(x)}{1+a} \right] dS_x. \tag{13}$$

If the solution exists, then it is unique up to a constant;

(3) if $m = 2$, and $\alpha \neq -1$, $\beta \neq -1$, then the necessary and sufficient condition for solvability of the problem (1)–(3) has the form

$$\frac{1}{1+b} \int\limits_{\partial\Omega} g_2(x)\, dS_x = \frac{n-1}{2} \int\limits_{\Omega} |x|^2 f(x)\, dx - \frac{n-3}{2} \int\limits_{\Omega} f(x)\, dx, \tag{14}$$

and

$$\int\limits_{\partial\Omega} \left(\left(E + \beta S^T\right)^{-1} x \right)_j g_2(x) - \left(\left(E + \alpha S^T\right)^{-1} x \right)_j g_1(x)\, dS_y$$

$$= \frac{n-1}{2} \int\limits_{\Omega} x_j |x|^2 f(x)\, dx - \frac{n-3}{2} \int\limits_{\Omega} x_j f(x)\, dx,\ j = 1, \ldots, n, \tag{15}$$

where $(x)_j = x_j$ is the jth element of a vector x. If a solution exists, then it is unique up to a first order polynomial.

Proof Consider the auxiliary Dirichlet problem

$$\Delta^2 v(x) = f(x), \quad x \in \Omega, \tag{16}$$

$$D_\nu^m v(x) = J_\alpha g_1(x), \quad x \in \partial\Omega, \tag{17}$$

$$D_\nu^{m+1} v(x) = J_\beta g_2(x), \quad x \in \partial\Omega, \tag{18}$$

where the operator J_α is defined in (5). We check that its solution $v(x)$ is also a solution of the considered problem (1)–(3). Indeed, the function $v(x)$ satisfies the Eq. (1). Applying the operator $1 + \alpha I_S$ to the condition (17) and using Lemmas 1 and 2, we get

$$g_1(x) = (1 + \alpha I_S) J_\alpha g_1(x) = (1 + \alpha I_S) D_\nu^m v(x)|_{\partial\Omega} = (1 + \alpha I_S) \Lambda^{[m]} v(x)|_{\partial\Omega}$$

$$= \Lambda^{[m]} (1 + \alpha I_S) v(x)|_{\partial\Omega} = D_\nu^m (1 + \alpha I_S) v(x)|_{\partial\Omega} = D_\nu^m v(x) + \alpha D_\nu^m v(x^*)|_{\partial\Omega},$$

where $x \in \partial\Omega$, i.e. condition (2) holds. Similarly, applying the operator $1 + \beta I_S$ to the condition (18) and using Lemmas 1 and 2, we get

$$g_2(x) = (1 + \beta I_S) J_\beta g_2(x) = (1 + \beta I_S) D_\nu^{m+1} v(x)|_{\partial\Omega} = (1 + \beta I_S) \Lambda^{[m+1]} v(x)|_{\partial\Omega}$$

$$= \Lambda^{[m+1]} (1 + \beta I_S) v(x)|_{\partial\Omega} = D_\nu^{m+1} (1 + \beta I_S) v(x)|_{\partial\Omega} = D_\nu^{m+1} v(x) + \beta D_\nu^{m+1} v(x^*)|_{\partial\Omega},$$

where $x \in \partial\Omega$, i.e. condition (3) holds also. So, the function $v(x)$ is a solution of the problem (1)–(3), and if $v(x)$ exists, then the problem (1)–(3) is solvable.

The case when the solution of the problem (16)–(18) does not exist, but $u(x)$ exists, is impossible. Indeed, let $u(x)$ be a solution of the Eq. (16). Applying the operator J_α to the condition (2) and using Lemmas 1 and 2, we have

$$J_\alpha g_1(x) = J_\alpha (D_\nu^m u(x) + \alpha D_\nu^m u(x^*))|_{\partial\Omega} = J_\alpha D_\nu^m (1 + \alpha I_S) u(x)|_{\partial\Omega}$$

$$= J_\alpha \Lambda^{[m]} (1 + \alpha I_S) u(x)|_{\partial\Omega} = \Lambda^{[m]} J_\alpha (1 + \alpha I_S) u(x)|_{\partial\Omega} = \Lambda^{[m]} u(x) = D_\nu^m u(x)|_{\partial\Omega},$$

where $x \in \partial\Omega$, i.e. condition (17) holds. Similarly, from (3) we get

$$J_\beta g_2(x) = J_\beta (D_\nu^{m+1} u(x) + \beta D_\nu^{m+1} u(x^*))|_{\partial\Omega} = J_\beta D_\nu^{m+1} (1 + \beta I_S) u(x)|_{\partial\Omega}$$

$$= J_\beta \Lambda^{[m+1]} (1 + \beta I_S) u(x)|_{\partial\Omega} = \Lambda^{[m+1]} J_\beta (1 + \beta I_S) u(x)|_{\partial\Omega} = \Lambda^{[m+1]} u(x)|_{\partial\Omega}$$

$$= D_\nu^{m+1} u(x)|_{\partial\Omega},$$

i.e. condition (18) holds also. Hence, $u(x)$ is a solution of the problem (16)–(18), which contradicts to the assumption. Problems (1)–(3) and (16)–(18) are solvable simultaneously. Smoothness of the functions $J_\alpha g_1(x)$ and $g_1(x)$ as well as $J_\beta g_2(x)$ and $g_2(x)$ are the same.

Using Theorems 1 and 2, we can find solvability conditions of the problem (16)–(18). Obviously these conditions will be the solvability conditions of the problem (1)–(3).

(1) Let $m = 0$. Then the problem (16)–(18) is the Dirichlet problem. For any functions on the right-hand sides of the problem with a given smoothness its solution exists and is unique.

(2) Let $m = 1$. In this case, by Theorem 1, the necessary and sufficient solvability condition of the problem (16)–18) is the integral equality

$$\frac{1}{2}\int\limits_{\Omega}(1-|x|^2)f(x)\,dx = \int\limits_{\partial\Omega}\left[J_\beta g_2(x) - J_\alpha g_1(x)\right]dS_x. \qquad (19)$$

Let us transform the integral on the right hand side of (19).

Lemma 3 *Let the function $\varphi(x)$ be continuous on $\partial\Omega$ and S be an orthogonal matrix, then*

$$\int\limits_{\partial\Omega}\varphi(Sx)\,dS_x = \int\limits_{\partial\Omega}\varphi(x)\,dS_x.$$

Proof Let the function $w(x)$ be a solution of the Dirichlet problem for the Laplace equation in Ω with condition $w(x)|_{\partial\Omega} = \varphi(x)$, $x \in \partial\Omega$. Then the function $w(Sx)$ is a solution of the Dirichlet problem for the Laplace equation in Ω with the condition $w(Sx)|_{\partial\Omega} = \varphi(Sx)$, $x \in \partial\Omega$. Therefore, due to the Poisson's formula, we have

$$\int\limits_{\partial\Omega}\varphi(Sx)\,dS_x = \int\limits_{\partial\Omega}w(Sx)\,dS_x = \omega_n w(0) = \int\limits_{\partial\Omega}\varphi(x)\,dS_x,$$

where ω_n is the area of the unit sphere. Lemma is proved.

Using the proved Lemma 3, the condition $\alpha \neq -1$, we find

$$\int\limits_{\partial\Omega}J_\alpha g_1(x)\,dS_x = \frac{1}{1-\alpha^2}\left(\int\limits_{\partial\Omega}g_1(x)\,dS_x - \alpha\int\limits_{\partial\Omega}I_S g_1(x)\,dS_x\right)$$

$$= \frac{1}{1-\alpha^2}\left[\int\limits_{\partial\Omega}g_1(x)\,dS_x - \alpha\int\limits_{\partial\Omega}g_1(Sx)\,dS_x\right] = \int\limits_{\partial\Omega}\frac{g_1(x)}{1+\alpha}\,dS_x. \qquad (20)$$

This implies that condition (19) can be transformed to the form (13)

$$\frac{1}{2}\int_{\Omega}(1-|x|^2)f(x)\,dx = \int_{\partial\Omega}\left[\frac{g_2(x)}{1+\beta}-\frac{g_1(x)}{1+\alpha}\right]dS_x.$$

(3) Let $m = 2$. By Theorem 2 the necessary and sufficient solvability conditions of the problem (16)–(18) take the form

$$\int_{\partial\Omega} J_\beta g_2(x)\,dS_x = \frac{n-1}{2}\int_{\Omega}|x|^2 f(x)\,dx - \frac{n-3}{2}\int_{\Omega} f(x)\,dx,$$

$$\int_{\partial\Omega} x_j[J_\beta g_2(x) - J_\alpha g_1(x)]\,dS_x = \frac{n-1}{2}\int_{\Omega} x_j|x|^2 f(x)dx - \frac{n-3}{2}\int_{\Omega} x_j f(x)dx,$$

where $j = 1, \ldots, n$. Using (20) the first condition can be easily transformed to the form (14). Further, taking into account orthogonality of powers of the matrices S, S^T, and using Lemma 3 we can write

$$\int_{\partial\Omega} x_j J_\alpha g_1(x)\,dS_x = \frac{1}{1-\alpha^2}\left(\int_{\partial\Omega} x_j g_1(x)\,dS_x - \alpha\int_{\partial\Omega} S^T\,(Sx)_j\,g_1(Sx)\,dS_x\right)$$

$$= \frac{1}{1-\alpha^2}\left(\int_{\partial\Omega} y_j g_1(y)\,dS_y - \alpha\int_{\partial\Omega}\left(S^T y\right)_j g_1(y)\,dS_y\right)$$

$$= \frac{1}{1-\alpha^2}\int_{\partial\Omega}\left(\sum_{k=0}^{1}\left(-\alpha S^T\right)^k y\right)_j g_1(y)\,dS_y = \frac{1-\alpha^2}{1-\alpha^2}\int_{\partial\Omega}\left(E+\alpha S^T\right)^{-1} g_1(y)\,dS_y$$

$$= \int_{\partial\Omega}\left(\left(E+\alpha S^T\right)^{-1} y\right)_j g_1(y)\,dS_y,$$

where $j = 1, \ldots, n$. Thus, the second condition can be rewritten in the form (15)

$$\int_{\partial\Omega}\left(\left(E+\beta S^T\right)^{-1} x\right)_j g_2(x) - \left(\left(E+\alpha S^T\right)^{-1} x\right)_j g_1(x)\,dS_y$$

$$= \frac{n-1}{2}\int_{\Omega} x_j|x|^2 f(x)\,dx - \frac{n-3}{2}\int_{\Omega} x_j f(x)\,dx,$$

for $j = 1, \ldots, n$. Theorem is proved.

Remark 2 If we consider the case of involution $S = -E$, then the solvability condition given by (15) can be simplified using the equality

$$\left(E + \alpha S^T\right)^{-1} = (E - \alpha E)^{-1} = \frac{1}{1-\alpha} E.$$

6 Representation of the Solution

In this section we give a method of constructing solutions of the problem (1)–(3) with homogeneous boundary conditions.

Theorem 5 *Let* $0 \le m \le 2$, $g_1(x) = g_2(x) = 0$. *Then*

(1) if $m = 0$, *then the solution of the problem (1)–(3) can be represented in the form*

$$u(x) = \int\limits_{\Omega} G_D(x, y) f(y)\, dy,$$

where $G_D(x, y)$ *is the Green's function of the Dirichlet problem for the biharmonic equation (1) in* Ω.

(2) if $m = 1$ *and (13) holds, then solution of the problem (1)–(3) can be represented in the form*

$$u(x) = \int\limits_0^1 \frac{v(sx)}{s}\, ds + C, \tag{21}$$

where C *is arbitrary constant and* $v(x)$ *is a solution of the following the Dirichlet problem*

$$\Delta^2 v(x) = (\Lambda + 4)\, f(x),\ x \in \Omega;\ v(x)|_{\partial\Omega} = 0,\ v(0) = 0,\ D_\nu^1 v(x)\big|_{\partial\Omega} = 0. \tag{22}$$

(3) if $m = 2$ *and (14), (15) hold, then solution of the problem (1)–(3) can be represented in the form*

$$u(x) = \int\limits_0^1 (1-s)\frac{v(sx)}{s^2}\, ds + \sum_{j=1}^{n} c_j x_j + c_0, \tag{23}$$

where c_j, $j = 0, \ldots, n$ *are arbitrary constants and* $v(x)$ *is a solution of the following Dirichlet problem*

$$\Delta^2 v(x) = (\Lambda + 4)\, (\Lambda + 3)\, f(x),\ x \in \Omega,\ v(x)|_{\partial\Omega} = 0,\ v(0) = 0,$$

$$D_\nu^1 v(x)\big|_{\partial\Omega} = 0,\ \frac{\partial v}{\partial x_j}(0) = 0,\ j = 1, \ldots, n. \tag{24}$$

Proof The auxiliary problem (16)–(18), whose solution coincides with the solution of the problem (1)–(3) (see the proof of Theorem 3), with the help of properties of the operator Λ takes the form

$$\Delta^2 v(x) = f(x),\ x \in \Omega,$$

$$\Lambda^{[m]} v(x)|_{\partial\Omega} = 0,\ \Lambda^{[m+1]} v(x)|_{\partial\Omega} = 0.$$

(1) Let $m = 0$, then in this case auxiliary problem is the Dirichlet problem and its solution coincides with a solution of the problem (1)–(3)

$$v(x) = u(x) = \int_\Omega G_D(x, y) f(y)\, dy. \tag{25}$$

(2) Let $m = 1$. Boundary conditions for the auxiliary problem take the form

$$\Lambda^{[1]} v(x)|_{\partial\Omega} \equiv \Lambda v(x)|_{\partial\Omega} = 0,\ \Lambda^{[2]} v(x)|_{\partial\Omega} \equiv (\Lambda^2 - \Lambda) v(x)|_{\partial\Omega} = \Lambda^2 v(x)|_{\partial\Omega} = 0.$$

Let us apply the operator $\Lambda + 4$ to the biharmonic equation of the problem. Due to the equality $\Delta^k \Lambda u = (\Lambda + 2k)\Delta^k u$ (see [13]) and denoting $w = \Lambda v$, for $w(x)$ we get the following Dirichlet problem (22)

$$\Delta^2 w(x) = (\Lambda + 4) f(x),\ x \in \Omega,$$

$$w(x)|_{\partial\Omega} = 0,\ \Lambda w(x)|_{\partial\Omega} = 0.$$

By the formula (25) we find

$$w(x) = \int_\Omega G_D(x, y)(\Lambda + 4) f(y)\, dy.$$

As in [13] equation $w = \Lambda v$ in the class of smooth functions $v(x)$ has a solution only if $w(0) = 0$, and this solution can be written in the form

$$u(x) = \int_0^1 \frac{w(sx)}{s}\, ds + C.$$

(3) Let $m = 2$. Boundary conditions for the auxiliary problem take the form

$$\Lambda^{[1]}v(x)|_{\partial\Omega} \equiv (\Lambda^2 - \Lambda)v(x)|_{\partial\Omega} = 0,\ \Lambda^{[2]}v(x)|_{\partial\Omega} \equiv (\Lambda^2 - \Lambda)(\Lambda - 2)v(x)|_{\partial\Omega}$$

$$= \Lambda(\Lambda^2 - \Lambda)v(x)|_{\partial\Omega} = 0.$$

Let us apply the operator $(\Lambda + 4)(\Lambda + 3)$ to the biharmonic equation of the problem. If we denote $w = (\Lambda^2 - \Lambda)v$, then due to the equality $\Delta^k \Lambda u = (\Lambda + 2k)\Delta^k u$ we get the Dirichlet problem (24)

$$\Delta^2 w(x) = (\Lambda + 4)(\Lambda + 3) f(x),\ x \in \Omega,$$

$$w(x)|_{\partial\Omega} = 0,\ \Lambda w(x)|_{\partial\Omega} = 0.$$

By the formula (25) we find

$$w(x) = \int_{\Omega} G_D(x, y)(\Lambda + 4)(\Lambda + 3) f(y)\, dy.$$

To find the function $v(x) = u(x)$ it is necessary to solve the equation $w = (\Lambda^2 - \Lambda)v$. As in [17] equation $\hat{w} = (\Lambda - \lambda)\hat{v}$ in the class of smooth functions $\hat{v}(x)$ has a solution only if $\lim\limits_{s\to 0} \frac{\hat{w}(sx)}{s^\lambda} = 0$, where $x \in \Omega$, and this solution can be written in the form

$$\hat{v}(x) = \int_0^1 \frac{\hat{w}(sx)}{s^{\lambda+1}}\, ds + H_\lambda(x),$$

where $H_\lambda(x)$ is arbitrary homogenous polynomial of order $\lambda \in N_0$. Using this result a smooth solution of the equation $w = (\Lambda^2 - \Lambda)u$ can be written in the form

$$u(x) = \int_0^1 \left(\int_0^1 \frac{w(stx)}{s^2}\, ds + H_1(tx) \right) \frac{dt}{t} + C = \int_0^1 \left(\int_0^t \frac{w(\tau x)}{\tau^2}\, d\tau \right) dt$$

$$+ H_1(x) + C - \int_0^1 \frac{w(\tau x)}{\tau^2} \int_\tau^1 dt\, d\tau + H_1(x) + C = \int_0^1 \frac{w(\tau x)}{\tau^2}(1 - \tau)\, d\tau + H_1(x) + C,$$

which coincides with (23). Conditions, under which this solution exists, are equivalent to the existence conditions of the last singular integral. It is easy to see that the integral converges if $w(0) = 0$ and $\frac{\partial w}{\partial x_j}(0) = 0$ for $j = 1, 2, \ldots, n$. Theorem is proved.

Example 1 Consider the following boundary value problem of the type (1)–(3)

$$\Delta^2 u(x) = x_i,\ x \in \Omega,$$

$$u(x) + 2u(x^*)|_{\partial\Omega} = 3x_j^2, \quad D_\nu^1 u(x) - 2D_\nu^1 u(x^*)|_{\partial\Omega} = -x_k^2, \ x \in \partial\Omega,$$

where $x^* = -Ex$ and $1 \le i, j, k \le n$. In this case $\alpha = 2$, $\beta = -2$ and because

$$J_\alpha = J_2 = -\frac{1}{3}(E - 2I_{-E}), \ J_\beta = J_{-2} = -\frac{1}{3}(E + 2I_{-E})$$

we have

$$J_2(3x_j^2) = -(1-2)x_j^2 = x_j^2$$

and

$$J_{-2}(-x_k^2) = \frac{1}{3}(1+2)x_k^2 = x_k^2.$$

The auxiliary problem (16)–(18) takes the form

$$\Delta^2 u(x) = x_i, \ x \in \Omega,$$

$$u(x)|_{\partial\Omega} = x_j^2, \quad D_\nu^1 u(x)|_{\partial\Omega} = x_k^2, \ x \in \partial\Omega.$$

Solution of this problem and also of the considered problem using Examples 2–4 of [16] has the form

$$u(x) = x_j^2 + (|x|^2 - 1)\Big(\frac{x_k^2}{2} - x_j^2\Big) + (|x|^2 - 1)^2\Big(\frac{x_i}{8(n+2)(n+4)} + \frac{1}{2n}\Big).$$

Boundary conditions and equation (see [16]) are fulfilled

$$u(x) + 2u(x^*)|_{\partial\Omega} = x_j^2 + 2(-x_j)^2 = 3x_j^2, \quad \Lambda u(x) - 2\Lambda$$

$$u(x^*)|_{\partial\Omega} = -\frac{1}{2}\Lambda(|x|^2 - 1)x_k^2|_{\partial\Omega} = -x_k^2.$$

Acknowledgements The work was supported by Act 211 of the Government of the Russian Federation, contract no.02.A03.21.0011, and by a grant from the Ministry of Science and Education of the Republic of Kazakhstan (grant no. AP05131268).

References

1. Agmon, S., Douglis, A., Nirenberg, L.: Estimates near the boundary for solutions of elliptic partial differential equations satisfying general boundary conditions. I. Commun. Pure Appl. Math. **12**, 623–727 (1959)
2. Andersson, L.-E., Elfving, T., Golub, G.H.: Solution of biharmonic equations with application to radar imaging. J. Comput. Appl. Math. **94**, 153–180 (1998)
3. Begerh, H., Vu, T.N.H., Zhang, Z.X.: Polyharmonic Dirichlet Problems. Proc. Steklov Inst. Math. **255**, 13–34 (2006)

4. Bitsadze, A.V.: On a class of conditionally solvable nonlocal boundary-value problems for harmonic functions. Sov. Phys. Doklad **280**, 521–524 (1985)
5. Bitsadze, A.V.: Some properties of polyharmonic functions. Differ. Equ. **24**, 825–831 (1988)
6. Bitsadze, A.V., Samarskii, A.A.: Some elementary generalizations of linear elliptic boundary value problems. Dokl. Akad. Nauk SSSR **185**, 739–740 (1969). (Russian)
7. Boggio, T.: Sulle funzioni di Green d'ordine m. Rendiconti del Circolo Matematico di Palermo. **20**, 97–135 (1905)
8. Criado, F., Criado, F.J., Odishelidze, N.: On the solution of some non-local problems. Czechoslov. Math. J. **54**, 487–498 (2004)
9. Ehrlich, L.N., Gupta, M.M.: Some difference schemes for the biharmonic equation. SIAM J. Numer. Anal. **12**, 773–790 (1975)
10. Kadirkulov, B.J., Kirane, M.: On solvability of a boundary value problem for the Poisson equation with a nonlocal boundary operator. Acta Math. Sci. **35**, 970–980 (2015)
11. Karachik, V.V.: Normalized system of functions with respect to the Laplace operator and its applications. J. Math. Anal. Appl. **287**, 577–592 (2003)
12. Karachik, V.V.: Solvability conditions for the Neumann problem for the Homogeneous Polyharmonic equation. Differ. Equ. **50**, 1449–1456 (2014)
13. Karachik, V.V.: On solvability conditions for the Neumann problem for a Polyharmonic equation in the unit ball. J. Appl. Ind. Math. **8**, 63–75 (2014)
14. Karachik, V.V.: Construction of polynomial solutions to some boundary value problems for Poisson's equation. Comput. Math. Math. Phys. **51**, 1567–1587 (2011)
15. Karachik, V.V.: A Neumann-type problem for the biharmonic equation. Sib. Adv. Math. **27**, 103–118 (2017)
16. Karachik, V.V., Antropova, N.A.: Polynomial solutions of the Dirichlet problem for the biharmonic equation in the ball. Differ. Equ. **49**, 251–256 (2013)
17. Karachik, V.V., Torebek, B.T.: On the Dirichlet-Riquier problem for Biharmonic equations. Math. Notes **102**, 31–42 (2017)
18. Karachik, V.V., Turmetov, BKh: On solvability of some Neumann-type boundary value problems for biharmonic equation. Electr. J. Differ. Equ. **2017**, 1–17 (2017)
19. Karachik, V.V., Turmetov, BKh, Bekaeva, A.E.: Solvability conditions of the biharmonic equation in the unit ball. Int. J. Pure Appl. Math. **81**, 487–495 (2012)
20. Kirane, M., Torebek, B.T.: On a nonlocal problem for the Laplace equation in the unit ball with fractional boundary conditions. Math. Method Appl. Sci. **39**, 1121–1128 (2016)
21. Kishkis, K.Y.: On some nonlocal problem for harmonic functions in multiply connected domain. Differ. Equ. **23**, 174–177 (1987)
22. Koshanova, M.D., Turmetov, BKh, Usmanov, K.I.: About solvability of some boundary value problems for Poisson equation with Hadamard type boundary operator. Electr. J. Differ. Equ. **2016**, 1–12 (2016)
23. Lai, M.-C., Liu, H.-C.: Fast direct solver for the biharmonic equation on a disk and its application to incompressible flows. Appl. Math. Comput. **164**, 679–695 (2005)
24. Love, A.E.H.: Biharmonic analysis, especially in a rectangle, and its application to the theory of elasticity. J. Lond. Math. Soc. **3**, 144–156 (1928)
25. Muratbekova, M.A., Shinaliyev, K.M., Turmetov, BKh: On solvability of a nonlocal problem for the Laplace equation with the fractional-order boundary operator. Bound. Value Probl. **2014**, 1–13 (2014)
26. Sadybekov, M.A., Turmetov, BKh: On analogues of periodic boundary value problems for the Laplace operator in ball. Eurasian Math. J. **3**, 143–146 (2012)
27. Sadybekov, M.A., Turmetov, BKh: On an analog of periodic boundary value problems for the Poisson equation in the disk. Differ. Equ. **50**, 268–273 (2014)
28. Skubachevskii, A.L.: Nonclassical boundary value problems I. J. Math. Sci. **155**, 199–334 (2008)
29. Skubachevskii, A.L.: Nonclassical boundary value problems II. J. Math. Sci. **166**, 377–561 (2010)

30. Turmetov, BKh, Ashurov, R.R.: On Solvability of the Neumann Boundary Value Problem for Non-homogeneous Biharmonic Equation. Br. J. Math. & Comput. Sci. **4**, 557–571 (2014)
31. Turmetov, BKh, Ashurov, R.R.: On solvability of the Neumann boundary value problem for a non-homogeneous polyharmonic equation in a ball. Bound. Value Probl. **2013**, 1–15 (2013)
32. Turmetov, BKh, Karachik, V.V.: On solvability of some boundary value problems for a biharmonic equation with periodic conditions. Filomat. **32**, 947–953 (2018)
33. Zaremba, S.: Sur l'integration de l'equation biharmonique. Bulletin international de l'Academie des sciences de Cracovie. 1–29 (1908)

Lyapunov-Type Inequality for Fractional Sub-Laplacians

Aidyn Kassymov and Durvudkhan Suragan

Abstract In the present paper, we prove the Lyapunov-type inequality for the fractional sub-Laplacian on the homogeneous Lie groups. We give some consequences of the obtained inequality. In addition, we also show the fractional L^2-Hardy inequality for the fractional sub-Laplacian on the half-space.

Keywords Lyapunov-type inequality · Sub-Laplacian · Homogeneous Lie group

1 Introduction

1.1 Homogenous Lie Group

A homogeneous group is a connected simply connected Lie group whose Lie algebra is equipped with dilations. For simplicity, we use the notation $\lambda x,\ \lambda > 0,\ x \in G$, for the dilation on G.

For any homogeneous group G there exists a homogeneous quasi-norm, which is a continuous non-negative function

$$G \ni x \mapsto q(x) \in [0, \infty), \tag{1}$$

with the properties $q(x) = q(x^{-1})$ for all $x \in G$, $q(\lambda x) = \lambda q(x)$ for all $x \in G$ and $\lambda > 0$, with $q(x) = 0$ if and only if $x = 0$.

A. Kassymov
Institute of Mathematics and Mathematical Modeling, 125 Pushkin str., Almaty, Kazakhstan
e-mail: kassymov@math.kz

Al-Farabi Kazakh National University, 71 Al-Farabi avenue, 050040 Almaty, Kazakhstan

D. Suragan (✉)
Department of Mathematics, School of Science and Technology, Nazarbayev University, 53 Kabanbay Batyr Ave, Nur-Sultan 010000, Kazakhstan
e-mail: durvudkhan.suragan@nu.edu.kz

A. Ashyralyev et al. (eds.), *Functional Analysis in Interdisciplinary Applications—II*, Springer Proceedings in Mathematics & Statistics 351,
https://doi.org/10.1007/978-3-030-69292-6_6

Note that for the simplicity of notation, here and after we assume that the origin 0 of R^N is the identity of G. This assumption is not restrictive due to properties of isomorphic Lie groups. We also use the following well-known polar decomposition on homogeneous Lie groups (see, e.g. [1, Sect. 3.1.7]): there is a (unique) positive Borel measure σ on the unit quasi-sphere $S := \{x \in G : q(x) = 1\}$, so that for every $f \in L^1(G)$ we have

$$\int_G f(x)dx = \int_0^\infty \int_S f(ry)r^{Q-1}d\sigma(y)dr. \tag{2}$$

Let $s \in (0, 1)$ and G be a homogeneous Lie group with homogeneous dimension Q and the group operation $\circ$. We define the fractional sub-Laplacian $(-\Delta_q)^s$ on G by the formula

$$(-\Delta_q)^s u(x) := 2 \lim_{\delta \searrow 0} \int_{G \setminus B_q(x,\delta)} \frac{u(x) - u(y)}{q^{Q+2s}(y^{-1} \circ x)} dy, \quad x \in G, \tag{3}$$

where q is a quasi-norm on G and $B_q(x, \delta)$ is a quasi-ball with respect to q, with radius δ centered at $x \in G$.

For a measurable function $u : G \to R$ we define the Gagliardo seminorm by

$$[u]_{s,q} = \left(\int_G \int_G \frac{|u(x) - u(y)|^2}{q^{Q+2s}(y^{-1} \circ x)} dxdy \right)^{1/2}. \tag{4}$$

Now we recall the definition of the fractional Sobolev spaces on homogeneous Lie groups denoted by H_q^s. The functional space

$$H_q^s = \{u \in L^2(G) : u \ \ is \ \ measurable, \ [u]_{s,q} < +\infty\}, \ s \in (0, 1), \tag{5}$$

endowed with the norm

$$\|u\|_{H_q^s} = (\|u\|^2_{L^2(G)} + [u]^2_{s,q})^{\frac{1}{2}}, \quad u \in H_q^s, \tag{6}$$

is called the fractional Sobolev spaces on G.

Let $\Omega \subset G$ is a Haar measurable set, we define the Sobolev space by

$$H^s_{q,\Omega} = \left\{ u \in L^2(\Omega) : u \ is \ measurable, \int_\Omega \int_\Omega \frac{|u(x) - u(y)|^2}{q^{Q+2s}(y^{-1} \circ x)} dxdy < \infty \right\}, \tag{7}$$

endowed with norm

$$\|u\|_{H^s_{q,\Omega}} = \left(\|u\|^2_{L^2(\Omega)} + \int_\Omega \int_\Omega \frac{|u(x) - u(y)|^2}{q^{Q+2s}(y^{-1} \circ x)} dxdy \right)^{\frac{1}{2}}, \quad u \in H^s_{q,\Omega}. \tag{8}$$

Let define Sobolev space $H_q^{s,0}(\Omega)$ as the completion of $C_0^\infty(\Omega)$ with respect to the norm $\|u\|_{H_q^s(\Omega)}$.

In the paper [2] the fractional Hardy inequality for the (standard) fractional Laplacian was proved in the form:

$$C\int_{R^N}\frac{|u(x)|^p}{|x|^{ps}}dx \leq \int_{R^N}\int_{R^N}\frac{|u(x)-u(y)|^p}{|x-y|^{N+ps}}dxdy, \tag{9}$$

where C is a positive constant (independent of u). Recently many different extentions of Hardy type inequalities have been obtained, namely, in [8] on stratified Lie groups, in [7] on homogeneous Lie groups and [9] on general vector fields. We refer a recent open access book [6] for further discussions in this direction.

In this paper, in order to prove the Lyapunov-type inequality, first we obtain an analogue of the fractional L^2-Hardy inequality on the homogeneous Lie groups (Theorem 2).

1.2 Lyapunov-Type Inequality

Nowadays, there are many extensions of above Lyapunov's inequality. In the paper [3] the authors obtained Lyapunov inequality for the multi-dimensional fractional p-Laplacian $(-\Delta_p)^s, 1 < p < \infty,\ s \in (0,1)$, with a homogeneous Dirichlet boundary condition, that is,

$$\{(-\Delta_p)^s u = \omega(x)|u|^{p-2}u,\ \ x \in \Omega, u(x) = 0,\ \ x \in R^N \setminus \Omega, \tag{10}$$

where $\Omega \subset R^N$ is a measurable set, $1 < p < \infty$, and $s \in (0,1)$. They considered two cases, when $N > sp$ and $N < sp$. Let us recall the following result of [3].

Theorem 1 *Let $\omega \in L^\theta(\Omega)$ with $N > sp,\ \frac{N}{sp} < \theta < \infty$, be a non-negative weight. Suppose that problem (10) has a non-trivial weak solution $u \in W_0^{s,p}(\Omega)$. Then*

$$\left(\int_\Omega \omega^\theta(x)\,dx\right)^{\frac{1}{\theta}} > \frac{C}{r_\Omega^{sp-\frac{N}{\theta}}}, \tag{11}$$

where $C > 0$ is a universal constant and r_Ω is the inner radius of Ω.

Similarly, in the work [4], the authors showed Lyapunov-type inequality for the system of the fractional p-Laplacian with homogeneous Dirichlet problem.

In this paper, we present an analogue of the Lyapunov-type inequality on G for the fractional sub-Laplacian with homogeneous Dirichlet boundary condition on the homogeneous Lie groups. This result is stated in Theorem 4.

Summarising our main results of this paper, we present the following facts:

- Analogue of the fractional L^2-Hardy inequality for the fractional sub-Laplacian on G;
- Lyapunov-type inequality for the fractional sub-Laplacian with Dirichlet boundary condition on G;
- Application of the Lyapunov-type inequality on G.
- Analogue of the fractional L^2-Hardy inequality for the fractional sub-Laplacian on G_+ on the half-space and with remainder.

The paper is organised as follows. In Sect. 2 we prove an analogue of the fractional L^2-Hardy inequality and Lyapunov-type inequality on G as well as present their applications. In Sect. 3 we obtain an analogue of the fractional L^2-Hardy inequality for the fractional sub-Laplacian on G_+ (on the half-space) and also we discuss a remainder term of the inequality.

2 Main Results

Firstly, we prove the fractional L^2-Hardy inequality on G. The proof is based on [2].

Assumption Let u be a positive measurable function on G. Let $k_\varepsilon(x, y)$, $\varepsilon > 0$, be a family of measurable functions on $G \times G$ such that $k_\varepsilon(x, y) = k_\varepsilon(y, x)$ and $0 < k_\varepsilon(x, y) \leq k(x, y)$,

$$\lim_{\varepsilon \to 0} k_\varepsilon(x, y) = k(x, y), \quad for\ a.e.\ x, y \in G. \tag{12}$$

Let a function $L_\varepsilon \in L^1_{loc}(G)$ be given by the formula

$$L_\varepsilon(x) = 2u^{-1}(x) \int_G (u(x) - u(y))k_\varepsilon(x, y)dy, \tag{13}$$

and this integral absolutely convergent,

$$\lim_{\varepsilon \to 0} L_\varepsilon(x) = L(x), \quad a.e. \tag{14}$$

and

$$\lim_{\varepsilon \to 0} \int_G L_\varepsilon(x)g(x)dx = \int_G L(x)g(x)dx, \tag{15}$$

for any bounded g with compact support in G.

Lemma 1 (Lemma 2.6. [2]) *Let $p \geq 1$. Then for all $0 \leq t \leq 1$ and $a \in C$ one has*

$$|a - t|^p \geq (1 - t)^{p-1}(|a|^p - t). \tag{16}$$

Lemma 2 *Under Assumption, suppose that $N(f)$ and $\int_G L|f|^2dx$ be finite for any f with compact support G, then*

$$N(f) \geq \int_G L(x)|f(x)|^2dx, \tag{17}$$

where $N(f) = \int_G \int_G |f(x) - f(y)|^2 k(x, y)dxdy$.

Proof Let $f(x) = u(x)v(x)$ and by (13) we have

$$L_\varepsilon(x)|f(x)|^2 = 2|v(x)|^2u(x) \int_G (u(x) - u(y))k_\varepsilon(x, y)dy, \tag{18}$$

and

$$L_\varepsilon(y)|f(y)|^2 = 2|v(y)|^2u(y) \int_G (u(y) - u(x))k_\varepsilon(y, x)dx, \tag{19}$$

By integrating over G and by $k_\varepsilon(x, y) = k_\varepsilon(y, x)$ we obtain

$$2\int_G L_\varepsilon(x)|f(x)|^2dx = \int_G L_\varepsilon(x)|f(x)|^2dx + \int_G L_\varepsilon(y)|f(y)|^2dy$$

$$= 2\int_G |v(x)|^2u(x)dx \int_G (u(x) - u(y))k_\varepsilon(x, y)dy - 2\int_G |v(y)|^2u(y)dy \int_G (u(x) - u(y))k_\varepsilon(y, x)dx$$

$$= 2\int_G \int_G (|v(x)|^2u(x) - |v(y)|^2u(y))(u(x) - u(y))k_\varepsilon(x, y)dxdy. \tag{20}$$

Thus, we have

$$\int_G L_\varepsilon(x)|f(x)|^2dx = \int_G \int_G (|v(x)|^2u(x) - |v(y)|^2u(y))(u(x) - u(y))k_\varepsilon(x, y)dxdy. \tag{21}$$

Let us rewrite it in the form

$$\int_G \int_G |f(x) - f(y)|^2k_\varepsilon(x, y)dxdy = \int_G L_\varepsilon(x)|f(x)|^2dx + \int_G \int_G M(x, y)k_\varepsilon(x, y)dxdy, \tag{22}$$

where

$$M(x, y) = |f(x) - f(y)|^2 - (|v(x)|^2u(x) - |v(y)|^2u(y))(u(x) - u(y))$$
$$= |u(x)v(x) - u(y)v(y)|^2 - (|v(x)|^2u(x) - |v(y)|^2u(y))(u(x) - u(y)). \tag{23}$$

From Assumption, we get

$$\int_G \int_G |f(x) - f(y)|^2 k(x, y) dx dy = \int_G L(x)|f(x)|^2 dx$$

$$+ \int_G \int_G M(x, y) k(x, y) dx dy, \tag{24}$$

By Lemma 1 with $p = 2$, $t = \frac{u(y)}{u(x)}$, $a = \frac{v(x)}{v(y)}$ and by symmetry we assume that $u(x) \geq u(y)$, so we have

$$M(x, y) = |u(x)v(x) - u(y)v(y)|^2 - (|v(x)|^2 u(x) - |v(y)|^2 u(y))(u(x) - u(y))$$

$$= |au(x)v(y) - tu(x)v(y)|^2 - (|a|^2 |v(y)|^2 u(x) - t|v(y)|^2 u(x))(u(x) - tu(x))$$

$$= |u(x)v(y)|^2 |a - t|^2 - |u(x)v(y)|^2 (|a|^2 - t)(1 - t)$$

$$= |u(x)v(y)|^2 (|a - t|^2 - (|a|^2 - t)(1 - t)) \geq 0. \tag{25}$$

That is, we have

$$M(x, y) \geq 0,$$

which implies

$$N(f) = \int_G \int_G |f(x) - f(y)|^2 k(x, y) dx dy = \int_G L(x)|f(x)|^2 dx$$

$$+ \int_G \int_G M(x, y) k(x, y) dx dy \geq \int_G L(x)|f(x)|^2 dx. \tag{26}$$

The proof of Lemma 2 is complete.

Theorem 2 *Let $0 < s < 1$ be such that $2s \neq 1$. Then for all $f \in H_q^s$, we have*

$$\int_G \frac{|f(x)|^2}{q^{2s}(x)} dx \leq C \int_G \int_G \frac{|f(x) - f(y)|^2}{q^{Q+2s}(y^{-1} \circ x)} dx dy, \quad Q \geq 2, \tag{27}$$

where $C = C(Q, s, q) > 0$.

Proof From Lemma 2 we have

$$\int_G \int_G |f(x) - f(y)|^2 k(x, y) dx dy = N(f) \geq \int_G L(x)|f(x)|^2 dx. \tag{28}$$

We take $k(x, y) = q^{-Q-2s}(y^{-1} \circ x)$ and $L(x) = \frac{1}{Cq^{2s}(x)}$. Moreover, from Assumption, $k(x, y)$ needs to be symmetric i.e., $k(x, y) = k(y, x)$. By using the properties of the homogeneous quasi-norm it can be checked directly

$$k(x,y) = q^{-Q-2s}(y^{-1} \circ x) = q^{-Q-2s}((y^{-1} \circ x)^{-1}) = q^{-Q-2s}(x^{-1} \circ y) = k(y,x). \tag{29}$$

Theorem 2 is proved.

We present the fractional Sobolev inequality on G:

Theorem 3 ([5]) *Let $s \in (0,1)$, $Q > 2s$ and q be a quasi-norm on G. For any measurable and compactly supported function $u : G \to R$ there exists a positive constant $C = C(Q,s,q) > 0$ such that*

$$||u||_{L^{2^*}(G)} \leq C[u]_{s,q}, \tag{30}$$

where $2^* = \frac{2Q}{Q-2s}$.

Now we give a Lyapunov-type inequality for the fractional sub-Laplacian with a homogeneous Dirichlet boundary problem on G. Let $s \in (0,1)$ be such that $Q > 2s$ and $\Omega \subset G$ be a Haar measurable set. We denote by $r_{\Omega,q}$ the inner quasi-radius of Ω, that is,

$$r_{\Omega,q} = \max\{q(x) : x \in \Omega\}. \tag{31}$$

Let us consider

$$\begin{cases} (-\Delta_q)^s u(x) = \omega(x)u(x), & x \in \Omega, \\ u(x) = 0, & x \in G \setminus \Omega, \end{cases} \tag{32}$$

where $\omega \in L^\infty(\Omega)$. A weak solution of problem (32) is called a function from $u \in H_q^{s,0}(\Omega)$ satisfying

$$\int_\Omega \int_\Omega \frac{(u(x)-u(y))(v(x)-v(y))}{q^{Q+2s}(y^{-1} \circ x)} dxdy = \int_\Omega \omega(x)u(x)v(x)dx \tag{33}$$

for all $v \in H^{s,0}_{q,\Omega}$.

Theorem 4 *Let $\Omega \subset G$ be a Haar measurable set. Let $\omega \in L^\theta(\Omega)$ be a non-negative weight with $\frac{Q}{2s} < \theta < \infty$. Suppose that problem (32) with $Q > 2s$ has a non-trivial weak solution $u \in H^{s,0}_{q,\Omega}$. Then, we have*

$$\|\omega\|_{L^\theta(\Omega)} \geq \frac{C}{r_{\Omega,q}^{2s-\frac{Q}{\theta}}}, \tag{34}$$

where $C = C(Q,s,q) > 0$.

Proof Let us introduce the notation

$$\beta = 2\alpha + 2^*(1-\alpha),$$

where $\alpha = \frac{\theta - Q/2s}{\theta - 1} \in (0, 1)$ and $2^* = \frac{2Q}{Q-2s}$. Let $\beta = 2\theta'$ with $1/\theta + 1/\theta' = 1$. Then, we obtain

$$\int_\Omega \frac{|u(x)|^\beta}{r^{2\alpha s}_{\Omega,q}} dx \leq \int_\Omega \frac{|u(x)|^\beta}{q^{2\alpha s}(x)} dx. \tag{35}$$

Now, by using Hölder's inequality we get

$$\int_\Omega \frac{|u(x)|^\beta}{q^{2\alpha s}(x)} dx \leq \int_\Omega \frac{|u(x)|^{2\alpha}|u(x)|^{2^*(1-\alpha)}}{q^{2\alpha s}(x)} dx \leq \left(\int_\Omega \frac{|u(x)|^2}{q^{2s}(x)dx}\right)^\alpha \left(\int_\Omega |u(x)|^{2^*} dx\right)^{1-\alpha}. \tag{36}$$

By using Theorems 3 and 2, we have

$$\int_\Omega \frac{|u(x)|^\beta}{q^{2\alpha s}(x)} dx \leq C_1^\alpha \left(\int_\Omega \int_\Omega \frac{|u(x)-u(y)|^2}{q^{Q+2s}(y^{-1}\circ x)} dxdy\right)^{\alpha/2} C_2^{2^*(1-\alpha)/2} [u]_{s,q}^{2^*(1-\alpha)/2} \tag{37}$$

$$\leq C_1^\alpha [u]^\alpha_{s,q} C_2^{(1-\alpha)2^*/2} [u]_{s,q}^{(1-\alpha)2^*/2} = C\left([u]^2_{s,q}\right)^{2(\alpha+2^*(1-\alpha))/2} = C\left(\int_\Omega \omega(x)|u(x)|^2 dx\right)^{\theta'} \tag{38}$$

$$\leq C\left(\int_\Omega \omega^\theta(x)dx\right)^{\theta'/\theta} \int_\Omega |u(x)|^{2\theta'} dx = C\|\omega\|^{\theta'}_{L^\theta(\Omega)} \int_\Omega |u(x)|^\beta dx. \tag{39}$$

Thus, we get

$$\int_\Omega \frac{|u(x)|^\beta}{q^{2\alpha s}(x)} dx \leq C\|\omega\|^{\theta'}_{L^\theta(\Omega)} \int_\Omega |u(x)|^\beta dx. \tag{40}$$

By inequality (35), we obtain

$$\frac{1}{r^{2\alpha s}_{\Omega,q}} \int_\Omega |u(x)|^\beta dx \leq \int_\Omega \frac{|u(x)|^\beta}{q^{2\alpha s}(x)} dx \leq C\|\omega\|^{\theta'}_{L^\theta(\Omega)} \int_\Omega |u(x)|^\beta dx. \tag{41}$$

Finally, we arrive at

$$\frac{C}{r^{2s-Q/\theta}_{\Omega,q}} \leq \|\omega\|_{L^\theta(\Omega)}. \tag{42}$$

Theorem 4 is proved.

Let us consider the spectral problem for the fractional sub-Laplacian with Dirichlet boundary condition:

$$\{(-\Delta_q)^s u = \lambda u, \ x \in \Omega, \ s \in (0,1), \ u(x) = 0, \ x \in G \setminus \Omega. \tag{43}$$

We have Rayleigh quotient for the fractional Dirichlet sub-Laplacian,

$$\lambda_1 = \inf_{u \in H_q^{s,0}(\Omega),\ u \neq 0} \frac{[u]_{s,q}^2}{\|u\|_{L^2(G)}^2}, \quad s \in (0,1). \tag{44}$$

As a consequence of Theorem 4 we obtain

Theorem 5 *Let λ_1 be the first eigenvalue of problem (43) given by (44). Let $Q > 2s,\ s \in (0,1)$. Then we have*

$$\lambda_1 \geq \sup_{\frac{Q}{2s} < \theta < \infty} \frac{C}{|\Omega|^{\frac{1}{\theta}} r_{\Omega,q}^{2s - Q/\theta}}, \tag{45}$$

where C is a positive constant given in Theorem 4, $|\cdot|$ is the Haar measure and $r_{\Omega,q}$ is the inner quasi-radius of Ω.

Proof From Theorem 4 taking $\omega(x) = \lambda$, we obtain

$$\|\omega\|_{L^\theta(\Omega)} = \|\lambda\|_{L^\theta(\Omega)} = \left(\int_\Omega \lambda^\theta dx \right)^{1/\theta} \geq \frac{C}{r_{\Omega,q}^{2s - Q/\theta}}. \tag{46}$$

For every $\theta > \frac{Q}{2s}$, we have

$$\lambda_1 \geq \frac{C}{|\Omega|^{\frac{1}{\theta}} r_{\Omega,q}^{sp - Q/\theta}}. \tag{47}$$

Thus, we get

$$\lambda_1 \geq \sup_{\frac{Q}{2s} < \theta < \infty} \frac{C}{|\Omega|^{\frac{1}{\theta}} r_{\Omega,q}^{2s - Q/\theta}}, \tag{48}$$

for all $\frac{Q}{2s} < \theta < \infty$.

Theorem 5 is proved.

3 Fractional Hardy Inequality on the Half-Space

In this section, we prove a version of the fractional Hardy inequality for the fractional sub-Laplacian on the half-space. The proof of this inequality is similar to the proof of Theorem 2. Let us consider $R_+^N := \{x = (x_1, \ldots, x_N) \in R^N : x_1, \ldots, x_{N-1} \in R, x_N > 0\}$, so we introduce $G_+ := (R_+^N, \circ)$.

Assumption Let u be a positive measurable function on G_+ and let $k_\varepsilon(x,y),\ \ \varepsilon > 0$ family of measurable function on $G_+ \times G_+$ such that $k_\varepsilon(x,y) = k_\varepsilon(y,x)$ and $0 < k_\varepsilon(x,y) \leq k(x,y)$,

$$\lim_{\varepsilon \to 0} k_\varepsilon(x,y) = k(x,y), \quad for\ a.e.\ x, y \in G_+. \tag{49}$$

Let

$$L_\varepsilon(x) = 2u^{-1}(x)\int_{G_+}(u(x)-u(y))k_\varepsilon(x,y)dy, \quad (50)$$

and this integral absolutely convergent,

$$\lim_{\varepsilon\to 0} L_\varepsilon(x) = L(x), \quad for\ a.e.\ x, y \in G_+, \quad (51)$$

and

$$\lim_{\varepsilon\to 0}\int_{G_+} L_\varepsilon(x)g(x)dx = \int_{G_+} L(x)g(x)dx, \quad (52)$$

for any bounded g with compact support G_+.

Lemma 3 *Under Assumption , suppose that $N(f)$ and $\int_{G_+} L|f|^2dx$ be a finite for any f with compact support G_+, then*

$$N(f) \geq \int_{G_+} L(x)|f(x)|^2dx, \quad (53)$$

where $N(f) = \int_{G_+}\int_{G_+}|f(x)-f(y)|^2k(x,y)dxdy$.

Proof Let $f = uv$ and by (13) we have

$$L_\varepsilon(x)|f(x)|^2 = 2|v(x)|^2u(x)\int_{G_+}(u(x)-u(y))k_\varepsilon(x,y)dy, \quad (54)$$

and

$$L_\varepsilon(y)|f(y)|^2 = 2|v(y)|^2u(y)\int_{G_+}(u(y)-u(x))k_\varepsilon(y,x)dx, \quad (55)$$

Then by integrating over G_+ and by $k_\varepsilon(x,y) = k_\varepsilon(y,x)$ we obtain

$$2\int_{G_+} L_\varepsilon(x)|f(x)|^pdx = \int_{G_+} L_\varepsilon(x)|f(x)|^2dx + \int_{G_+} L_\varepsilon(y)|f(y)|^2dy$$

$$= 2\int_{G_+}|v(x)|^2u(x)dx\int_{G_+}(u(x)-u(y))k_\varepsilon(x,y)dy$$

$$-2\int_{G_+}|v(y)|^2u(y)dy\int_{G_+}(u(x)-u(y))k_\varepsilon(y,x)dx$$

$$= 2\int_{G_+}\int_{G_+}(|v(x)|^2u(x)-|v(y)|^pu(y))(u(x)-u(y))k_\varepsilon(x,y)dxdy, \quad (56)$$

that is,

$$\int_{G_+}\int_{G_+}(|v(x)|^2u(x)-|v(y)|^2u(y))(u(x)-u(y))k_\varepsilon(x,y)dxdy$$

$$=\int_{G_+}L_\varepsilon(x)|f(x)|^2dx. \tag{57}$$

Let us restate is as

$$\int_{G_+}\int_{G_+}|f(x)-f(y)|^2k_\varepsilon(x,y)dxdy=\int_{G_+}L_\varepsilon(x)|f(x)|^2dx$$

$$+\int_{G_+}\int_{G_+}M(x,y)k_\varepsilon(x,y)dxdy, \tag{58}$$

where

$$M(x,y)=|f(x)-f(y)|^2-(|v(x)|^pu(x)-|v(y)|^2u(y))(u(x)-u(y))$$

$$=|u(x)v(x)-u(y)v(y)|^2-(|v(x)|^2u(x)-|v(y)|^2u(y))(u(x)-u(y)). \tag{59}$$

From Assumption, we have

$$\int_{G_+}\int_{G_+}|f(x)-f(y)|^2k(x,y)dxdy=\int_{G_+}L(x)|f(x)|^2dx$$

$$+\int_{G_+}\int_{G_+}M(x,y)k(x,y)dxdy, \tag{60}$$

By using Lemma 1 with $t=\frac{u(y)}{u(x)}$, $a=\frac{v(x)}{v(y)}$ and by the symmetry we assume that $u(x)\geq u(y)$, so we obtain that

$$M(x,y)\geq 0.$$

By using this fact in (60), we get

$$N(f)=\int_{G_+}\int_{G_+}|f(x)-f(y)|^2k(x,y)dxdy=\int_{G_+}L(x)|f(x)|^2dx$$

$$+\int_{G_+}\int_{G_+}M(x,y)k(x,y)dxdy\geq\int_{G_+}L(x)|f(x)|^2dx. \tag{61}$$

The proof of Lemma 3 is complete.

Theorem 6 (Fractional Hardy inequality on the half-space) *Let $0<s<1$ be such that $2s\neq 1$ and $Q\geq 2$. Then for all $f\in H^s_q(G_+)$, we have*

$$\int_{G_+} \frac{|f|^2}{x_N^{2s}} dx \leq C \int_{G_+} \int_{G_+} \frac{|f(x) - f(y)|^2}{q^{Q+2s}(y^{-1} \circ x)} dx dy, \tag{62}$$

where $C = C(Q, s, p, q) > 0$.

Proof From Lemma 3 we get

$$\int_{G_+} \int_{G_+} |f(x) - f(y)|^2 k(x, y) dx dy = N(f) \geq \int_{G_+} L(x) |f(x)|^2 dx. \tag{63}$$

By taking $k(x, y) = q^{-Q-2s}(y^{-1} \circ x)$ and $L(x) = \frac{1}{C x_N^{2s}}$ with

$$k(x, y) = q^{-Q-2s}(y^{-1} \circ x) = q^{-Q-2s}((y^{-1} \circ x)^{-1}) = q^{-Q-2s}(x^{-1} \circ y) = k(y, x), \tag{64}$$

we complete the proof of Theorem 6.

Futhermore, to show the fractional Hardy inequality with a remainder on G_+ we need to recall the following results.

Lemma 4 (Lemma 2.6. [2]) *Let* $p \geq 2$ *then for all* $0 \leq t \leq 1$ *and all* $a \in C$ *one has*

$$|a - t|^p \geq (1 - t)^{p-1}(|a|^p - t) + c_p t^{\frac{p}{2}} |a - 1|^p, \tag{65}$$

where

$$c_p := \min_{0<\beta<1/2} ((1 - \beta)p - \beta^p + p\beta^{p-1}), \tag{66}$$

and $0 < c_p < 1$. *If* $p = 2$, *(65) is an equality with* $c_2 = 1$.

Lemma 5 *Under Assumption, suppose that* $N(f)$, $N_u(v)$ *and* $\int_{G_+} L|f|^2 dx$ *be finite for any* f *with compact support with* $f = uv$ *on* G_+, *then we have*

$$N(f) \geq \int_{G_+} L(x)|f(x)|^2 dx + N_u(v), \tag{67}$$

where $N_u(v) = \int_{G_+} \int_{G_+} |v(x) - v(y)|^2 u(x) u(y) k(x, y) dx dy$.

Proof The proof is the same as the proof of Lemma 3, but only difference is that now we need to use inequality (65).

Now we ready to state the Hardy inequality with remainder on G_+.

Theorem 7 (Hardy inequality with remainder on G_+) *Let* $0 < s < 1$ *be such that* $2s \neq 1$ *and* $Q \geq 2$. *Then for all* $f \in H_q^s(G_+)$ *and* $v := x_N^{\frac{1-2s}{2}} f$, *we have*

$$C \int_{G_+} \frac{|f|^2}{x_N^{2s}} dx + \int_{G_+} \int_{G_+} \frac{|v(x) - v(y)|^2 x_N^{\frac{2s-1}{2}} y_N^{\frac{2s-1}{2}}}{q^{Q+2s}(y^{-1} \circ x)} dx dy$$

$$\leq \int_{G_+} \int_{G_+} \frac{|f(x) - f(y)|^2}{q^{Q+2s}(y^{-1} \circ x)} dxdy, \tag{68}$$

where $C = C(Q, s, p, q) > 0$.

Proof The proof is the same as the proof of Theorem 6, but only difference is now we need to use Lemma 5 with $u = x_N^{\frac{2s-1}{2}}$.

Acknowledgements The authors were supported in parts by the MESRK grant AP05130981.

References

1. Fischer, V., Ruzhansky, M.: Quantization on nilpotent Lie groups. In: Progress in Mathematics, vol. 314. Birkhäuser, Basel (2016)
2. Frank, R.L., Seiringer, R.: Non-linear ground state representations and sharp Hardy inequalities. J. Funct. Anal. **255**(12), 3407–3430 (2008)
3. Jleli, M., Kirane, M., Samet, B.: Lyapunov-type inequalities for fractional partial differential equations. Appl. Math. Lett. **66**, 30–39 (2017)
4. Jleli, M., Kirane, M., Samet, B.: Lyapunov-type inequalities for a fractional p-Laplacian system. Fract. Calc. Appl. Anal. **20**(6), 1485–1506 (2017)
5. Kassymov, A., Suragan, D.: Some functional inequalities for the fractional p-sub-Laplacian (2018). arXiv:1804.01415v2
6. Ruzhansky, M., Suragan, D.: Hardy Inequalities on Homogeneous Groups. In: Progress in Mathematics, vol. 327, 588 p. Birkhäuser, Basel (2019)
7. Ruzhansky, M., Suragan, D.: Hardy and Rellich inequalities, identities, and sharp remainders on homogeneous groups. Adv. Math. **317**, 799–822 (2017)
8. Ruzhansky, M., Suragan, D.: On horizontal Hardy, Rellich, Caffarelli-Kohn-Nirenberg and p-sub-Laplacian inequalities on stratified groups. J. Differ. Equ. **262**, 1799–1821 (2017)
9. Ruzhansky, M., Suragan, D.: Local Hardy and Rellich inequalities for sums of squares of vector fields. Adv. Differ. Equ. **22**, 505–540 (2017)

Differential Equations and Boundary Value Problems

Cone Rectangular Metric Spaces over Banach Algebras and Fixed Point Results of T-Contraction Mappings

Abba Auwalu and Ali Denker

Abstract In this paper, we introduce the notion of T-contraction mappings on cone rectangular metric spaces over Banach algebras. Further, we establish the existence and uniqueness of fixed point for such mappings. Our results extend and generalize the Banach contraction principle and many recent results in the literature. Moreover, an example to illustrate the main result is also presented.

Keywords Cone rectangular metric space · T-Contraction mappings · Banach contraction principle · Fixed point

1 Introduction

In 2007, Long-Guang and Xian [4] introduced the concept of a cone metric space. They replaced the set of real numbers which forms the domain of distance function by an ordered Banach space. Further, they proved some fixed point theorems for contractive type conditions in such a space. In 2009 Beiranvand et al. [2] introduced new classes of contractive mappings defined on metric spaces called T-contractions and established the extended Banach contractive principle. Recently, Morales and Rojas [5] extended the work of [2] to cone metric metric spaces. In 2013, Hao and Shaoyuan [3], introduced the notion of cone metric spaces over Banach algebras and show that they are not equivalent to metric spaces in terms of existence of the fixed points of mappings. Further, they proved Banach contraction principle in such a space by replacing usual real contraction constant with a vector constant. In 2016, Shukla et al., [7] introduced the notion of a cone rectangular metric space over Banach

A. Auwalu · A. Denker (✉)
Department of Mathematics, Near East University, Mersin 10, Nicosia, TRNC, Turkey
e-mail: ali.denker@neu.edu.tr; adenker63@gmail.com

A. Auwalu
e-mail: abba.auwalu@neu.edu.tr; abbaauwalu@yahoo.com

A. Ashyralyev et al. (eds.), *Functional Analysis in Interdisciplinary Applications—II*, Springer Proceedings in Mathematics & Statistics 351,
https://doi.org/10.1007/978-3-030-69292-6_7

algebras and proved Banach contraction principle in such a space. In this work we introduce the notion of a T-contraction mapping on cone rectangular metric spaces over Banach algebras and give an example to show that this concept is more general than that of contraction. Furthermore, we establish the existence and uniqueness of fixed point for such mapping.

2 Preliminaries

In this section, we shall give some definitions and results needed in the sequel.

Definition 1 ([6]) A Banach algebra $\mathscr{A}$ is a real Banach space in which an operation of multiplication is defined subject to the following properties for all $x, y, z \in \mathscr{A}$, $\lambda \in \mathbb{R}$:

(1) $(xy)z = x(yz)$;
(2) $x(y+z) = xy + xz$ and $(x+y)z = xz + yz$;
(3) $\lambda(xy) = (\lambda x)y = x(\lambda y)$;
(4) $\|xy\| \leq \|x\| \, \|y\|$.

Definition 2 ([3]) A subset $\mathscr{P}$ of Banach algebra $\mathscr{A}$ is called a cone if:

(1) $\mathscr{P}$ is nonempty closed and $\{\theta, e\} \subset \mathscr{P}$;
(2) $k_1\mathscr{P} + k_2\mathscr{P} \subset \mathscr{P}$ for all nonnegative real numbers k_1, k_2;
(3) $\mathscr{P}^2 = \mathscr{P}\mathscr{P} \subset \mathscr{P}$;
(4) $\mathscr{P} \cap (-\mathscr{P}) = \{\theta\}$, where θ and e denote the zero and unit elements of the Banach algebra $\mathscr{A}$ respectively. For a given cone $\mathscr{P} \subset \mathscr{A}$ we write $y \preccurlyeq z$ if and only if $z - y \in \mathscr{P}$ where $\preccurlyeq$ is a partial order relation defined on $\mathscr{P}$. We shall write $y \prec z$ to indicate that $y \preccurlyeq z$ but $y \neq z$ while $y \ll z$ will stand for $z - y \in int\mathscr{P}$ where $int\mathscr{P}$ denotes the interior of $\mathscr{P}$. If $int\mathscr{P} \neq \varnothing$ then $\mathscr{P}$ is called a solid cone.

Definition 3 ([8]) Let $\mathscr{P}$ be a solid cone in a Banach algebra $\mathscr{A}$. A sequence $\{z_i\} \subset \mathscr{P}$ is said to be a c-sequence if for each $\theta \ll c$ there exists $n_0 \in \mathbb{N}$ such that $z_i \ll c$ for all $i > n_0$.

Lemma 1 ([8]) *Let $\mathscr{P}$ be a solid cone in a Banach algebra $\mathscr{A}$ and $\{z_i\} \subset \mathscr{P}$ be a sequence with $\|z_i\| \to 0$ $(i \to \infty)$ then for each $\theta \ll c$ there exists $n_0 \in \mathbb{N}$ such that for all $i > n_0$ we have $z_i \ll c$.*

Lemma 2 ([6]) *Let $\mathscr{A}$ be a Banach algebra with a unit e and $\lambda \in \mathscr{A}$. If the spectral radius $\delta(\lambda)$ of λ is less than one i.e. $\delta(\lambda) = \lim_{n\to\infty} \|\lambda^n\|^{\frac{1}{n}} = inf_{n\in\mathbb{N}}\|\lambda^n\|^{\frac{1}{n}} < 1$ then $(e - \lambda)$ is invertible in $\mathscr{A}$. Moreover, $(e-\lambda)^{-1} = \sum_{j=0}^{\infty} \lambda^j$.*

Remark 1 ([8]) If the spectral radius $\delta(\lambda) < 1$ then $\|\lambda^i\| \to 0$ $(i \to \infty)$

Lemma 3 ([7]) *Let $\mathscr{A}$ be a real Banach algebra with a solid cone $\mathscr{P}$. For $a, b, c, \lambda \in \mathscr{P}$ if:*

(1) $a \preccurlyeq b \ll c$ *then* $a \ll c$.
(2) $\theta \preccurlyeq a \ll c$ *for each* $\theta \ll c$ *then* $a = \theta$.
(3) $a \preccurlyeq \lambda a$ *and* $\delta(\lambda) < 1$ *then* $a = \theta$.

Definition 4 ([3]) Let $\mathscr{X}$ be a nonempty set and $\mathscr{A}$ be a Banach algebra. Suppose that $\forall x, y, z \in \mathscr{X}$, $\rho : \mathscr{X} \times \mathscr{X} \to \mathscr{A}$ is a mapping satisfying the following conditions:
(C_1) $\theta \preccurlyeq \rho(x, y)$ and $\rho(x, y) = \theta \Leftrightarrow x = y$;
(C_2) $\rho(x, y) = \rho(y, x)$;
(C_3) $\rho(x, y) \preccurlyeq \rho(x, z) + \rho(z, y)$ (Triangle inequality).
Then ρ is called a cone metric on $\mathscr{X}$ and $(\mathscr{X}, \rho)$ is a cone metric space over $\mathscr{A}$.

Remark 2 A metric space is a cone metric space over Banach algebra $\mathscr{A}$ where by $\mathscr{A} = \mathbb{R}$ and $\mathscr{P} = [0, +\infty)$. The converse is not necessarily true (e.g. see [3, 4]).

Definition 5 ([7]) Let $\mathscr{X}$ be a nonempty set and $\mathscr{A}$ be a Banach algebra. Suppose that $\rho : \mathscr{X} \times \mathscr{X} \to \mathscr{A}$ is a mapping such that $\forall x, y, u, v \in \mathscr{X}, x \neq u, u \neq v, v \neq y$ the following conditions are satisfied:
(R_1) $\theta \preccurlyeq \rho(x, y)$ and $\rho(x, y) = \theta \Leftrightarrow x = y$;
(R_2) $\rho(x, y) = \rho(y, x)$;
(R_3) $\rho(x, y) \preccurlyeq \rho(x, u) + \rho(u, v) + \rho(v, y)$ (Rectangular inequality).
Then ρ is called a cone rectangular metric on $\mathscr{X}$ and $(\mathscr{X}, \rho)$ is called a cone rectangular metric space over $\mathscr{A}$.

Remark 3 A cone metric space over Banach algebra $\mathscr{A}$ is a cone rectangular metric space over Banach algebra $\mathscr{A}$. The converse is not necessarily true (e.g. see [7]).

Definition 6 ([7]) Let $(\mathscr{X}, \rho)$ be a cone rectangular metric space over Banach algebra $\mathscr{A}$, $z \in \mathscr{X}$ and $\{z_i\}$ be a sequence in $(\mathscr{X}, \rho)$. Then we say
(1) $\{z_i\}$ converges to z if for each $c \in \mathscr{A}$ with $\theta \ll c$, there is a natural number n_0 such that $\rho(z_i, z) \ll c$ for all $i \geq n_0$. We denote this by $z_i \to z$ $(i \to \infty)$.
(2) $\{z_i\}$ is a Cauchy sequence if, for each $c \in \mathscr{A}$ with $\theta \ll c$, there is a natural number n_0 which is independent of n such that $\rho(z_i, z_{i+n}) \ll c$ for all $i \geq n_0$.
(3) $(\mathscr{X}, \rho)$ is said to be complete if every Cauchy sequence in $(\mathscr{X}, \rho)$ is convergent.

Lemma 4 *Let* $(\mathscr{X}, \rho)$ *be a complete cone rectangular metric space over Banach algebra* $\mathscr{A}$, $\mathscr{P}$ *be the underlying solid cone and* $\{z_i\}$ *be a sequence in* $(\mathscr{X}, \rho)$. *If* $\{z_i\}$ *converges to* $z \in \mathscr{X}$, *then*
(1) $\{\rho(z_i, z)\}$ *is a c-sequence.* (2) *for any* $j \in \mathbb{N}$, $\{\rho(z_i, z_{i+j})\}$ *is a c-sequence.*

Proof Follows from Definitions 3, 5 and 6 (1).

Lemma 5 ([8]) *Let* $\mathscr{A}$ *be a Banach algebra with a solid cone* $\mathscr{P}$ *and let* $\{\alpha_n\}$ *and* $\{\beta_n\}$ *be sequences in* $\mathscr{P}$. *If* $\{\alpha_n\}$ *and* $\{\beta_n\}$ *are c-sequences and* $k_1, k_2 \in \mathscr{P}$, *then* $\{k_1\alpha_n + k_2\beta_n\}$ *is also a c-sequence.*

3 Main Results

In this section, we introduce the notion of a T-contraction mapping on cone rectangular metric spaces over Banach algebras and give an example to show that this concept is more general than that of contraction. Furthermore, we establish the existence and uniqueness of fixed point for such mapping.

Definition 7 Let $(\mathscr{X}, \rho)$ be a complete cone rectangular metric space over Banach algebra $\mathscr{A}$ and $\mathscr{P}$ be a solid cone in $\mathscr{A}$. Let $T, F : \mathscr{X} \to \mathscr{X}$ be two mappings. Then F is said to be a T-contraction if there exists $\lambda \in \mathscr{P}$ with $\delta(\lambda) < 1$ such that

$$\rho(TFz, TFy) \preccurlyeq \lambda\rho(Tz, Ty), \text{ for all } z, y \in \mathscr{X}. \tag{1}$$

Remark 4 A contraction is T-contraction by taking $T = I$ where I is the identity mapping. The converse is not necessarily true as can be seen the following example.

Example 1 Let $\mathscr{A} = C^1_{\mathbb{R}}[0, 1]$ and define a norm on $\mathscr{A}$ by $\|z\| = \|z\|_\infty + \|z'\|_\infty$, for $z \in \mathscr{A}$. Define multiplication in $\mathscr{A}$ as just pointwise multiplication. Then $\mathscr{A}$ is a real Banach algebra with a unit element $e = 1$ (e.g. see [7]). The set $\mathscr{P} = \{z \in \mathscr{A} : z(t) \geq 0, t \in [0, 1]\}$ is a non-normal cone in $\mathscr{A}$. Let $\mathscr{X} = \{1, 2, 3, 4\}$. Define a mapping $\rho : \mathscr{X} \times \mathscr{X} \to \mathscr{A}$ by

$$\rho(z, y) = \begin{cases} 0, & \text{if } z = y, \\ 3e^t, & \text{if } z, y \in \{1, 2\} \text{ and } z \neq y, \\ e^t, & \text{otherwise.} \end{cases}$$

Then $(\mathscr{X}, \rho)$ is a cone rectangular metric space over Banach algebra $\mathscr{A}$. Consider the mappings $T, F : \mathscr{X} \to \mathscr{X}$ defined by

$$Tz = \begin{cases} 3, & \text{if } z \neq 4, \\ 1, & \text{if } z = 4, \end{cases}$$

and

$$Fz = \begin{cases} 2, & \text{if } z = 1, \\ 1, & \text{if } z = 2, \\ z, & \text{if } z = 3. \end{cases}$$

Then the mapping F is a T-contraction on $(\mathscr{X}, \rho)$ with $\lambda = \frac{1}{3}$ but not a contraction.

Theorem 1 *Let $(\mathscr{X}, \rho)$ be a complete cone rectangular metric space over Banach algebra $\mathscr{A}$ and $\mathscr{P}$ be a solid cone in $\mathscr{A}$. Suppose that $T : \mathscr{X} \to \mathscr{X}$ is a bijective mapping and $F : \mathscr{X} \to \mathscr{X}$ a T-contraction mapping. Then F has a unique fixed point z in $\mathscr{X}$.*

Proof Let z_0 be arbitrary point in $\mathscr{X}$. We define a sequence $\{z_i\}$ in $(\mathscr{X}, \rho)$ by

$$z_{i+1} = Fz_i, \text{ for all } i \in \mathbb{N}. \tag{2}$$

Suppose $z_j = z_{j+1}$ for some $j \in \mathbb{N}$, then $z = z_j = Fz_j$ is a fixed point of F and the result is proved. Hence, we assume that $z_i \neq z_{i+1}$ for all $i \in \mathbb{N}$. We shall show that $\{Tz_i\}$ is a Cauchy sequence in $(\mathscr{X}, \rho)$. By using (1) and (2), we have

$$\begin{aligned} \rho(Tz_i, Tz_{i+1}) &= \rho(TFz_{i-1}, TFz_i) \preccurlyeq \lambda\rho(Tz_{i-1}, Tz_i), \\ \therefore \quad \rho(Tz_i, Tz_{i+1}) &\preccurlyeq \lambda\rho(Tz_{i-1}, Tz_i) \prec \rho(Tz_{i-1}, Tz_i). \end{aligned} \tag{3}$$

That is, the sequence $\{\rho(Tz_i, Tz_{i+1})\}$ is strictly decreasing and from this it follows that $\rho(Tz_i, Tz_{i+1}) \neq \rho(Tz_j, Tz_{j+1})$ whenever $i \neq j$. Hence, from (1), (2) and (3) we obtain

$$\rho(Tz_i, Tz_{i+1}) \preccurlyeq \lambda\rho(Tz_{i-1}, Tz_i) = \lambda\rho(TFz_{i-2}, TFz_{i-1})$$

$$\preccurlyeq \lambda^2\rho(Tz_{i-2}, Tz_{i-1}) = \lambda^2\rho(TFz_{i-3}, TFz_{i-2})$$

$$\preccurlyeq \lambda^3\rho(Tz_{i-3}, Tz_{i-2}) \preccurlyeq \cdots \preccurlyeq \lambda^i\rho(Tz_0, Tz_1)$$

$$\therefore \quad \rho(Tz_i, Tz_{i+1}) \preccurlyeq \lambda^i\rho(Tz_0, Tz_1) \text{ for all } i \in \mathbb{N}. \tag{4}$$

Similarly, by using (R_3), (1), (2), (4) and the fact that $\rho(Tz_i, Tz_{i+1}) \neq \rho(Tz_j, Tz_{j+1})$ whenever $i \neq j$, we have that

$$\rho(Tz_i, Tz_{i+2}) = \rho(TFz_{i-1}, TFz_{i+1}) \preccurlyeq \lambda\rho(Tz_{i-1}, Tz_{i+1})$$

$$\preccurlyeq \lambda[\rho(Tz_{i-1}, Tz_i) + \rho(Tz_i, Tz_{i+2}) + \rho(Tz_{i+2}, Tz_{i+1})]$$

$$(e - \lambda)\rho(Tz_i, Tz_{i+2}) \preccurlyeq \lambda[\rho(Tz_{i-1}, Tz_i) + \rho(Tz_{i+1}, Tz_{i+2})]$$

$$\preccurlyeq \lambda[\lambda^{i-1}\rho(Tz_0, Tz_1) + \lambda^{i+1}\rho(Tz_0, Tz_1)] = \lambda(e + \lambda^2)\lambda^{i-1}\rho(Tz_0, Tz_1)$$

$$\therefore \quad \rho(Tz_i, Tz_{i+2}) \preccurlyeq (e - \lambda)^{-1}\lambda^i(e + \lambda^2)\rho(Tz_0, Tz_1), \text{ for all } i \in \mathbb{N}. \tag{5}$$

Since $\delta(\lambda) < 1$ then by Lemma 2, it follows that $(e - \lambda)$ is invertible in $\mathscr{A}$. Moreover,

$$(e-\lambda)^{-1}=\sum_{j=0}^{\infty}\lambda^j. \tag{6}$$

Also by Remark 1, we obtain that $\|\lambda^i\| \to 0\ (i \to \infty)$. (7)

Now, for the sequence $\{Tz_i\}$ we consider $\rho(Tz_i, Tz_{i+n})$ in two cases:
Case 1. If n is odd, say $n = 2j+1\ (j \in \mathbb{N})$, then by using (R_3), (4), (6) and the fact that $\rho(Tz_i, Tz_{i+1}) \neq \rho(Tz_j, Tz_{j+1})$ whenever $i \neq j$ we have that

$$\rho(Tz_i, Tz_{i+2j+1}) \preccurlyeq \rho(Tz_{i+2j}, Tz_{i+2j+1}) + \rho(Tz_{i+2j-1}, Tz_{i+2j}) + \rho(Tz_i, Tz_{i+2j-1})$$

$$\preccurlyeq \rho(Tz_{i+2j}, Tz_{i+2j+1}) + \rho(Tz_{i+2j-1}, Tz_{i+2j}) + \rho(Tz_{i+2j-2}, Tz_{i+2j-1})$$

$$+\rho(Tz_{i+2j-3}, Tz_{i+2j-2}) + \rho(Tz_i, Tz_{i+2j-3})$$

$$\preccurlyeq \rho(Tz_{i+2j}, Tz_{i+2j+1}) + \rho(Tz_{i+2j-1}, Tz_{i+2j}) + \rho(Tz_{i+2j-2}, Tz_{i+2j-1})$$

$$+\cdots+\rho(Tz_{i+1}, Tz_{i+2}) + \rho(Tz_i, Tz_{i+1})$$

$$\preccurlyeq \lambda^{i+2j}\rho(Tz_0, Tz_1) + \lambda^{i+2j-1}\rho(Tz_0, Tz_1) + \lambda^{i+2j-2}\rho(Tz_0, Tz_1)$$

$$+\cdots+\lambda^{i+1}\rho(Tz_0, Tz_1) + \lambda^i\rho(Tz_0, Tz_1)$$

$$\preccurlyeq \lambda^i\left(\sum_{j=0}^{\infty}\lambda^j\right)\rho(Tz_0, Tz_1) = \lambda^i(e-\lambda)^{-1}\rho(Tz_0, Tz_1)$$

$$\therefore\ \ \rho(Tz_i, Tz_{i+2j+1}) \preccurlyeq \lambda^i(e-\lambda)^{-1}\rho(Tz_0, Tz_1).$$

Using (7) and part (4) of Definition 1, we have $\|\lambda^i(e-\lambda)^{-1}\rho(Tz_0, Tz_1)\| \leq \|\lambda^i\|\ \|(e-\lambda)^{-1}\rho(Tz_0, Tz_1)\| \to 0\ (i \to \infty)$, and it follows by Lemma 1, that for any $c \in \mathscr{A}$ with $\theta \ll c$, there exists $n_1 \in \mathbb{N}$ such that

$$\rho(Tz_i, Tz_{i+2j+1}) \preccurlyeq \lambda^i(e-\lambda)^{-1}\rho(Tz_0, Tz_1) \ll c, \text{ for all } i > n_1.$$

Therefore, by part (1) of Lemma 3 we have that

$$\rho(Tz_i, Tz_{i+2j+1}) \ll c, \text{ for all } i > n_1. \tag{8}$$

Case 2. If n is even, say $n = 2j$ ($j \in \mathbb{N}$), then by using (R_3), (4), (5), (6) and the fact that $\rho(Tz_i, Tz_{i+1}) \neq \rho(Tz_j, Tz_{j+1})$ whenever $i \neq j$ we obtain that

$$\rho(Tz_i, Tz_{i+2j}) \preccurlyeq \rho(Tz_{i+2j-1}, Tz_{i+2j}) + \rho(Tz_{i+2j-1}, Tz_{i+2j-2}) + \rho(Tz_i, Tz_{i+2j-2})$$

$$\preccurlyeq \rho(Tz_{i+2j-1}, Tz_{i+2j}) + \rho(Tz_{i+2j-2}, Tz_{i+2j-1}) + \rho(Tz_{i+2j-3}, Tz_{i+2j-2})$$

$$+\rho(Tz_{i+2j-3}, Tz_{i+2j-4}) + \rho(Tz_i, Tz_{i+2j-4})$$

$$\preccurlyeq \rho(Tz_{i+2j-1}, Tz_{i+2j}) + \rho(Tz_{i+2j-2}, Tz_{i+2j-1}) + \rho(Tz_{i+2j-3}, Tz_{i+2j-2})$$

$$+ \cdots + \rho(Tz_{i+3}, Tz_{i+2}) + \rho(Tz_i, Tz_{i+2})$$

$$\preccurlyeq \lambda^{i+2j-1}\rho(Tz_0, Tz_1) + \lambda^{i+2j-2}\rho(Tz_0, Tz_1) + \lambda^{i+2j-3}\rho(Tz_0, Tz_1)$$

$$+ \cdots + \lambda^{i+2}\rho(Tz_0, Tz_1) + (e - \lambda)^{-1}\lambda^i(e + \lambda^2)\rho(Tz_0, Tz_1)$$

$$\preccurlyeq \lambda^{i+2}\Big(\sum_{j=0}^{\infty}\lambda^j\Big)\rho(Tz_0, Tz_1) + (e - \lambda)^{-1}\lambda^i(e + \lambda^2)\rho(Tz_0, Tz_1)$$

$$\therefore\ \rho(Tz_i, Tz_{i+2j+1}) \preccurlyeq \lambda^i(e - \lambda)^{-1}\left(e + 2\lambda^2\right)\rho(Tz_0, Tz_1).$$

Using (7) and part (4) of Definition 1, we have $\|\lambda^i(e - \lambda)^{-1}\left(e + 2\lambda^2\right)\rho(Tz_0, Tz_1)$ $\| \to 0$ ($i \to \infty$) and it follows by Lemma 1, that for any $c \in \mathscr{A}$ with $\theta \ll c$, there exists $n_2 \in \mathbb{N}$ such that

$$\rho(Tz_i, Tz_{i+2j}) \preccurlyeq \lambda^i(e - \lambda)^{-1}\left(e + 2\lambda^2\right)\rho(Tz_0, Tz_1) \ll c, \text{ for all } i > n_2.$$

Therefore, by part (1) of Lemma 3 we have that

$$\rho(Tz_i, Tz_{i+2j}) \ll c, \text{ for all } i > n_2. \tag{9}$$

Let $n_0 := \max\{n_1, n_2\}$, hence from (8) and (9), we have that

$$\rho(Tz_i, Tz_{i+n}) \ll c, \text{ for all } i > n_0.$$

Which implies by Definition 6, that $\{Tz_i\}$ is a Cauchy sequence in $(\mathscr{X}, \rho)$ and since $(\mathscr{X}, \rho)$ is complete, the sequence $\{Tz_i\}$ converges i.e. there exists z^* in $\mathscr{X}$ such that $Tz_i \to z^*$ $(i \to \infty)$. Since T is a surjection mapping, there exists a point z in $\mathscr{X}$ such that $Tz = z^*$. Now, using (R_3), (1) and (2) we have

$$\rho(z^*, TFz) \preccurlyeq \rho(z^*, Tz_i) + \rho(Tz_i, TFz_i) + \rho(TFz_i, TFz)$$

$$\preccurlyeq \rho(z^*, Tz_i) + \rho(Tz_i, Tz_{i+1}) + \lambda\rho(Tz_i, Tz)$$

$$= \rho(z^*, Tz_i) + \rho(Tz_i, Tz_{i+1}) + \lambda\rho(Tz_i, z^*)$$

$$\therefore \quad \rho(z^*, TFz) \preccurlyeq \alpha\rho(Tz_i, z^*) + \rho(Tz_i, Tz_{i+1}),$$

where $\alpha = (e + \lambda) \in \mathscr{P}$. Using Lemmas 4 and 5; $\{\rho(Tz_i, z^*)\}$, $\{\rho(Tz_i, Tz_{i+1})\}$ and $\{\alpha\rho(Ty_i, z^*) + \rho(Tz_i, Tz_{i+1})\}$ are c-sequences. Hence, for any $c \in \mathscr{A}$ with $\theta \ll c$, there exists $n_0 \in \mathbb{N}$ such that

$$\rho(z^*, TFz) \preccurlyeq \alpha\rho(Tz_i, z^*) + \rho(Tz_i, Tz_{i+1}) \ll c, \text{ for all } i > n_0, \tag{10}$$

which implies by part (1) and (2) of Lemma 3, that $\rho(z^*, TFz) = \theta$. Hence, $TFz = z^* = Tz$. Since the mapping T is injective, we have $Fz = z$. Thus, z is a fixed point of F. Next we shall show that the fixed point is unique. Assume that y is another fixed point of F, i.e. $Fy = y$, then using (1) we have that

$$\rho(Tz, Ty) = \rho(TFz, TFy) \preccurlyeq \lambda\rho(Tz, Ty).$$

Since $\delta(\lambda) < 1$, using part (3) of Lemma 3 we have that $\rho(Tz, Ty) = \theta$ i.e. $Tz = Ty$ and since T is injective, so $z = y$, i.e. the fixed point of F is unique. This completes the proof.

Corollary 1 *Let $(\mathscr{X}, \rho)$ be a complete cone rectangular metric space over Banach algebra $\mathscr{A}$ and $\mathscr{P}$ be a solid cone in $\mathscr{A}$. Let $F : \mathscr{X} \to \mathscr{X}$ be mapping satisfying the following condition*

$$\rho(Fz, Fy) \preccurlyeq \lambda\rho(z, y), \text{ for all, } z, y \in \mathscr{X},$$

where $\lambda \in \mathscr{P}$ *such that* $\delta(\lambda) < 1$. *Then* F *has a unique fixed point* z *in* $\mathscr{X}$.

Proof The result follows from Theorem 1 where $T = I$ (identity mapping).

Corollary 2 *Let* $(\mathscr{X}, \rho)$ *be a complete cone metric space over Banach algebra* $\mathscr{A}$ *and* $\mathscr{P}$ *be a solid cone in* $\mathscr{A}$. *Suppose that* $T : \mathscr{X} \to \mathscr{X}$ *is a bijective mapping and* $F : \mathscr{X} \to \mathscr{X}$ *a T-contraction mapping. Then* F *has a unique fixed point* z *in* $\mathscr{X}$.

Proof The result follows from Remark 3 and Theorem 1.

Corollary 3 *Let* $(\mathscr{X}, \rho)$ *be a complete cone metric space over Banach algebra* $\mathscr{A}$ *and* $\mathscr{P}$ *be a solid cone in* $\mathscr{A}$. *Let* $F : \mathscr{X} \to \mathscr{X}$ *be mapping satisfying the following condition*

$$\rho(Fz, Fy) \preccurlyeq \lambda\rho(z, y), \text{ for all, } z, y \in \mathscr{X},$$

where $\lambda \in \mathscr{P}$ *such that* $\delta(\lambda) < 1$. *Then* F *has a unique fixed point* z *in* $\mathscr{X}$.

Proof The result follows from Remark 3 and Corollary 1.

Corollary 4 *Let* $(\mathscr{X}, \rho)$ *be a complete metric space. Suppose that* $T : \mathscr{X} \to \mathscr{X}$ *is a bijective mapping and* $F : \mathscr{X} \to \mathscr{X}$ *a T-contraction mapping. Then* F *has a unique fixed point* z *in* $\mathscr{X}$.

Proof The result follows from Remark 2 and Corollary 2.

Corollary 5 *Let* $(\mathscr{X}, \rho)$ *be a complete metric space. Let* $F : \mathscr{X} \to \mathscr{X}$ *be mapping satisfying the following condition*

$$\rho(Fz, Fy) \preccurlyeq \lambda\rho(z, y), \text{ for all, } z, y \in \mathscr{X},$$

where $\lambda \in [0, 1)$. *Then* F *has a unique fixed point* z *in* $\mathscr{X}$.

Proof The result follows from Remark 2 and Corollary 3.

Example 2 Let $\mathscr{A} = C^1_{\mathbb{R}}[0, 1]$ and define a norm on $\mathscr{A}$ by $\|z\| = \|z\|_\infty + \|z'\|_\infty$, for $z \in \mathscr{A}$. Define multiplication in $\mathscr{A}$ as just pointwise multiplication. Then $\mathscr{A}$ is a real Banach algebra with a unit element $e = 1$ (e.g. see [7]). The set $\mathscr{P} = \{z \in \mathscr{A} : z(t) \geq 0, t \in [0, 1]\}$ is a non-normal cone in $\mathscr{A}$. Let $\mathscr{X} = \{1, 2, 3, 4\}$. Define $\rho : \mathscr{X} \times \mathscr{X} \to \mathscr{A}$ by

$$\rho(z, y) = \begin{cases} 0, & \text{if } z = y, \\ 3e^t, & \text{if } z, y \in \{1, 2\} \text{ and } z \neq y, \\ e^t, & \text{otherwise.} \end{cases}$$

Then $(\mathscr{X}, \rho)$ is a complete cone rectangular metric space over Banach algebra $\mathscr{A}$. Let the mappings $T, F : \mathscr{X} \to \mathscr{X}$ be defined by

$$Tz = \begin{cases} 3, & \text{if } z \neq 4, \\ 1, & \text{if } z = 4, \end{cases}$$

and

$$Fz = \begin{cases} 2, & \text{if } z = 1, \\ 1, & \text{if } z = 2, \\ z, & \text{if } z = 3. \end{cases}$$

Then the mapping F is a T-contraction with $\lambda = \frac{1}{3}$ but not contraction. Clearly, T is bijective mapping. Moreover, F satisfies all the conditions of Theorem 1 and $z = 3$ is the unique fixed point of F

4 Conclusion

The aim of this paper is to introduce the notion of T-contraction mappings on cone rectangular metric spaces over Banach algebras and establishing the existence and uniqueness of fixed point for such mappings. Our results extend and generalize the Banach contraction principle in [1–4], and many recent results in the literature. Moreover, an example to illustrate the main result is also presented.

References

1. Auwalu, A.: Synchronal algorithm for a countable family of strict psedocontractions in q-uniformly smooth Banach spaces. Int. J. Math. Anal. **8**, 727–745 (2014)
2. Beiranvand, A., Moradi, S., Omid, M., Pazandeh, H.: Two fixed-point theorems for special mappings, vol. 1 (2009). arXiv:0903.1504v1 [math.FA]
3. Hao, L., Shaoyuan, X.: Cone metric spaces with Banach algebras and fixed point theorems of generalized Lipschitz mappings. Fixed Point Theory Appl. **320**, 1–10 (2013). https://doi.org/10.1186/1687-1812-2013-320
4. Huang, L.G., Zhang, X.: Cone metric spaces and fixed point theorems of contractive mappings. J. Math. Anal. Appl. **332**, 1468–1476 (2007)
5. Morales, R., Rojas, E.: Cone metric spaces and fixed point theorems of T contractive mappings. Revista Notas de Matemática **4**, 66–78 (2008)
6. Rudin, W.: Functional Analysis. McGraw-Hill Inc., New York (1991)
7. Shukla, S., Balasubramanian, S., Pavlović, M.: A generalized banach fixed point theorem. Bull. Mal. Math. Sci. Soc. **39** (2016). https://doi.org/10.1007/s40840-015-0255-5
8. Xu, S., Radenović, S.: Fixed point theorems of generalized Lipschitz mappings on cone metric spaces over Banach algebras without assumption of normality. Fixed Point Theory Appl. (2014). https://doi.org/10.1186/1687-1812-2014-102(2014)

Trends and Risk of HIV/AIDS in Turkey and Its Cities

Evren Hincal, Murat Sayan, Bilgen Kaymakamzade, Tamer Şanlidağ, Farouk Tijjani Sa'ad, and Isa Abdullahi Baba

Abstract We developed a mathematical model to study the dynamics of HIV/AIDS in Turkey and its environs. Two equilibrium points were found and local stability analysis of the equilibria was conducted. It was found that the stability of the equilibria depend on a threshold quantity; the basic reproduction ratio R_0. When $R_0 \leq 1$, the disease free equilibrium is locally asymptotically stable and disease dies out. When $R_0 \geq 1$, the endemic equilibrium is locally asymptotically stable and disease persist. We considered the dynamics of the disease in 25 cities of Turkey and found out that the basic reproduction ratio for 17 cities is greater than one, while 8 of them have less than one. This indicates there is going to be epidemics in these 17 cities and the disease will die out in the remaining cities. The results also indicated that some of the cities (with bigger basic reproduction ratio) are at bigger risk than the others. There is need for quick intervention by the relevant authorities, especially in the cities with greater risk of the disease.

E. Hincal (✉) · B. Kaymakamzade · F. T. Sa'ad · I. A. Baba
Department of Mathematics, Near East University, Mersin 10 Nicosia, Northern Cyprus, Turkey
e-mail: evren.hincal@neu.edu.tr

B. Kaymakamzade
e-mail: bilgen.kaymakamzade@neu.edu.tr

F. T. Sa'ad
e-mail: farouksaaad@yahoo.co.uk

I. A. Baba
e-mail: isababa7@yahoo.com

M. Sayan
Faculty of Medicine, Kocaeli University, Clinical Laboratory, PCR Unit, Kocaeli, Turkey
e-mail: sayanmurat@hotmail.com

M. Sayan · T. Şanlidağ
Near East University, Research Center of Experimental Health Sciences, Nicosia, Northern Cyprus, Turkey
e-mail: trsanlidag@yahoo.com

T. Şanlidağ
Faculty of Medicine, Department of Medical Microbiology, Celal Bayar University, Manisa, Turkey

A. Ashyralyev et al. (eds.), *Functional Analysis in Interdisciplinary Applications—II*,
Springer Proceedings in Mathematics & Statistics 351,
https://doi.org/10.1007/978-3-030-69292-6_8

Keywords HIV · Mathematical model · Basic reproduction number · Stability analysis

1 Introduction

Human Immunodeficiency Virus (HIV) is a lent virus that causes HIV infection. With time especially if left untreated progressed to AIDS (Acquired Immune Deficiency Syndrome). The progression came as a result of the virus infecting the CD4+ T-cells; an integral part of the human immune system which fights and kill foreign cells or any infection. For a healthy person, the level of CD4+ T-cells is in the range 800 to 1200 $cells/mm^3$. When once this number falls below $200 cells/mm^3$ in a HIV patient, then the patient is considered to progress to AIDS [14]. AIDS highly-developed into a world pandemic since 1980s; the first time it was discovered. According to reports, about 38 million people were living with HIV infection, around 4 million have been newly infected, and almost 3 million deaths due to AIDS occurred in 2005. The year with the highest epidemic was 2003 [11].

HIV can be transmitted via direct contact with contaminated blood products, needle or syringes, sexual intercourse, during birth or through breastfeeding (mother to child). Despite the fact that AIDS is always as a result of a HIV infection, not every HIV infected individual has AIDS. As a matter of fact, HIV infected individuals may live a healthy life for many years before they progressed to AIDS [11]. If left untreated, HIV-1 infection is generally disastrous within 5–10 years. With ART (antiretroviral therapy), infected individuals can live longer with less symptoms. It was reported that between 250,000 and 350,000 deaths were avoided in 2005 due to the increase in treatment strategies (WHO/UNAIDS 2005) [5]. It was reported that 2.7 million people were newly infected, 1.8 million people died of AIDS-related causes worldwide. By the end of 2010, at least 34 million people were living with HIV/AIDS across the globe. In 2011, China stated that about 2.8 million people died of AIDS-related diseases and about 7.8 million people are HIV-infected by the end of the year [12]. The HIV epidemic has become one of the major causes of death across the globe.

Today, HIV/AIDS should be regarded as a rising disease in Turkey. With a population of 80,191,582, AIDS was discovered in the year 1985 from two patients [3]. The progression of the disease continues as time passes by. In 1990, there were 34 cases, 91 cases in 1995, 119 cases in 1999, and 210 new cases in 2004 [3]. The statistics given by the Ministry of Health was that, a total of 1800 to 1992 people were reported to have HIV/AIDS from the first discovery to December 2004. Among the patients reported, 800 have CD4+ T-cell count less than $200 cells/mm^3$ as such they were classified as AIDS patients. This figure, might not give the actual number of people infected due to the setbacks of the current surveillance system. The ministry also reported that, about 2.5 million people were tested for HIV in 2003, unfortunately; only 12,500 of those tested were commercial sex workers (almost 0.5% of those tested) [11]. Moreover, the primary mode of transmission in Turkey is through Heterosexual sexual intercourse, which was mainly commercial sex workers, followed by homosexual sex and intravenous drug use. Only those tested in the above exercise, are approximate number of registered sex workers in Turkey which by far, is so small compared to the actual sex workers' population. The estimated number of unregistered sex workers in Istanbul only is almost eight times the number of registered commercial sex workers in the whole of Turkey [14].

In 2011, the Ministry of Health gave the total number of HIV/AIDS infected people as 5224 [3]. Additionally, in 2013, the Ministry stated that at least 6000 individuals were infected with HIV, and the numbers of newly reported cases in 2010, 2011, and first six months of 2013 were 589, 1068, and 587 respectively [2]. It was reported that increase in number of unregistered sex workers, inadequate knowledge and awareness of sexually transmitted diseases, increase in the number of homosexuals, and intravenous drug use are the major cause of epidemic of HIV infection in Turkey [16]. Moreover, the main route of transmitting the virus in Turkey is via heterosexual sex (53%), homosexual sex (9%), and intravenous drug users (IDU at 3%). The increase in number of HIV/AIDS patients signals a major public health issue in Turkey in the coming years, as such; it must be regarded as an emerging disease [12].

Epidemiology is the scientific study of epidemics and epidemic diseases, specifically the factors that facilitate the incidence, spread, and control of infectious diseases prevalence in the human population [5]. Mathematical models help to study the progress, spread and control of infectious diseases. It entails how to know the possible result of an epidemic and helps in managing different control programs [1, 5, 13, 15]. In the early 20th century, mathematical methods were introduced into epidemiology by McKendrick [McKendrick and Kermack, 1927] and others [6]. The mathematical modeling presented in this paper is more of a deterministic form that is to aid in the understanding of the disease. Dynamical systems, either system of ordinary or partial differential equations, are lending new insights into HIV/AIDS dynamics. Population models are most commonly used, and given hypotheses about the interactions of those populations, models can be created, analyzed, and refined.

Many researchers worked in modeling HIV/AIDS dynamics. Different phenomena are explained by different models, but none of the models exhibit all that is observed clinically. This can be attributed to the fact that much about the disease mechanism is still unknown. But once a model is tested and believed to behave well both analytically and numerically as compared with clinical data, the model can then be used to test such things as treatment strategies and the addition of secondary infections.

Basic reproductive ratio (R_0), is an important concept in epidemiology, and is one of the most valuable ideas that mathematical thinking has brought to epidemic theory [7]. It was originally developed for the study of demographics [8–11], it was studied independently for vector—borne diseases and directly transmitted human infections [6, 12]. There are many researches on (R_0) and its significance in determining epidemics of infectious disease in literature [12].

R_0 is defined as the expected number of secondary infections caused by one individual in a completely susceptible population. From the definition above, it can be understood that when $R_0 < 1$, it means each infected individual will produce on average, less than one new infected individual, and we therefore predict that the infection will die out from the population. If $R_0 > 1$, the infection will spread there by causing epidemics. This threshold behavior is the most important and useful aspect of the R_0 concept. In this paper we will use R_0 to determine the occurrence of HIV/AIDS epidemics or otherwise in Turkey and its environs, using the raw data obtained from Turkey.

The paper is organized as follows: In Sect. 2, the model is presented and the basic reproduction number is obtained. In Sect. 3, stability of the equilibria was investigated. In Sect. 4, results are obtained and epidemics or otherwise of HIV/AIDS are determined for Turkey and its environs using the raw data obtained from Turkey in 2016. Finally Sect. 5 is the discussion and conclusion of the research.

2 Construction of the Model

The model is a system of differential equations as follows; where the meaning of parameters and variables are given by Table 1.

$$\begin{aligned}\frac{dS}{dt} &= \Lambda - \beta\frac{SH}{N} - \mu S,\\ \frac{dH}{dt} &= \beta\frac{SH}{N} - (\mu + v)H,\\ \frac{dR}{dt} &= vH + \mu S + \mu H,\end{aligned} \tag{1}$$

$$S(t) > 0, \quad H(t) \geq 0, \quad and \quad R(t) \geq 0. \tag{2}$$

Since $N = S + H + R$, then equation (1) can be reduced to

Table 1 Meaning of parameters and variables of the model (1)

Parameters/variables	Meaning
$S(t)$	Population of susceptibles
$H(t)$	Infected with HIV/AIDS
$R(t)$	Removed by natural death or death due to HIV/AIDS
Λ	Birth rate of Turkish people by states
$\frac{1}{v}$	Average life expectancy of HIV/AIDS patients under treatment
$\frac{1}{\mu}$	Life expectancy of Turkish people
β	Incidence rate

$$\frac{dS}{dt} = \Lambda - \beta\frac{SH}{N} - \mu S, \; \frac{dH}{dt} = \beta\frac{SH}{N} - (\mu + v)H. \tag{3}$$

It follows from (3) that,

$$\frac{dS}{dt} + \frac{dH}{dt} + \frac{dR}{dt} = \Lambda - \beta\frac{SH}{N} - \mu S + \beta\frac{SH}{N} - (\mu + v)H + vH + \mu S + \mu H \leq \Lambda. \tag{4}$$

Then, $limsup(S + H + R) \leq \Lambda$ as t approaches infinity. Thus the feasible region for (1) is

$$\pi = \{(S + H + R) : S + H + R \leq \Lambda, \;\; S(t) > 0, \;\; H(t) \geq 0, \;\; R(t) \geq 0\}. \tag{5}$$

2.1 *Assumptions*

The following assumptions are made.

1. $S(0)$ is considered to be 2016 population for each state.
2. We assumed homogeneous mixing in the population.
3. $v = 0.0149$ is considered (That is averagely 67 years is the life span of HIV+ people).

2.2 *Equilibrium Points*

We find two equilibrium points: Disease free and endemic equilibrium points.

$$E_0 = (S_0 = \frac{\Lambda}{\mu}, H_0 = 0) \;\; and \;\; E_1 = (S_1 = \frac{\Lambda + v}{\mu}, H_1 = \frac{\Lambda\alpha - \mu v - \mu^2}{\alpha(v + \mu)}). \tag{6}$$

The equilibrium points E_1 exists only if

$$\frac{\Lambda\alpha}{\Lambda(v + \mu)} \geq 1. \tag{7}$$

2.3 *Basic Reproduction Ratio* (R_0)

This is defined as the number of secondary infections caused by an infective individual in a wholly susceptible population. Here we used the next generation matrix to find it as,

$$F = [\alpha SH], V = [(\mu + v)\ ], M = \partial F = [\alpha S\ ], M(E_0) = [\alpha S_0\ ]K = \partial V, V = [\alpha + v\ ], \tag{8}$$

$$K^{-1} = \frac{1}{v+\mu}, MK^{-1} = \frac{\alpha S_0}{v+\mu}.$$

The spectral radius, which is the dominant eigenvalue, is $\frac{\alpha S_0}{v+\mu}$. Hence the basic reproduction ratio is;

$$R_0 = \frac{\alpha S_0}{(v+\mu)}. \tag{9}$$

3 Stability Analysis of the Equilibria

In this section stability analysis of the equilibria is obtained. The conditions for the stability of the equilibria in each case depends on the magnitude of the basic reproduction ratio R_0. We have the following theorems and their proofs.

Theorem 1 *E_0 is locally asymptotically stable if $R_0 < 1$.*

Proof The Jacobian matrix for equation (1) at the disease free equilibrium is

$$J(E_0) = \begin{Bmatrix} -\mu & -\beta S_0 \\ 0 & -(v+\mu) + \alpha S_0 \end{Bmatrix}. \tag{10}$$

The eigenvalues are $\lambda_1 = -\mu$ and $\lambda_2 = -(v+\mu) + \alpha S_0$. Since μ is nonnegative, then the first eigenvalue is negative. To show the local stability of E_0 we need to show $\lambda_2 < 0$.
$\lambda_2 < 0 \;\; if \;\; \alpha S_0 < (v+\mu)$. This implies $\lambda_2 < 0 \;\; when \;\; \frac{\alpha S_0}{(v+\mu)} < 1$. Therefore, E_0 is locally asymptotically stable if $R_0 < 1$. □

Theorem 2 *E_1 is locally asymptotically stable if $R_0 > 1$.*

Proof The Jacobian matrix for equation (1) at the disease free equilibrium is

$$J(E_1) = \begin{Bmatrix} \frac{\mu(v+\mu)-\Lambda\alpha}{(v+\mu)} - \mu & -\mu - v \\ \frac{\Lambda\alpha-\mu(v+\mu)}{(v+\mu)} & 0 \end{Bmatrix}. \tag{11}$$

To show the local stability of the endemic equilibrium, it is enough to show the trace of the above matrix is less than zero and the determinant is greater than zero

$$Trace(J(E_1)) = \frac{\Lambda\alpha}{(v+\mu)} \;\; and \;\; Det(J(E_1)) = \Lambda\alpha - \mu(v+\mu). \tag{12}$$

Since μ, Λ, v and μ are all positive then $Trace(J(E_1)) < 0$, and $Det(J(E_1)) > 0$ if $\frac{\alpha S_0}{(v+\mu)} > 1$. Therefore, E_1 is locally asymptotically stable if $R_0 > 1$.

4 Results

In this section, results are calculated by simulating the model using the real data obtained from Turkey in 2016. We considered data from 25 cities in Turkey, and compared the HIV/AIDS epidemics in the cities using the basic reproduction ratios. Table 2, presents the number of cases and the values of basic reproduction ratios for 25 cities as calculated from the model.

Table 2 Meaning of parameters and variables of the model (1)

States	HIV cases	R_0
Ankara	468	6.1093
Antalya	340	12.4188
Adana	215	7.4001
Bolu	20	1.3967
Bursa	202	3.2656
Canakkale	12	0.3343
Denizli	65	2.2824
Diyarkabir	60	9.3288
Elazig	40	3.5748
Gaziantep	171	14.4311
Isparta	10	0.5751
Istanbul	4447	28.4510
Izmir	355	3.1323
Kayseri	83	3.9013
Kocaeli	233	11.0001
Mardin	2	0.5914
Mersin	150	5.3210
Ordu	6	0.2005
Sakaraya	75	3.4082
Samsun	200	4.9347
Sivas	30	1.4282
Urfa	1	0.1577
Tokat	1	0.0393
Duzce	2	0.1874
Edirne	7	0.2323

5 Discussions and Conclusions

A mathematical model for HIV/AIDS is constructed. Two equilibrium points (Disease free and endemic) are found and stability of each of the equilibrium point was shown to depend on the magnitude of the basic reproduction ratio. It was shown that if

$$R_0 = \frac{\alpha S_0}{(v + \mu)}. \tag{13}$$

is less than one, the disease free equilibrium is locally asymptotically stable. Also, if the value is greater than or equals to one the endemic equilibrium is locally asymptotically stable.

The basic reproduction ratios for Ankara, Antalya, Adana, Bolu, Bursa, Denizli, Diyarbakır, Elazğ, Gaziantep, Istanbul, Izmir, Kayseri, Kocaeli, Mersin, Sakarya, Samsun, and Sivas are respectively as 6.1093, 12.4188, 7.4001, 1.3967, 3.2656, 2.2824, 9.3288, 3.5748, 14.4311, 28.4510, 3.1323, 3.9013, 11.0001, 5.3210, 3.4082, 4.9347, and 1.4282. Though the endemic equilibrium for the above 17 cities is stable, but the threat is more in some cities than the others. For example, Antalya, Gaziantep, Istanbul and Kocaeli have $R_0 > 10$, which means one HIV/AIDS positive individual can infect more than ten susceptible individuals. Ankara, Adana, Dıyarbakır and Mersin have basic reproduction ratio greater than 5 which means one HIV/AIDS positive individual can infect more than five susceptible individuals. Lastly, Bolu, Bursa, Denizli, Elazığ, Izmir, Kayseri, Sakarya, Samsun, and Sivas have basic reproduction ratio greater than 1 which means one HIV/AIDS positive individual can infect more than one susceptible individual.

It was also shown that the following cities; Çanakkale, Isparta, Mardin, Ordu, Urfa, Tokat, Düzce and Edirne have basical reproduction ratios less than 1 which shows there will be no epidemic in these cities. Although the values of the R_0 show no epidemics in the cities but it is evident from the magnitude of the R0 that some cities are at greater threat than others. For example Isparta and Mardin have 0.5751 and 0.5914 as their basic reproduction ratios respectively. Çanakkale, Ordu, Urfa, Tokat, Düzce, and Edirne have 0.3343, 0.2005, 0.1577, 0.0393, 0.1874, and 0.2323 respectively.

The magnitude of the basic reproduction ratio was shown to depend directly on the incidence rate and S_0 and inversely on average life span of HIV/AIDS patients (v) and life expectancy of the population (μ). In general from Fig. 1, it can be observed that cities with higher number of cases have higher basic reproduction ratios. In a nutshell to overcome the epidemics the concerned authorities should reduce the basic reproduction ratio R_0 to less than one. This can be achieved by reducing both the incidence rate and number of susceptibles in the population. The result also shows the need for quick intervention by the relevant authorities in converting HIV/AIDS in Turkey.

In our analysis we did not account for the effect of behavioral change arising both from number of AIDS cases in the community as well as awareness by governmental and nongovernmental organizations. We also did not consider the possible effects of

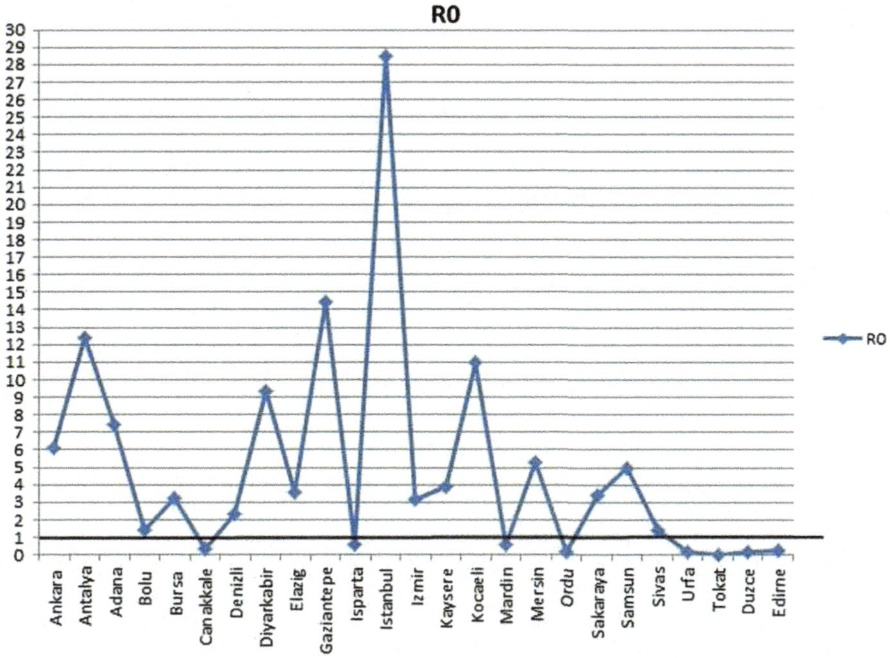

Fig. 1 Basic Reproduction ratio for HIV/AIDS in Turkey and some of its cities

extensive use of ARVs in terms of method of distributing drugs through public or private health institutions or a combination of both could determine whether patients on ARVs revert back to the infective class. This together with reduced infectiousness due to lower viral loads for those on treatment was not accounted for. These can be considered as draw backs to our model.

There are several implications of our results to public health. The endemic equilibrium should be brought as low as possible especially during the first wave of the epidemic. This model suggests that this can be achieved by prolonging the lifetime of the AIDS patients for as long as possible. Second, HIV prevalence at low prevalence levels become less sensitive to changes in the dynamics of HIV epidemic because it is overpowered by demographic changes especially the recruitment of susceptibles. At low prevalence levels, there is hence need to track trends in number of persons infected with HIV than tracking HIV prevalence.

References

1. Al-Sheikh, S., Musali, F., Alsolami, S.M.: Stability analysis of an HIV/AIDS epidemic model with screening. Int. Math. Forum **6**, 3251–3273 (2001)
2. Anderson, R.M., Medley, G. May, F.R. M., Johnson, A.M.: A preliminary study of the transmission dynamics of the human immunodeficiency virus (HIV), the causative agent of AIDS. IMA J. Math. Appl. Med. Biol. **3**, 229–263 (1986)
3. Diekmann, O, Heesterbeek, J.A.P.: Mathematical Epidemiology of Infectious Diseases, Model Building, Analysis and Interpretation. 3rd edn. Wiley, New York (2000)
4. Dietz, K.: The estimation of the basic reproduction number for infectious diseases. Stat. Methods Med. Res. **2**, 23–41 (1975)
5. Alspar, Dilek., et al.: Molecular epidemiology of HIV in a cohort of men having sex with men from Istanbul. Med. Microbiol. Immunol. (2013). https://doi.org/10.1007/500430-0120285-7
6. Dublin, L.I., Lotka, A.J.: On the true rate of natural increase of a population. J. Am. Stat. Assoc. **20**, 305–339 (1925)
7. Guvenc, K., et al.: Analysis of the treatment costs of HIV/AIDS in Turkey. Farm Econ. Health Econ. Therapeutic Pathways **17**, 13–17 (2016)
8. Heesterbeek, J.A.P.: A brief history of R_0 and a recipe for its calculation. Acta Biotheret. **50**, 189–204 (2002)
9. Heesterbeek, J.A.P., Dietz, K.: The concept of R_0 in epidemic theory. Stat. Neerlandica **50**, 89–110 (1996)
10. Liming, C., Xuenzhi, L., Mini, G.B.: Stability analysis of an HIV/Aids epidemic model with treatment. J. Comput. Appl. Math. **229**, 313–323 (2009)
11. Pinar, A., Selma, K.: Is there a hidden HIV/AIDS epidemic in Turkey?: The gap between the numbers and the facts. Marmara Med. J. **2**, 90–97 (2007)
12. Qianqian, L, et. al: Stability analysis of an HIV/AIDS dynamics model with drug resistance. Discret. Dyn. Nat. Soc. (2012). https://doi.org/10.1155/2012/162527
13. Sharp, F.R., Lotka, A.J.: A problem in age distribution. Philos. Mag. **6**, 435–438 (1911)
14. Tugrul, E., Nuketc, P.E.: Status HIV/AIDS epidemic In Turkey. Acta Med. **1**, 19–24 (2012)
15. Unal, A.: Aids knowledge and attitudes in a Turkish population. An epidemiology Study. BMC Public Health (2005). https://doi.org/10.1186/1471-2458-5-95
16. Zindoga, M., Prasenjit, D., Christinah, C., Farai, N.G.l.: Global analysis of an HIV/AIDS epidemic model. World J. Model. Simlul. **6**, 231–240 (2010)

On the Stability of Schrödinger Type Involutory Differential Equations

Allaberen Ashyralyev, Twana Abbas Hidayat, and Abdisalam A. Sarsenbi

Abstract In the present paper, the Schrödinger type involutory differential equation is considered which is stated as

$$i\frac{dv(t)}{dt} + Av(t) + bAv(-t) = f(t), t \in I = (-\infty, \infty), v(0) = \varphi$$

in a Hilbert space H with a self-adjoint positive definite operator A. Here, operator approach enables us to apply the results on abstract problem on multi-dimensional or nonlocal problems which deserve a studious treatment. Throughout the paper, the main theorem on stability estimates for the solution of the abstract problem under the condition $|b| < 1$ is established. Furthermore, the main theorem is applied to a one-dimensional problem with nonlocal condition and involution and a multi-dimensional problem with Dirichlet and Neumann conditions on the boundary.

Keywords Involutory differential equation · Stability · Hilbert space · Positive operator

---

A. Ashyralyev
Department of Mathematics, Near East University, Nicosia, TRNC, 10 Mersin, Turkey
e-mail: allaberen.ashyralyev@neu.edu.tr

Peoples' Friendship University of Russia, (RUDN University)6 Miklukho-Maklaya St, Moscow 117198, Russian Federation

Institute of Mathematics and Mathematical Modeling, 050010 Almaty, Kazakhstan

T. A. Hidayat (✉)
Department of Mathematics, College of Education, Sulaymaniyah, Iraq
e-mail: twana.margaret@gmail.com

A. A. Sarsenbi
Department of Mathematical Methods and Modeling, M.Auezov South Kazakhstan State University, Shymkent, Kazakhstan
e-mail: abdisalam@mail.ru

A. Ashyralyev et al. (eds.), *Functional Analysis in Interdisciplinary Applications—II*,
Springer Proceedings in Mathematics & Statistics 351,
https://doi.org/10.1007/978-3-030-69292-6_9

1 Introduction

It is known that various problems in physics lead to the Schrödinger equation. Methods of solutions of the problems for Schrödinger equation have been studied extensively by many researchers (see, e.g., [6–8, 11, 16, 19, 23], and the references given therein).

Time delay is one of the most common phenomena occurring in many engineering applications. In control theory, the process of sampled-data control is a typical example where time delay happens in the transmission from measurement to controller.

Theory and applications of delay linear and nonlinear Schrödinger equations with the delay term is an operator of lower order with respect to the operator term were widely investigated (see, e.g., [14, 17, 18, 26–28], and the references given therein).

For example, in the article [27], the boundary stabilization of a Schrödinger equation with variable coefficient where the boundary observation suffers from a fixed time delay was studied. This is a generalization of the similar work for the Schrödinger equation in [18] by using the separation principle [17] for constant coefficients. The variable coefficients make the system too complicated to estimate the solution, which relies on the estimation of the eigenvalues and eigenfunctions by asymptotic analysis. In [26], existence and uniqueness of local solutions of nonlinear Schrödinger equation with delay was investigated. In [14], the existence and upper semi-continuity of the global attractor for discrete nonlinear delay Schrödinger equation was established. The paper [28] was devoted to study of traveling waves of nonlinear Schrödinger equation with distributed delay by applying geometric singular perturbation theory, differential manifold theory and the regular perturbation analysis for a Hamiltonian system. Under the assumptions that the distributed delay kernel was strong general delay kernel and the average delay was small, the existence of solitary wave solutions was investigated by differential manifold theory. Then by utilizing the regular perturbation analysis for a Hamiltonian system, the periodic traveling wave solutions were explored. Finally, theory and applications of partial differential equations with the delay term is an operator of same order with respect to the other operator term were widely investigated for delay parabolic differential equations (see, e.g., [1, 3–5, 12, 13, 24], and the references given therein).

In the paper [2], the stability of the initial value problem for the Schrödinger equation with time delay in a Hilbert space with self-adjoint positive definite operator was investigated. Theorems on stability estimates for the solution of this problem were established. The application of theorems for three types of Schrödinger problems were provided.

In an experiment measuring the population growth of a species of water fleas, Nesbit (1997), used a DDE model in his study. In simplified form his population equation was

$$N'(t) = aN(t-d) + bN(t).$$

He got into a difficulty with this model because he did not have a reasonable history function to carry out the solution of this equation. To overcome this roadblock he proposed to solve a "time reversal" problem in which he sought the solution to an IDE that is neither a DDE, nor a FDE. He used a "time reversal" equation to get the juvenile population prior to the beginning time $t = 0$. The time reversal problem is a special case of a type of equation called an involutory differential equation. These are defined as equations of the form

$$y'(t) = f(t; y(t); y(u(t))), y(t_0) = y_0. \quad (1)$$

Here $u(t)$ is involution function, that is $u(u(t)) = t$, and t_0 is a fixed point of u. For the "time reversal" problem, we have the simplest IDE, one in which the deviating argument is $u(t) = -t$. This function is involution since

$$u(u(t)) = u(-t) = -(-t) = t.$$

Our goal in this paper is to investigate the stability of the Schrödinger type involutory differential equation

$$i\frac{dv(t)}{dt} + Av(t) + bAv(-t) = f(t), t \in I = (-\infty, \infty), v(0) = \varphi \quad (2)$$

in a Hilbert space H with a self-adjoint positive definite operator A. The main theorem on stability estimates for the solution of problem (2) under the assumption $|b| < 1$ is established. The application of the theorem for four types of Schrödinger type involutory problems are provided. Therefore, this article has great significance for obtaining stability estimates of the Schrödinger type of involutory differential equations.

The paper is organized as follows. The Sect. 1 is introduction. In Sect. 2, the main theorem on stability of problem (2) is established. In Sect. 3, theorems on the stability estimates for the solution of four problems for the Schrödinger type involutory equation are proved. Finally, Sect. 4 is conclusion.

2 The Main Theorem on Stability

A function $v(t)$ is called a solution of problem (2), if the following conditions are satisfied:

i. $v(t)$ is continuously differentiable function on the interval $(-\infty, \infty)$. The derivative at the endpoint $t = 0$ is understood as the appropriate unilateral derivative.
ii. The element $v(t)$ belongs to $D(A)$ for all $t \in (-\infty, \infty)$, and the function $Av(t)$ is continuous on the interval $(-\infty, \infty)$.
iii. $v(t)$ satisfies the main equation and initial condition in (2).

Now, we will obtain the initial value problem for the second order differential equation equivalent to problem (2) under smoothness solution. Using initial condition and equation in problem (2), we get

$$v\left(0\right)=\varphi, v'\left(0\right)=-if(0)+i(1+b)A\varphi. \tag{3}$$

Differentiating equation (2), we get

$$iv^{''}\left(t\right)=bAv'(-t)-Av'(t)+f'(t). \tag{4}$$

Substituting $-t$ for t into Eq. (2), we get

$$iv'(-t)=-bAv(t)-Av(-t)+f(-t). \tag{5}$$

Using these equations, we can eliminate $v(-t)$ and $v^{'}(-t)$ terms. Actually, using Eqs. (4) and (5), we get

$$v^{''}\left(t\right)=b^2A^2v(t)+bA^2v(-t)-bAf(-t)+iAv'(t)-if'(t).$$

Using that and Eq. (2), we get

$$v^{''}\left(t\right)=b^2A^2v(t)+A\left[-iv'(t)-Av(t)+f(t)\right]-bAf(-t)+iAv'(t)-if'(t)$$

or

$$v^{''}(t)+\left(1-b^2\right)A^2v(t)=Af(t)-bAf(-t)-if'(t). \tag{6}$$

So, we get the initial value problem (3) and (6) for the second order differential equation in a Hilbert space H. It is easy to see that $\left(1-b^2\right)A^2$ is the self-adjoint positive definite operator under the assumption $|b|<1$. We have that (see [12])

$$v\left(t\right)=c\left(t\right)v(0)+s\left(t\right)v'(0)+\int_0^t s\left(t-y\right)\left[Af(y)-bAf(-y)-if'(y)\right]dy, \tag{7}$$

where

$$c\left(t\right)=\frac{e^{it\mu A}+e^{-it\mu A}}{2}, s\left(t\right)=\left(\mu A\right)^{-1}\frac{e^{it\mu A}-e^{-it\mu A}}{2i}, \mu=\sqrt{1-b^2}.$$

Since

$$-i\int_0^t s\left(t-y\right)f'(y)dy=is(t)f(0)-i\int_0^t c\left(t-y\right)f(y)dy,$$

$$-\int_0^t s\left(t-y\right)bAf(-y)dy=-b\int_{-t}^0 s\left(t+y\right)Af(y)dy,$$

we have the following formula

$$v(t) = c(t)\varphi + is(t)(1+b)A\varphi + \int_0^t s(t-y)\,Af(y)dy \tag{8}$$

$$-b\int_{-t}^{0} s(t+y)\,Af(y)dy - i\int_0^t c(t-y)\,f(y)dy,\ t \in I.$$

For the self-adjoint positive definite operator A, we have the following estimates

$$\|c(t)\|_{H\to H} \le 1,\ \|\mu As(t)\|_{H\to H} \le 1. \tag{9}$$

Theorem 2.1 *Assume that $\varphi \in H$ and $f(t)$ is a continuous function on I. For the solution of problem (2) the following stability estimate holds*

$$\sup_{t\in I}\|v(t)\|_H \le \left(1+\frac{1+|b|}{\mu}\right)\left[\|\varphi\|_H + \int_{-\infty}^{\infty}\|f(y)\|_H\,dy\right]. \tag{10}$$

The proof of Theorem 2.1 is based on triangle inequality and formula (8), on estimates (9).

Theorem 2.2 *Assume that $\varphi \in D(A)$ and $f(t)$ is a continuously differentiable function on I. Then, for the solution of problem (2) the following stability estimates hold*

$$\sup_{t\in I}\left\|v'(t)\right\|_H,\ \sup_{t\in I}\|Av(t)\|_H$$

$$\le \left(1+\tfrac{(1+|b|)}{\mu}\right)\|A\varphi\|_H + \left[\tfrac{2}{\mu^2}(1+|b|)+\tfrac{1}{\mu}\right]\left[\|f(0)\|_H + \int_{-\infty}^{\infty}\left\|f'(y)\right\|_H dy\right]. \tag{11}$$

Proof Applying formula (8), we get

$$Av(t) = c(t)\,A\varphi + i\,As(t)(1+b)A\varphi + \int_0^t As(t-y)\,Af(y)dy$$

$$-b\int_{-t}^{0} As(t+y)\,Af(y)dy - i\int_0^t Ac(t-y)\,f(y)dy.$$

Since

$$\int_0^t As(t-y)\,Af(y)dy = \frac{1}{\mu^2}\left[f(t)-c(t)f(0)\right] - \frac{1}{\mu^2}\int_0^t c(t-y)\,f'(y)dy,$$

$$\int_{-t}^{0} As\,(t+y)\,Af(y)dy = \frac{1}{\mu^2}\,[f(-t) - c(t)f(0)] + \frac{1}{\mu^2}\int_{-t}^{0} c\,(t+y)\,f'(y)dy,$$

$$\int_{0}^{t} Ac\,(t-y)\,f(y)dy = As(t)f(0) + \int_{0}^{t} As\,(t-y)\,f'(y)dy,$$

we have that

$$Av(t) = c\,(t)\,A\varphi + iAs\,(t)\,(1+b)A\varphi$$

$$+\frac{1}{\mu^2}\,[f(t) - c(t)f(0)] - \frac{b}{\mu^2}\,[f(-t) - c(t)f(0)] - iAs(t)f(0)$$

$$-\frac{1}{\mu^2}\int_{0}^{t} c\,(t-y)\,f'(y)dy - \frac{b}{\mu^2}\int_{-t}^{0} c\,(t+y)\,f'(y)dy$$

$$-i\int_{0}^{t} As\,(t-y)\,f'(y)dy = \sum_{i=1}^{4} G_i\,(t),$$

where

$$G_1\,(t) = c\,(t)\,A\varphi + iAs\,(t)\,(1+b)A\varphi,$$

$$G_2\,(t) = \frac{1}{\mu^2}\,[f(t) - c(t)f(0)] - \frac{b}{\mu^2}\,[f(-t) - c(t)f(0)] - iAs(t)f(0),$$

$$G_3\,(t) = -\frac{1}{\mu^2}\int_{0}^{t} c\,(t-y)\,f'(y)dy - \frac{b}{\mu^2}\int_{-t}^{0} c\,(t+y)\,f'(y)dy,$$

$$G_4\,(t) = -i\int_{0}^{t} As\,(t-y)\,f'(y)dy.$$

Now, applying the triangle inequality, we obtain

$$\|Av(t)\|_H \le \sum_{i=1}^{4} \|G_i\,(t)\|_H\,,$$

for any $t \in I$. Therefore, we will estimate $\|G_i\,(t)\|_H$, $i = 1, 2, 3, 4$, separately. First, using estimates (9),we obtain

$$\|G_1\,(t)\|_H \le \left(1 + \frac{(1+|b|)}{\mu}\right)\|A\varphi\|_H$$

for any $t \in T$. Second, using the triangle inequality and estimates (9), we get

$$\|G_2(t)\|_H \le \left(\frac{2}{\mu^2}(1+|b|)+\frac{1}{\mu}\right)\|f(0)\|_H + \frac{1}{\mu^2}(1+|b|)\int_{-\infty}^{\infty}\left\|f'(y)\right\|_H dy$$

for any $t \in I$. Third, using the triangle inequality and estimates (9), we get

$$\|G_3(t)\|_H \le \frac{1}{\mu^2}(1+|b|)\int_{-\infty}^{\infty}\left\|f'(y)\right\|_H dy$$

for any $t \in I$. Fourth, using the triangle inequality and estimates (9), we get

$$\|G_4(t)\|_H \le \frac{1}{\mu}\int_{-\infty}^{\infty}\left\|f'(y)\right\|_H dy$$

for any $t \in I$. Combining these estimates, we can write

$$\|Av(t)\|_H \le \left(1+\frac{(1+|b|)}{\mu}\right)\|A\varphi\|_H + \left(\frac{2}{\mu^2}(1+|b|)+\frac{1}{\mu}\right)\|f(0)\|_H$$

$$+\left[\frac{2}{\mu^2}(1+|b|)+\frac{1}{\mu}\right]\int_{-\infty}^{\infty}\left\|f'(y)\right\|_H dy$$

for any $t \in I$. This completes the proof of Theorem 2.2.

Applying similar approach one can obtain same stability estimates for the solution of the initial value problem for the involutory Schrödinger differential equation

$$i\frac{dv(t)}{dt} + Av(t) + bAv(t_0 - t) = f(t), t \in I = (-\infty, \infty), v\left(\frac{t_0}{2}\right) = \varphi \qquad (12)$$

in a Hilbert space H with a self-adjoint positive definite operator A.

3 Applications

In this section, we consider the applications of Theorems 2.1–2.2.

First, the boundary value problem for the Schrödinger type involutory differential equation with nonlocal conditions

$$\begin{cases} iu_t(t,x) - (a(x)u_x(t,x))_x + \delta u(t,x) \\ = b\left(-\left(a(x)u_x(-t,x)\right)_x + \delta u(-t,x)\right) \\ +f(t,x), t \in I, 0 < x < 1, \\ u(0,x) = \varphi(x), 0 \le x \le 1, \\ u(t,0) = u(t,1), u_x(t,0) = u_x(t,1), \ t \in I \end{cases} \tag{13}$$

is considered. Under compatibility conditions problem (13) has a unique solution $u(t,x)$ for the smooth functions $a(x) \ge a > 0,\ x \in (0,1),\ \delta > 0,\ a(1) = a(0)$, $\varphi(x)\ (x \in [0,1]),\ f(t,x)(t \in I,\ 0 < x < 1)$ and $|b| < 1$. This allows us to reduce the boundary value problem (13) to the boundary value problem (2) in a Hilbert space $H = L_2[0,1]$ with a self-adjoint positive definite operator A^x defined by formula

$$A^x u(x) = -(a(x)u_x)_x + \delta u \tag{14}$$

with domain

$$D(A^x) = \{u(x) : u(x), u_x(x), (a(x)u_x)_x \in L_2[0,1], u(1) = u(0), u_x(1) = u_x(0)\}.$$

Applying the symmetry property of the space operator A^x with the domain $D(A^x) \subset W_2^2[0,1]$ and estimates (9) in $H = L_2[0,1]$, we can obtain the following theorem on stability of problem (13).

Theorem 3.1 *For solutions of problem (13) we have following stability estimates*

$$\sup_{t\in I} \|u(t,\cdot)\|_{L_2[0,1]} \le M_1 \left[\|\varphi\|_{L_2[0,1]} + \int_{-\infty}^{\infty} \|f(y,\cdot)\|_{L_2[0,1]} dy \right],$$

$$\sup_{t\in I} \|u_t(t,\cdot)\|_{L_2[0,1]} + \sup_{t\in I} \|u(t,\cdot)\|_{W_2^2[0,1]}$$

$$\le M_1 \left[\|\varphi\|_{W_2^2[0,1]} + \|f(0,\cdot)\|_{L_2[0,1]} + \int_{-\infty}^{\infty} \left\|f'(y,\cdot)\right\|_{L_2[0,1]} dy \right],$$

where M_1 does not depend on $\varphi(x)$ and $f(t,x)$. Here, $W_2^2\,[0,1]$ is the Sobolev space of all square integrable functions $\psi\,(x)$ defined on $[0,1]$ equipped with the norm

$$\|\psi\|_{W_2^2[0,1]} = \left\{ \int_0^1 \left[\psi^2(x) + \psi_{xx}^2(x)\right] dx \right\}^{1/2}.$$

Second, the boundary value problem for the Schrödinger type involutory differential equation with the involution

$$\begin{cases} iu_t(t,x) - (a(x)u_x(t,x))_x + \beta\,(a(-x)u_x\,(t,-x))_x + \sigma u(t,x) \\ = b\,(-\,(a(x)u_x(-t,x))_x + \beta\,(a(-x)u_x\,(-t,-x))_x + \sigma u(-t,x)) \\ + f(t,x), t \in I, -l < x < l, \\ u(0,x) = \varphi(x), -l \le x \le l, \varphi\,(-l) = \varphi\,(l) = 0, \\ u(t,-l) = u(t,l) = 0, \ t \in I \end{cases} \tag{15}$$

is considered. Under compatibility conditions problem (15) has a unique solution $u(t,x)$ for the smooth functions
$a \ge a(x) = a\,(-x) \ge \delta > 0$ and $\sigma > 0$ is a sufficiently large number,
$\varphi(x)$ $(x \in [-l,l])$, $f(t,x)(t \in I, -l < x < l)$ and $|b| < 1$. This allows us to reduce the boundary value problem (15) to the boundary value problem (2) in a Hilbert space $H = L_2[-l,l]$ with a self-adjoint positive definite operator A^x defined by formula (see [9])

$$A^x u(x) = -(a(x)u_x)_x + \beta\,(a(-x)u_x\,(t,-x))_x + \sigma u(t,x) \tag{16}$$

with domain

$$D(A^x) = \{u(x) : u(x), u_x(x), (a(x)u_x)_x \in L_2[-l,l], u(-l) = u(l) = 0\}.$$

Applying the symmetry property of the space operator A^x with the domain $D(A^x) \subset W_2^2\,[-l,l]$ and estimates (9) in $H = L_2[-l,l]$, we can obtain the following theorem on stability of problem (15).

Theorem 3.2 *Assume that $\delta - a\,|\beta| \ge 0$. For solutions of problem (15) we have following stability estimates*

$$\sup_{t\in I} \|u(t,\cdot)\|_{L_2[-l,l]} \le M_3 \left[\|\varphi\|_{L_2[-l,l]} + \int_{-\infty}^{\infty} \|f(y,\cdot)\|_{L_2[-l,l]}\,dy \right],$$

$$\sup_{t\in I} \|u_t(t,\cdot)\|_{L_2[-l,l]} + \sup_{t\in I} \|u(t,\cdot)\|_{W_2^2[-l,l]}$$

$$\le M_3 \left[\|\varphi\|_{W_2^2[-l,l]} + \|f(0,\cdot)\|_{L_2[-l,l]} + \int_{-\infty}^{\infty} \left\|f'(y,\cdot)\right\|_{L_2[-l,l]} dy \right],$$

where M_3 does not depend on $\varphi(x)$ and $f(t,x)$. Here, $W_2^2[-l,l]$ is the Sobolev space of all square integrable functions $\psi(x)$ defined on $[-l,l]$ equipped with the norm

$$\|\psi\|_{W_2^2[0,1]} = \left\{ \int_{-l}^{l} \left[\psi^2(x) + \psi_{xx}^2(x)\right] dx \right\}^{1/2}.$$

Third, let Ω be the unit open cube in the n-dimensional Euclidean space $R^n (x = (x_1, \cdots, x_n) : 0 < x_k < 1, k = 1, \cdots, n)$ with boundary S, $\overline{\Omega} = \Omega \cup S$. In $I \times \Omega$, the boundary value problem for the multi-dimensional Schrödinger type involutory differential equation with the Dirichlet condition

$$\begin{cases} i\frac{\partial u(t,x)}{\partial t} - \sum\limits_{r=1}^{n} \left(a_r(x)u_{x_r}(t,x)\right)_{x_r} \\ = -b\sum\limits_{r=1}^{n} \left(a_r(x)u_{x_r}(-t,x)\right)_{x_r} \\ +f(t,x), t \in I, x \in \Omega, \\ u(0,x) = \varphi(x), x \in \overline{\Omega}, \\ u(t,x) = 0, \ x \in S, \ t \in I \end{cases} \tag{17}$$

is considered. Here $a_r(x) \geq a > 0$, $(x \in \Omega)$, $\varphi(x) (x \in \overline{\Omega})$ and $f(t,x)$ $(t \in I, x \in \Omega)$ are given smooth functions and $|b| < 1$.

We consider the Hilbert space $L_2(\overline{\Omega})$ of the all square integrable functions defined on $\overline{\Omega}$, equipped with the norm

$$\| f \|_{L_2(\overline{\Omega})} = \left(\int \cdots \int_{x \in \overline{\Omega}} |f(x)|^2 dx_1 \cdots dx_n \right)^{\frac{1}{2}}.$$

Under compatibility conditions problem (17) has a unique solution $u(t,x)$ for the smooth functions $\varphi(x)$, $a_r(x)$ and $f(t,x)$. This allows us to reduce the problem (17) to the boundary value problem (2) in the Hilbert space $H = L_2(\overline{\Omega})$ with a self-adjoint positive definite operator A^x defined by formula

$$A^x u(x) = -\sum_{r=1}^{n} (a_r(x) u_{x_r})_{x_r} \tag{18}$$

with domain

$$D(A^x) = \left\{ u(x) : u(x), u_{x_r}(x), (a_r(x)u_{x_r})_{x_r} \in L_2(\overline{\Omega}), 1 \leq r \leq n, u(x) = 0, x \in S \right\}.$$

Therefore, estimates (9) in $H = L_2(\overline{\Omega})$ permit us to get the following theorem on stability of problem (17).

Theorem 3.3 *For the solutions of problem (17), we have following stability estimates*

$$\sup_{t\in I} \|u(t,\cdot)\|_{L_2(\overline{\Omega})} \leq M_4 \left[\|\varphi\|_{L_2(\overline{\Omega})} + \int_{-\infty}^{\infty} \|f(y,\cdot)\|_{L_2(\overline{\Omega})} dy \right],$$

$$\sup_{t\in I} \|u_t(t,\cdot)\|_{L_2(\overline{\Omega})} + \sup_{t\in I} \|u(t,\cdot)\|_{W_2^2(\overline{\Omega})}$$

$$\leq M_4 \left[\|\varphi\|_{W_2^2(\overline{\Omega})} + \|f(0,\cdot)\|_{L_2(\overline{\Omega})} + \int_{-\infty}^{\infty} \left\|f'(y,\cdot)\right\|_{L_2(\overline{\Omega})} dy \right],$$

where M_4 does not depend on $\varphi(x)$ and $f(t,x)$. Here and in the future, $W_2^2(\overline{\Omega})$ is the Sobolev space of all square integrable functions $\psi(x)$ defined on $\overline{\Omega}$ equipped with the norm

$$\|\psi\|_{W_2^2(\overline{\Omega})} = \left(\int \cdots \int_{x\in\overline{\Omega}} \left[|\psi(x)|^2 + \sum_{r=1}^{n} \left|\psi_{x_r x_r}(x)\right|^2 \right] dx_1 \cdots dx_n \right)^{\frac{1}{2}}.$$

The proof of Theorem 3.3 is based on estimates (9) in $H = L_2(\overline{\Omega})$ and the symmetry property of the operator A^x defined by formula (18) and the following theorem on the coercivity inequality for the solution of the elliptic differential problem in $L_2(\overline{\Omega})$.

Theorem 3.4 *For the solution of the elliptic differential problem [25]*

$$\begin{cases} A^x u(x) = \mu(x), x \in \Omega, \\ u(x) = 0, x \in S, \end{cases}$$

the following coercivity inequality holds

$$\sum_{r=1}^{n} \left\|u_{x_r x_r}\right\|_{L_2(\overline{\Omega})} \leq M_5 ||\mu||_{L_2(\overline{\Omega})}.$$

Here M_5 does not depend on $\mu(x)$.

Fourth, in $I \times \Omega$, the boundary value problem for the multi-dimensional Schrödinger type involutory differential equation with the Neumann condition

$$\begin{cases} i\frac{\partial u(t,x)}{\partial t}-\sum\limits_{r=1}^{n}\left(a_r(x)u_{x_r}(t,x)\right)_{x_r}+\delta u(t,x) \\ =b\left(-\sum\limits_{r=1}^{n}\left(a_r(x)u_{x_r}(-t,x)\right)_{x_r}+\delta u(-t,x)\right) \\ +f(t,x), t\in I, x\in\Omega, \\ u(0,x)=\varphi(x), x\in\overline{\Omega}, \\ \frac{\partial u(t,x)}{\partial \mathbf{m}}=0,\ \ x\in S,\ rt\in I \end{cases} \tag{19}$$

is considered. Here, $\mathbf{m}$ is the normal vector to S, $a_r(x)\geq a>0$, $(x\in\Omega)$, $\varphi(x)$ $(x\in\overline{\Omega})$ and $f(t,x)$ $(t\in I, x\in\Omega)$ are given smooth functions and $\delta>0$, $|b|<1$.

Problem (19) has a unique solution $u(t,x)$ for the smooth functions $\varphi(x)$ and $a_r(x)$, $f(t,x)$. This allows us to reduce the problem (19) to the boundary value problem (2) in the Hilbert space $H=L_2(\overline{\Omega})$ with a self-adjoint positive definite operator A^x defined by formula

$$A^x u(x)=-\sum_{r=1}^{n}(a_r(x)u_{x_r})_{x_r}+\delta u \tag{20}$$

with domain

$$D(A^x)=\left\{u(x):u(x),u_{x_r}(x),(a_r(x)u_{x_r})_{x_r}\in L_2(\overline{\Omega}),1\leq r\leq n,\frac{\partial u\,(x)}{\partial \mathbf{m}}=0,x\in S\right\}.$$

Therefore, estimates (9) in $H=L_2(\overline{\Omega})$ permit us to get the following theorem on stability of problem (19).

Theorem 3.5 *For the solutions of problem (19), we have following stability estimates*

$$\sup_{t\in I}\|u(t,\cdot)\|_{L_2(\overline{\Omega})}\leq M_6\left[\|\varphi\|_{L_2(\overline{\Omega})}+\int_{-\infty}^{\infty}\|f(y,\cdot)\|_{L_2(\overline{\Omega})}dy\right],$$

$$\sup_{t\in I}\|u_t(t,\cdot)\|_{L_2(\overline{\Omega})}+\sup_{t\in I}\|u(t,\cdot)\|_{W_2^2(\overline{\Omega})}$$

$$\leq M_6\left[\|\varphi\|_{W_2^2(\overline{\Omega})}+\|f(0,\cdot)\|_{L_2(\overline{\Omega})}+\int_{-\infty}^{\infty}\left\|f'(y,\cdot)\right\|_{L_2(\overline{\Omega})}dy\right],$$

where M_6 does not depend on $\varphi(x)$ and $f(t,x)$.

The proof of Theorem 3.4 is based on estimates (9) in $H = L_2(\overline{\Omega})$ and the symmetry property of the operator A^x defined by formula (19) and the following theorem on the coercivity inequality for the solution of the elliptic differential problem in $L_2(\overline{\Omega})$.

Theorem 3.6 *For the solution of the elliptic differential problem [25]*

$$\begin{cases} A^x u(x) = \mu(x), x \in \Omega, \\ \frac{\partial u(x)}{\partial \mathbf{m}} = 0, x \in S, \end{cases}$$

the following coercivity inequality holds

$$\sum_{r=1}^{n} \left\| u_{x_r x_r} \right\|_{L_2(\overline{\Omega})} \leq M_7 ||\mu||_{L_2(\overline{\Omega})}.$$

Here M_7 does not depend on $\mu(x)$.

4 Conclusion

In the present paper, the initial value problem for the Schrödinger type involutory differential equation in a Hilbert space is investigated. Theorems on stability estimates for the solution of the problem are established. The applications of these theorems for four types of Schrö dinger type involutory differential problems are provided. Moreover, applying the result of the monograph [12], the single-step difference schemes for the numerical solution of boundary value problem (2) can be constructed.

Acknowledgements The publication has been prepared with the support of the "RUDN University Program 5-100" and funded by the Science Committee of the Ministry of Education and Science of the Republic of Kazakhstan (Grant No. AP05131225).

References

1. Agirseven, D.: Approximate solutions of delay parabolic equations with the Drichlet condition. Abstr. Appl. Anal. **2012** (2012). https://doi.org/10.1155/2012/682752
2. Agirseven, D.: On the stability of the Schrodinger equation with time delay. Filomat **32**(3), 759–766 (2018)
3. Ardito, A., Ricciardi, P.: Existence and regularity for linear delay partial differential equations. Nonlinear Anal. **4**, 411–414 (1980)
4. Ashyralyev, A., Agirseven, D.: On convergence of difference schemes for delay parabolic equations. Comput. Math. Appl. **66**(7), 1232–1244 (2013)
5. Ashyralyev, A., Agirseven, D.: Well-posedness of delay parabolic difference equations. Adv. Differ. Equ. **2014**, 18 (2014). https://doi.org/10.1186/1687-1847-2014-18

6. Ashyralyev, A., Hicdurmaz, B.: A note on the fractional Schrödinger differential equations. Kybernetes **40**(5–6), 736–750 (2011)
7. Ashyralyev, A., Hicdurmaz, B.: On the numerical solution of fractional Schrödinger differential equations with the Dirichlet condition. Int. J. Comput. Math. **89**, (13–14) (2012). Special Issue: SI: 1927–1936. https://doi.org/10.1080/00207160.2012.698841.
8. Ashyralyev, A., Hicdurmaz, B.: A stable second order of accuracy difference scheme for a fractional Schrödinger differential equation. Appl. Comput. Math. **17**(1), 10–21 (2018)
9. Ashyralyev, A., Sarsenbi, A.M.: Well-posedness of an elliptic equation with involution.Electron. J. Differential Equ. Art. Num. **284** (2015)
10. Ashyralyev, A., Sarsenbi, A.M.: Well-posedness of a parabolic equation with the involution. Numer. Funct. Anal. Optim. **38**(10), 1295–1304 (2017)
11. Ashyralyev, A., Sirma, A.: Nonlocal boundary value problems for the Schrödinger equation. Comput. Math. Appl. **55**(3), 392–407 (2008). https://doi.org/10.1016/j.camwa.2007.04.021
12. Ashyralyev, A., Sobolevskii, P.E.: New Difference Schemes for Partial Differential Equations, Operator Theory Advances and Applications. Birkhäuser Verlag, Basel (2004)
13. Blasio, G.D.: Delay differential equations with unbounded operators acting on delay terms. Nonlinear Anal. Theory Methods Appl. **52**(1), 1–18 (2003)
14. Chen, T., Zhou, S.F., Zhao, C.D.: Attractors for discrete nonlinear Schrödinger equation with delay. Acta Mathematicae Applicatae Sinica, English Series **26**(4), 633–642 (2010). https://doi.org/10.1007/s10255-007-7101-y
15. Falbo, C.E.: Idempotent differential equations. J. Interdiscip. Math. **6**(3), 279–289 (2003)
16. Gordeziani, D.G., Avalishvili, G.A.: Time-nonlocal problems for Schrödinger type equations: I. Prob. Abstr. Spaces, Differ. Equ. **41**(5), 703–711 (2005)
17. Guo, B.Z., Shao, Z.C.: Regularity of a Schrödinger equation with Dirichlet control and colocated observation. Syst. Control Lett. **54**, 1135–1142 (2005)
18. Guo, B.Z., Yang, K.Y. Output feedback stabilization of a one-dimensional Schrödinger equation by boundary observation with time delay. IEEE Trans. Autom. Control **55** 1226–1232 (2010)
19. Mayfield, M.E.: Non-Reflective Boundary Conditions for Schrödinger's Equation, Ph.D. Thesis, University of Rhode Island (1989)
20. Nakatsuji, H.: Inverse Schrödinger equation and the exact wave function. Phys. Rev. A. **65**, 1–15 (2002)
21. Nesbit, R.: Delay Differential Equations for Structured Populations, Structured Population Models in Marine, Terrestrial and Freshwater Systems, pp. 89–118. Tuljapurkar & Caswell, ITP (1997)
22. Sadybekov, M.A., Sarsenbi, A.M.: Criterion for the basis property of the eigenfunction system of a multiple differentiation operator with an involution. Differ. Equ. **48** 1112–1118 (2012)
23. Serov, V., äivärinta, L.P: Inverse scattering problem for two-dimensional Schrödinger operator. J. Inverse Ill-Posed Probl. **14**(3), 295–305 (2006)
24. Sinestrari, E.: On a class of retarded partial differential equations. Math. Z. **186**, 223–224 (1984)
25. Sobolevskii, P.E.: Difference Methods for the Approximate Solution of Differential Equations. Izdat. Voronezh. Gosud. University, Voronezh (1975). (Russian)
26. Wu, J.: Theory and Applications of Partial Functional-Differential Equations. Applied Mathematical Sciences, vol. 119. Springer, New York (1996)
27. Yang, K.Y., Yao, C.Z.: Stabilization of one-dimensional Schrödinger equation with variable coefficient under delayed boundary output feedback. Asian J. Control **15**(5), 1531–1537 (2013)
28. Zhao, Z., Ge, W.: Traveling wave solutions for Schrödinger equation with distributed delay. Appl. Math. Model. **35**(2), 675–687 (2011)

On the Ternary Semigroups of Continuous Mappings

Firudin Kh. Muradov

Abstract A ternary semigroup is a nonempty set with a ternary operation which is associative. In this paper, some properties of ternary semigroups are investigated and an abstract characterization of ternary semigroups of continuous mappings defined on ternary separated topological spaces is given.

Keywords Ternary semigroup · Continuous mapping · Minimal ideal

1 Introduction

Semigroups of full and partial endomorphisms permit the description of ties between the original relations of the system and algebraic properties of its semigroups of endomorphisms. Gluskin [2] revealed deep ties between the theory of semigroups and other mathematical theories: graph theory, topology, linear algebra, and others. It turned out that semigroups of endomorphisms of a quasi-ordered set define this set exactly up to isomorphism. Semigroups of endomorphisms of a linear space of rank not less than 2 define this space exactly up to isomorphism. Some semigroups of homeomorphic transformations of a topological space, considered as an algebraic semigroup define this space exactly up to homeomorphism. Many researchers have focused their efforts on the characterization of topological spaces by semigroups of continuous, open, closed, quasi-open mappings defined on these spaces [3, 5, 10].

The theory of ternary algebraic systems is introduced by Lehmer in [1]. He investigated algebraic systems called triplexes. In [8], Dutta and Kar introduced and studied the notion of regular ternary semirings. Ternary semigroups are universal algebras with one associative ternary operation. The notion of ternary semigroup is known to Banach who is credited with example of a ternary semigroup which can not be reduced to a semigroup. Sioson [4] studied ternary semigroups with special reference to ideals and radicals . Santiago [9] developed regular and completely regular

F. Kh. Muradov (✉)
Department of Mathematics, Near East University, 10 Mersin, North Cyprus, Turkey
e-mail: firudin.muradov@neu.edu.tr

A. Ashyralyev et al. (eds.), *Functional Analysis in Interdisciplinary Applications—II*, Springer Proceedings in Mathematics & Statistics 351,
https://doi.org/10.1007/978-3-030-69292-6_10

ternary semigroups. In [6], Wagner studied semiheaps. Regularity, Greens equivalences (Mustafaev and Babaev [6]) and complete simplicity (Gluskin [2]) have been studied for semiheaps. In this paper, some properties of ternary semigroups are investigated and an abstract characterization of ternary semigroups of continuous mappings defined on ternary separated topological spaces is given.

2 Ternary Semigroups

A *ternary semigroup* is a non-empty set T together with a ternary operation $[abc]$ satisfying the associative law $[[abc]\,de] = [a\,[bcd]\,e] = [ab\,[cde]], \forall a, b, c, d, e \in T$. For any subsets A, B, C of T, we will denote by $[ABC]$ the set of all elements $[abc]$, where $a \in A$, $b \in B$, $c \in C$. The set $[AAA]$ is often denoted by $A^{[3]}$. As an example, any semigroup can be made into a ternary semigroup by defining the ternary product to be $[abc] = abc$. The set of all odd permutations under composition is a ternary semigroup. A non-empty subset A of a ternary semigroup T is called a *ternary subsemigroup* of T, if $A^{[3]} \subseteq A$. A non-empty subset L of a ternary semigroup T is called a *left (right, lateral) ideal* of T, if $[TTL] \subseteq L$ $([LTT] \subseteq L, [TLT] \subseteq L)$. A non-empty subset A of a ternary semigroup T is called a *two sided ideal* of T, if it is a left and right ideal of T and A is called an *ideal* of T, if it is a left, right and lateral ideal of T. A ternary semigroup T is called a *left commutative ternary semigroup*, if $[stu] = [tsu]$ for all $s, t, u \in T$.

Let S and T be two ternary semigroups and let $S \times T$ be their cartesian product. A ternary operation introduced in $S \times T$ is as follows:

$$[(s_1, t_1)\,(s_2, t_2)\,(s_3, t_3)] = ([s_1 s_2 s_3]\,, [t_1 t_2 t_3])\,.$$

It is easy to show that the so defined ternary operation is associative on the pairs $(s, t) \in S \times T$. Hence $S \times T$ is a ternary semigroup, which is called *direct product* of S and T. A mapping $f : S \to T$ is a *homomorphism,* if $f\,([xyz]) = [f\,(x)\,f\,(y)\,f\,(z)]$ for all $x, y, z \in S$.

Let Ω_1 and Ω_2 be two non-empty sets and let $\Re\,(\Omega_1, \Omega_2)$ be the set of all pairs of binary relations (ρ, σ), where $\rho \in \Omega_1 \times \Omega_2$ and $\sigma \in \Omega_2 \times \Omega_1$. The set $\Re\,(\Omega_1, \Omega_2)$ is a ternary semigroup with respect to the ternary operation

$$[(\rho_1, \sigma_1)\,(\rho_2, \sigma_2)\,(\rho_3, \sigma_3)] = (\rho_1 \circ \sigma_2 \circ \rho_3, \sigma_1 \circ \rho_2 \circ \sigma_3)\,.$$

The set of all functions from Ω_1 to Ω_2 is denoted by $F\,(\Omega_1, \Omega_2)$. The set $\aleph\,(\Omega_1, \Omega_2) = F\,(\Omega_1, \Omega_2) \times F\,(\Omega_2, \Omega_1)$ is a ternary subsemigroup of $\Re\,(\Omega_1, \Omega_2)$.

Let S be a ternary semigroup. A mapping $\lambda : S \to S$ such that $\lambda\,([stu]) = [\lambda\,(s)\,tu]$ is called a left multiplication operator on S. Let $\Lambda\,(S)$ denote the set of all left multiplication operators on S. Then, $\Lambda\,(S)$ is a ternary semigroup with the usual composition of mappings, i.e., $[\lambda_1 \lambda_2 \lambda_3] = \lambda_1 \circ \lambda_2 \circ \lambda_3$. For each pair (a, b) of elements of the ternary semigroup T, it is denoted by $\lambda_{a,b}$, $\mu_{a,b}$ and $\delta_{a,b}$ the *inner left*, *right* and *lateral multiplication* operators on T defined as follows:

$$\lambda_{a,b}(x) = [abx], \ \mu_{a,b}(x) = [xba], \delta_{a,b}(x) = [axb], \ \forall x \in T.$$

We denote by $\Lambda_i(S)$ the set of all inner left multiplication operators on S. Since $\lambda_{a,b}([stu]) = [ab[stu]] = [[abs]tu] = \left[\lambda_{a,b}(s)\, tu\right]$, $\lambda_{a,b} \in \Lambda(S)$ $\forall a, b \in S$, and so $\Lambda_i(S)$ is a ternary subsemigroup of $\Lambda(S)$. Moreover, $\Lambda_i(S)$ is a left ideal of $\Lambda(S)$, since $\left[\lambda_1\lambda_2\lambda_{s,t}\right](u) = \lambda_1\lambda_2([stu]) = \lambda_1([\lambda_2(s)\,tu]) = [\lambda_1\lambda_2(s)\,tu] = \lambda_{\lambda_1\lambda_2(s),t}(u)$ for all $\lambda_1\lambda_2 \in \Lambda(S)$ and $s, t \in S$. Pairs $(s_1, t_1)(s_2, t_2) \in S$, $(s_1, t_1) \neq (s_2, t_2)$ are called *left equi-acting* if $[s_1t_1u] = [s_2t_2u]$ for all $u \in S$.

Theorem 1 *Let S be a left commutative ternary semigroup containing L as a left ideal. If S does not contain left equi-acting pairs on L, then the mapping $\theta(s, t) = \lambda_{s,t}$ is an isomorphism from $S \times S$ to $\Lambda(L)$.*

Let T be a ternary semigroup and let Λ and M be the sets of all left and right multiplication operators on T, respectively. Consider the sets $\Omega_1 = T \cup M$ and $\Omega_2 = T \cup \Lambda$. Assign a pair $(\varphi_a, \phi_a) \in \aleph(\Omega_1, \Omega_2)$ to each element $a \in T$ by

$$\varphi_a(x) = \lambda_{a,x}, \varphi_a\left(\mu_{x,y}\right) = \mu_{x,y}(a) \ \ \textit{and} \ \ \phi_a(x) = \mu_{x,a}, \phi_a\left(\lambda_{x,y}\right) = \lambda_{a,x}(y),$$

for all $x, y \in T, \lambda \in \Lambda, \mu \in M$. Elements $a, b \in T$ are called *equi-acting* if $[axy] = [bxy]$, $[xya] = [xyb]$, and $[xay] = [xby]$ for all $x, y \in T$.

Theorem 2 *Let T be a ternary semigroup with no equi-acting elements. The mapping $f(a) = (\varphi_a, \phi_a)$ is an isomorphism from T to $\aleph(\Omega_1, \Omega_2)$.*

Proof Let $f(a) = f(b)$ for some $a, b \in T$ such that $a \neq b$. Then, $(\varphi_a, \phi_a) = (\varphi_b, \phi_b)$, equivalently $\varphi_a = \varphi_b$ and $\phi_a = \phi_b$. It follows from $\varphi_a(x) = \varphi_b(x)$ that $\lambda_{a,x} = \lambda_{b,x}$ and $\varphi_a\left(\mu_{x,y}\right) = \varphi_b\left(\mu_{x,y}\right)$ for all $x, y \in T$. Therefore, $[axy] = [bxy]$ and $[ayx] = [byx]$ for all $x, y \in T$. Similarly, it follows from $\phi_a(x) = \phi_b(x)$ that $\mu_{x,a} = \mu_{x,b}$ and $\phi_a\left(\lambda_{x,y}\right) = \phi_b\left(\lambda_{x,y}\right)$, which are equivalent to $[yax] = [ybx]$ and $[xya] = [xyb]$ for all $x, y \in T$. Consequently, a and b in T are equi-acting elements, which contradicts the assumption. It follows directly from the definition that $f([abc]) = [f(a)\, f(b)\, f(c)]$. □

Let T be a ternary semigroup with no equi-acting elements and let $\Omega_1 = T \cup M$ and $\Omega_2 = T \cup \Lambda$. Suppose that S is a ternary semigroup containing T as an ideal. Let us also consider a in S and t in T. To each pair (a, t) assign the mappings $\lambda'_{a,t}$ and $\mu'_{a,t}$ of T by

$$\lambda'_{a,t}(x) = [atx], \ \ \mu'_{a,t}(x) = [xta], \ \ x \in T.$$

Now assign a pair $(\varphi_a, \phi_a) \in \aleph(\Omega_1, \Omega_2)$ to each element $a \in S$ as

$$\varphi_a(x) = \lambda'_{a,x}, \ \varphi_a\left(\mu_{x,y}\right) = [ayx] \ \ \textit{and} \ \ \phi_a(x) = \mu'_{x,a}, \ \ \phi_a\left(\lambda_{x,y}\right) = [xya],$$

for all $x, y \in T$.

Theorem 3 *The map* $f'(a) = (\varphi_a, \phi_a)$, $\forall a \in S$ *is a homomorphism from* S *to* $\aleph(\Omega_1, \Omega_2)$, *which induces an isomorphism on* T.

Clearly, if S is a ternary semigroup which does not contain equi-acting elements to its ideal T, then the homomorphism $f' : S \to \aleph(\Omega_1, \Omega_2)$ is an isomorphism.

Let T be a ternary semigroup. An element $e \in T$ is said to be selfpotent if $e^{[3]} = e$. T is said to be selfpotent if every element of T is selfpotent. An element $a \in T$ is said to be regular if there exists an element $x \in T$ such that $[axa] = a$. If a and b are $[aba] = a$ and $[bab] = b$, then a and b are said to be the inverses of one another. Every regular element has an inverse.

Proposition 1 *Every pair of elements of the ternary semigroup* T *are the inverses of one another if and only if* T *is selfpotent and* $[xyz] = [xzz] = [xxz]$ *for every* $x, y, z \in T$.

The following Proposition states that the class of ternary semigroups for which every pair of elements are inverses of one another has the property opposite to the commutativity, that is why they can be called anticommutative.

Proposition 2 *Every pair of elements of the ternary semigroup* T *are inverses of one another if and only if* $[xty] \neq [ytx]$ *for every* $x, y, z \in T$ *such that* $x \neq y$.

3 Ternary Semigroups of continuous Mappings of Topological Spaces

A ternary semigroup T which is simultaneously a topological space with a topology σ, such that the ternary semigroup operation is continuous in the given topology is called a topological ternary semigroup and is denoted by T_σ. An isomorphism of topological ternary semigroups is a ternary semigroup isomorphism which is also a homeomorphism of the underlying topological spaces.

Let Ω_1 and Ω_2 be two topological T_1-spaces and $\varphi : \Omega_1 \to \Omega_2$, $\phi : \Omega_2 \to \Omega_1$. The set of all pairs (φ, ϕ) of continuous maps between Ω_1 and Ω_2 is denoted by $C(\Omega_1, \Omega_2)$. The set $C(\Omega_1, \Omega_2)$ is a ternary semigroup with respect to the ternary operation $[(\varphi_1, \phi_1)(\varphi_2, \phi_2)(\varphi_3, \phi_3)] = (\varphi_1\phi_2\varphi_3, \phi_1\varphi_2\phi_3)$. For every $x \in \Omega_1$ and $y \in \Omega_2$ denote by h_y and g_x the constant maps $h_y : \Omega_1 \to \Omega_2$ and $g_x : \Omega_2 \to \Omega_1$ such that $h_y(\Omega_1) = y$ and $g_x(\Omega_2) = x$. Let $P(\Omega_1, \Omega_2)$ denote the subset of $C(\Omega_1, \Omega_2)$ consisting of all pairs $\left(h_y, g_x\right)$ of constant maps.

Theorem 4 $P(\Omega_1, \Omega_2)$ *is a minimal ideal of the ternary semigroup* $C(\Omega_1, \Omega_2)$.

Assigning to any point (x, y) of the topological product $\Omega_1 \times \Omega_2$ the element $\left(h_y, g_x\right) \in P(\Omega_1, \Omega_2)$ defines a one-to-one mapping from $\Omega_1 \times \Omega_2$ onto $P(\Omega_1, \Omega_2)$. We denote the topology on $P(\Omega_1, \Omega_2)$ induced by this mapping as ρ_0. Clearly, the topological spaces $\Omega_1 \times \Omega_2$ and $P_{\rho_0}(\Omega_1, \Omega_2)$ are homeomorphic. A ternary semigroup $S(\Omega_1, \Omega_2)$ of continuous mappings between Ω_1 and Ω_2 containing $P(\Omega_1, \Omega_2)$ is called a separating if for every non-empty closed sets $A_1 \subseteq \Omega_1$ and $A_2 \subseteq \Omega_2$ and any points $\xi \in \Omega_1 \setminus A_1$, $\eta \in \Omega_2 \setminus A_2$ there exist $\alpha \in \Omega_1$, $\beta \in \Omega_2$ and $(\varphi, \phi) \in S(\Omega_1, \Omega_2)$ such that $\varphi(A_1) = \beta, \varphi(\xi) \neq \beta, \phi(A_2) = \alpha, \phi(\eta) \neq \alpha$. Note that not all pairs of topological spaces have separating ternary semigroups. Clearly, if the topological spaces Ω_1 and Ω_2 are completely regular spaces containing a simple arc, then they have separating ternary semigroups of continuous mappings. Now, let Ω_1 and Ω_2 be the compact totally disconnected spaces. Note that each component of a compact is the intersection of all open-closed sets containing this component. Therefore, in a compact totally disconnected space Ω the intersection of all open-closed sets of Ω containing a point $\xi \in \Omega$ coincides with ξ. If $\eta \in \Omega$, $\eta \neq \xi$, then there exists an open-closed set A_ξ containing ξ but not η. Now suppose that A_1, A_2 are closed sets of Ω_1, Ω_2, respectively and $\alpha \in \Omega_1 \setminus A_1$, $\beta \in \Omega_2 \setminus A_2$. Assign to every $\xi \in A_1$ the open-closed set $T_\xi \subset \Omega_1$ containing ξ but not containing α. Analogously, we can assign to every $\eta \in A_2$ the open-closed set $K_\eta \subset \Omega_2$ containing η but not containing β. We can select a finite cover $T_{\xi_1}, T_{\xi_2}, \ldots, T_{\xi_k}$ from $\{T_\xi\}_{\xi \in A_1}$ and a finite cover $K_{\eta_1}, K_{\eta_2}, \ldots, K_{\eta_n}$ from $\{K_\eta\}_{\eta \in A_2}$. The set $T = \bigcup_{i=1}^{k} T_{\xi_i}$ is the open-closed set of Ω_1 containing A_1 but not containing the point α. The set $K = \bigcup_{i=1}^{n} K_{\eta_i}$ is the open-closed set of Ω_2 containing A_2 but not containing the point β. The mappings

$$\varphi(x) = \begin{cases} \eta_1 & , if \;\; x \in T, \\ \beta & if \;\; x \in \Omega_1/T, \end{cases} \qquad \phi(y) = \begin{cases} \xi_1 & , if \;\; y \in K, \\ \alpha & , if \;\; y \in \Omega_2/K \end{cases}$$

are continuous and $C(\Omega_1, \Omega_2)$ is therefore separating ternary semigroup of Ω_1 and Ω_2.

We say that the topological T_1-spaces Ω_1 and Ω_2 are ternary separated if they have a separating ternary semigroup.

Let A be an ideal of a ternary semigroup T. Define a family Γ of subsets of A as

$$F \in \Gamma \leftrightarrow \exists_{x,y \in T, a \in A} F = \delta_{x,y}^{-1}(a),$$

where $\delta_{x,y}$ is an inner lateral multiplication generated by $x, y \in T$. Let's take the family Γ as a subbasis of closed sets for some topology on A. Throughout this paper, τ denotes this topology.

Theorem 5 *Let Ω_1 and Ω_2 be two topological T_1 -spaces and let $S(\Omega_1, \Omega_2)$ be a ternary semigroup of continuous mappings of Ω_1 and Ω_2 containing $P(\Omega_1, \Omega_2)$. Then, the topology ρ_0 is finer than the topology τ on $P(\Omega_1, \Omega_2)$.*

Proof Let F be a closed set in $P_\tau(\Omega_1, \Omega_2)$. Clearly, $F = \bigcap_t \left(\bigcup_{i_t=1}^{n_t} F_{i_t} \right)$, where $F_{i_t} \in \Gamma$. By the definition of Γ, there exist $(\varphi_1, \phi_1), (\varphi_2, \phi_2) \in S(\Omega_1, \Omega_2)$ and $(h_\beta, g_\alpha) \in P(\Omega_1, \Omega_2)$ such that

$$\delta^{-1}_{(\varphi_1,\phi_1),(\varphi_2,\phi_2)}\left(h_\beta, g_\alpha\right) = F_{i_t}.$$

Suppose that $F_{i_t} \subset P(\Omega_1, \Omega_2)$ corresponds to $A \subset \Omega_1 \times \Omega_2$ under the homeomorphism between $\Omega_1 \times \Omega_2$ and $P_{\rho_0}(\Omega_1, \Omega_2)$. Then, $\left(h_\beta, g_\alpha\right) \in F_{i_t}$ if and only if $(\alpha, \beta) \in A$. If A_1 denotes the projection of A onto Ω_1 and A_2 denotes the projection of A onto Ω_2, then $A = A_1 \times A_2$. Here A_1 is the inverse image of $\beta \in \Omega_2$ under φ_1 and A_2 is the inverse image of $\alpha \in \Omega_1$ under ϕ_1, i.e., $A_1 \times A_2$ is the closed set of $\Omega_1 \times \Omega_2$. Thus, F_{i_t} is the closed subset of the space $P_{\rho_0}(\Omega_1, \Omega_2)$. □

The one-to-one mapping existing between the elements of the space $P_\tau(\Omega_1, \Omega_2)$ and the elements of the set $\Omega_1 \times \Omega_2$ induces a topology on $\Omega_1 \times \Omega_2$ that we will denote by τ.

Theorem 6 *Let Ω_1 and Ω_2 be two ternary separated topological spaces. For every closed sets $F_1 \subseteq \Omega_1$ and $F_2 \subseteq \Omega_2$, the set $F_1 \times F_2$ is closed in the space $(\Omega_1 \times \Omega_2)_\tau$.*

Proof If at least one of the sets F_1 and F_2 is empty, then the proof is trivial. Suppose that F_1 and F_2 are non-empty closed sets. Since the spaces $\Omega_1 \times \Omega_2$ and $P_{\rho_0}(\Omega_1, \Omega_2)$ are homeomorphic, there is a closed set F in $P_{\rho_0}(\Omega_1, \Omega_2)$ corresponding to $F_1 \times F_2$. It is sufficient to show that F is also closed in $P_\tau(\Omega_1, \Omega_2)$. Since Ω_1 and Ω_2 are ternary separated topological spaces, for every $\left(h_\beta, g_\alpha\right) \in P(\Omega_1, \Omega_2) \setminus F$ there exist $\left(h_{\beta^*}, g_{\alpha^*}\right) \in P(\Omega_1, \Omega_2)$ and $(\varphi_1, \phi_1), (\varphi_2, \phi_2) \in \Re$ such that $[(\varphi_1, \phi_1)\, F\, (\varphi_2, \phi_2)] = \left(h_{\beta^*}, g_{\alpha^*}\right)$ and $\left[(\varphi_1, \phi_1)\left(h_\beta, g_\alpha\right)(\varphi_2, \phi_2)\right] \neq \left(h_{\beta^*}, g_{\alpha^*}\right)$. Here $\Re$ is the separating ternary semigroup between Ω_1 and Ω_2. Let $F_{\left(h_\beta, g_\alpha\right)} = \delta^{-1}_{(\varphi_1,\phi_1),(\varphi_2,\phi_2)}\left(h_{\beta^*}, g_{\alpha^*}\right)$. Clearly, $F \subseteq F_{\left(h_\beta, g_\alpha\right)}$ and $\left(h_\beta, g_\alpha\right) \notin F_{\left(h_\beta, g_\alpha\right)}$. Since $F_{\left(h_\beta, g_\alpha\right)} \in \Gamma$, it suffices to note that

$$F = \bigcap_{\left(h_\beta, g_\alpha\right) \in P(\Omega_1, \Omega_2) \setminus F} F_{\left(h_\beta, g_\alpha\right)}.$$

□

Theorem 7 *Let Ω_1 and Ω_2 be ternary separated topological spaces and let $\Re$ be their separating ternary semigroup containing $P(\Omega_1, \Omega_2)$. Then, the topology τ is finer than the topology ρ_0 on the minimal ideal $P(\Omega_1, \Omega_2)$.*

Proof Let V_1 and V_2 be arbitrary open sets of Ω_1 and Ω_2, respectively. By Theorem 6 the sets

$$\Omega_1 \times V_2 = (\Omega_1 \times \Omega_2) \setminus \Omega_1 \times (\Omega_2 \setminus V_2),$$

$$V_1 \times \Omega_2 = (\Omega_1 \times \Omega_2) \setminus (\Omega_1 \setminus V_1) \times \Omega_2$$

are open in $(\Omega_1 \times \Omega_2)_\tau$ and hence $V_1 \times V_2 = (\Omega_1 \times V_2) \cap (V_1 \times \Omega_2)$ is open in $(\Omega_1 \times \Omega_2)_\tau$. □

We can summarize the above results as follows.

Theorem 8 *Let Ω_1 and Ω_2 be two ternary separated topological spaces and $\Re$ be their separating ternary semigroup containing $P(\Omega_1, \Omega_2)$. Then, the topologies τ and ρ_0 coincide on the minimal ideal $P(\Omega_1, \Omega_2)$.*

Theorem 9 *Let Ω_1 and Ω_2, Ω_1' and Ω_2' be two pairs of ternary separated topological spaces and let $\Re$ and $\Re'$ be their separating ternary semigroups containing $P(\Omega_1, \Omega_2)$. If $\Re$ and $\Re'$ are isomorphic, then the spaces Ω_i and Ω_i' are homeomorphic $(i = 1, 2)$.*

Proof An isomorphism f_1 of $\Re$ and $\Re'$ induces a one-to-one mapping ξ_i between the sets Ω_i and Ω_i', $(i = 1, 2)$ defined by

$$\left\{\xi_1(\alpha) = \alpha', \xi_2(\beta) = \beta'\right\} \leftrightarrow f_1\left(h_\beta, g_\alpha\right) = \left(h'_{\beta'}, g'_{\alpha'}\right)$$

for every $(\alpha, \beta) \in (\Omega_1 \times \Omega_2)$ and $\left(\alpha', \beta'\right) \in \left(\Omega_1' \times \Omega_2'\right)$. Note that the isomorphism f_1 of the ternary semigroups $\Re$ and $\Re'$ induces a homeomorphism of the spaces $P_\tau(\Omega_1, \Omega_2)$ and $P_\tau\left(\Omega_1', \Omega_2'\right)$. According to Theorem 8, on $P(\Omega_1, \Omega_2)$ and $P\left(\Omega_1', \Omega_2'\right)$ the topology τ coincides with ρ_0. So, the spaces $P_{\rho_0}(\Omega_1, \Omega_2)$ and $P_{\rho_0}\left(\Omega_1', \Omega_2'\right)$ are homeomorphic. Let f_2 be a homeomorphism between $\Omega_1 \times \Omega_2$ and $P_{\rho_0}(\Omega_1, \Omega_2)$ and let f_3 be a homeomorphism between $P_{\rho_0}\left(\Omega_1', \Omega_2'\right)$ and $\Omega_1' \times \Omega_2'$. Then, $f = f_3 \circ f_1 \circ f_2$ is the homeomorphism between the spaces $\Omega_1 \times \Omega_2$ and $\Omega_1' \times \Omega_2'$ such that $f(\alpha, \beta) = (\xi_1(\alpha), \xi_2(\beta))$ for every $(\alpha, \beta) \in \Omega_1 \times \Omega_2$. From this, it follows that ξ_i is a homeomorphism between the spaces Ω_i and Ω_i' for $i = 1, 2$.
□

Since $C(\Omega_1, \Omega_2)$ is the separating ternary semigroup of the ternary separated topological spaces Ω_1 and Ω_2, we obtain the following.

Theorem 10 *Let Ω_1 and Ω_2, Ω_1' and Ω_2' be pairs of ternary separated topological spaces. The ternary semigroups $C(\Omega_1, \Omega_2)$ and $C\left(\Omega_1', \Omega_2'\right)$ are isomorphic if and only if the spaces Ω_i and Ω_i' are homeomorphic, $(i = 1, 2)$.*

Corollary 1 *Let Ω_1 and Ω_2, Ω_1' and Ω_2' be pairs of completely regular spaces containing a simple arc. The ternary semigroups $C(\Omega_1, \Omega_2)$ and $C\left(\Omega_1', \Omega_2'\right)$ are isomorphic if and only if the spaces Ω_i and Ω_i' are homeomorphic, $(i = 1, 2)$.*

Corollary 2 *Let Ω_1 and Ω_2, Ω_1' and Ω_2' be pairs of compact totally disconnected spaces. The ternary semigroups $C(\Omega_1, \Omega_2)$ and $C\left(\Omega_1', \Omega_2'\right)$ are isomorphic if and only if the spaces Ω_i and Ω_i' are homeomorphic, $(i = 1, 2)$.*

References

1. Lehmer, D.H.: A ternary analogue of Abelian groups. Am. J. Math. **54**, 329–338 (1932)
2. Gluskin, L.M.: Semigroups. Itogi Nauki. **1**, 33–58 (1965)

3. Shneperman, L.B.: Semigroups of continuous mappings of topological spaces. Sib. Math. J. **1**, 221–229 (1965)
4. Sioson, F.M.: Ideal theory in ternary semigroups. Math. Jpn. **10**, 63–84 (1965)
5. Yusufov, V.S.: Open mappings. U.M.N. **41**, pp. 185–186 (1965)
6. Wagner, V.V.: The theory of generalized heaps and generalized groups. Mat. Sb. NS. **32**, 545–632 (1953)
7. Mustafaev, L.G., Babaev, E.A.: Commutative bands of semiheaps with cancellation. Izv. Akad. Nauk Azerb. SSR, Ser. Fiz. Teh. Mat. Nauk **5–6**, 80–83 (1971)
8. Dutta, T.K., Kar, S.: On regular ternary semirings. In: Advances in Algebra: Proceedings of the ICM Satellite Conference in Algebra and Related Topics, vol. 2003, pp. 343–355. World Scientific, Singapore (2003)
9. Santiago, M.L., Sri Bala, S.: Ternary semigroups. Semigroup Forum. **81**, 380–388 (2010)
10. Muradov, FKh: Semigroups of Quasi-open mappings and Lattice-Equivalence. Int. J. Algebra **6**, 1443–1447 (2012)

A Unified Numerical Method for Solving System of Nonlinear Wave Equations

Ozgur Yildirim and Meltem Uzun

Abstract In this study a composite numerical method is implemented for solving the nonlinear coupled system of sine Gordon equations. The first and second order of accuracy unconditionally stable difference schemes are used in fixed point iteration for the approximate solution of the nonlinear coupled system. The unique solvability of the system is considered. Some numerical results are presented in tables and error analysis is given. Presented numerical results are carried out using the MATLAB software.

1 Introduction

Linear and nonlinear wave equations have wide range of applications in physics and engineering, including fluid dynamics, nonlinear optics, plazma physics, biophysics, and condensed-matter physics, etc. (see [1–14] and the references given therein).

It is, in general, not easy to obtain exact solution for nonlinear problems. Therefore, these problems have been solved numerically by many methods such as finite element method [20–22], finite difference method [2–19], predictor–corrector schemes [23], and radial basis function (RBF) collocation method [24], etc.

In this study, a unified numerical method which combines first and second order of accuracy unconditionally stable finite difference schemes with fixed point iteration is studied for the nonlinear system of coupled sine Gordon equations

$$\begin{cases} \frac{\partial^2 u}{\partial t^2} - \frac{\partial^2 u}{\partial x^2} + \gamma_1 \sin(\delta_{11} u + \delta_{12} v) = f_1 \text{ in } [0, T] \times \Omega, \\ \frac{\partial^2 v}{\partial t^2} - \frac{\partial^2 v}{\partial x^2} + \gamma_2 \sin(\delta_{21} u + \delta_{22} v) = f_2 \text{ in } [0, T] \times \Omega \end{cases} \tag{1}$$

O. Yildirim (✉) · M. Uzun
Department of Mathematics, Yildiz Technical University, Istanbul, Turkey
e-mail: ozgury@yildiz.edu.tr

M. Uzun
e-mail: meltemu@yildiz.edu.tr

A. Ashyralyev et al. (eds.), *Functional Analysis in Interdisciplinary Applications—II*, Springer Proceedings in Mathematics & Statistics 351,
https://doi.org/10.1007/978-3-030-69292-6_11

under the boundary conditions

$$u_i = 0 \text{ on } \Gamma, i = 1, 2 \tag{2}$$

and the initial conditions

$$u_i(0, x) = u_0^i(x) \text{ in } \Omega \text{ and } \frac{\partial u_i}{\partial t}(0, x) = u^i(x) \text{ in } \Omega,\ i = 1, 2. \tag{3}$$

Here, Ω is an open bounded set of $\mathbb{R}$ with boundary $\Gamma = \partial\Omega$. The physical constants γ_i, δ_{ij} are real numbers and f_i are forcing functions for $i, j = 1, 2$. Note that some results of this work, without theoretical statements, are presented in [10].

Let $A = -\Delta$ be an unbounded self-adjoint positive definite operator in a Hilbert space $H = L^2(\Omega)$. Problem (1)–(3) is reduced to a system of Cauchy problems in H

$$\begin{cases} \frac{\partial^2 u}{\partial t^2} + Au + \gamma_1 \sin(\delta_{11} u + \delta_{12} v) = f_1 \text{ in } [0, T], \\ \frac{\partial^2 v}{\partial t^2} + Av + \gamma_2 \sin(\delta_{21} u + \delta_{22} v) = f_2 \text{ in } [0, T], \\ u(0) = u_0,\ \frac{du}{dt}(0) = u_1, \\ v(0) = v_0,\ \frac{dv}{dt}(0) = v_1. \end{cases} \tag{4}$$

In this study, the first order of accuracy stable difference scheme

$$\begin{cases} \tau^{-2}(u_{k+1} - 2u_k + u_{k-1}) + Au_{k+1} + \gamma_1 \sin(\delta_{11} u_k + \delta_{12} v_k) = f_{1k}, \\ f_{1k} = f_1(t_{k+1}, u(t_k), v(t_k)),\ t_{k+1} = (k+1)\tau,\ 1 \le k \le N-1,\ N\tau = 1, \\ \tau^{-2}(v_{k+1} - 2v_k + v_{k-1}) + Av_{k+1} + \gamma_2 \sin(\delta_{21} u_k + \delta_{22} v_k) = f_{2k}, \\ f_{2k} = f_2(t_{k+1}, u(t_k), v(t_k)),\ t_{k+1} = (k+1)\tau,\ 1 \le k \le N-1,\ N\tau = 1, \\ u_0 = \varphi_1,\ \tau^{-1}(u_1 - u_0) = \psi_1, \\ v_0 = \varphi_2,\ \tau^{-1}(u_1 - u_0) = \psi_2 \end{cases} \tag{5}$$

and the second order of accuracy stable difference scheme

$$\begin{cases} \tau^{-2}(u_{k+1} - 2u_k + u_{k-1}) + \frac{1}{2}Au_k + \frac{1}{4}A(u_{k+1} + u_{k-1}) \\ +\gamma_1 \sin(\delta_{11}u_k + \delta_{12}v_k) = f_{1k}, \\ f_{1k} = f_1(t_k, u_k, v_k), \quad t_k = k\tau, \ 1 \le k \le N-1, N\tau = 1, \\ \tau^{-2}(v_{k+1} - 2v_k + v_{k-1}) + \frac{1}{2}Av_k + \frac{1}{4}A(v_{k+1} + v_{k-1}) \\ +\gamma_2 \sin(\delta_{21}u_k + \delta_{22}v_k) = f_{2k}, \\ f_{2k} = f_2(t_k, u_k, v_k), \quad t_k = k\tau, \ 1 \le k \le N-1, N\tau = 1, \\ u_0 = \varphi_1, v_0 = \varphi_2, \\ (I + \tau^2 A)\tau^{-1}(u_1 - u_0) = \frac{\tau}{2}(f_{10} - Au_0) + \psi_1, \\ (I + \tau^2 A)\tau^{-1}(v_1 - v_0) = \frac{\tau}{2}(f_{20} - Av_0) + \psi_2 \end{cases} \tag{6}$$

which was presented in [4, 6] are used for the iterations, with a modification for nonlinear system. Here, the set of a family of grid points

$$\Omega_h = [0,1]_\tau \times [0,\pi]_h = \{(t_k, x_n) : t_k = k\tau, \quad 0 \le k \le N,$$
$$N\tau = 1, x_n = nh, 0 \le n \le M, Mh = \pi\} \tag{7}$$

with parameters τ and h is considered.

The theoretical result on unique solvability will be given in the subsequent section. Some results of numerical experiments are presented with error analysis, in the last section.

2 Unique Solvability of Abstract Problem

In this section we present theoretical statements for unique solvability of problem (1)–(3). Let us define the Hilbert space H as $H = L^2(\Omega)$ equipped with the inner product and norm

$$(\psi, \phi) = \int_\Omega \psi(x)\phi(x)dx, \quad |\psi| = (\psi, \psi)^{1/2}, \forall \phi, \psi \in L^2(\Omega).$$

We consider system (4) in the following vector form

$$\begin{cases} \mathbf{u}'' + \mathbf{Au} + \gamma \sin \delta \mathbf{u} = \mathbf{f} \text{ in } [0, T], \\ \mathbf{u(0)} = \mathbf{u}_0, \ \mathbf{u}'(\mathbf{0}) = \mathbf{u}_1, \end{cases} \tag{8}$$

where

$$\mathbf{u}=\begin{bmatrix} u \\ v \end{bmatrix}, \mathbf{f}=\begin{bmatrix} f_1 \\ f_2 \end{bmatrix}, \ \mathbf{u}''=\begin{bmatrix} \frac{d^2u}{dt^2} \\ \frac{d^2v}{dt^2} \end{bmatrix}, \sin\mathbf{u}=\begin{bmatrix} \sin u \\ \sin v \end{bmatrix},$$

$$\mathbf{A}=\begin{bmatrix} A & 0 \\ 0 & A \end{bmatrix}, \ \delta=\begin{bmatrix} \delta_{11} & \delta_{12} \\ \delta_{21} & \delta_{22} \end{bmatrix}, \ \gamma=\begin{bmatrix} \gamma_1 & 0 \\ 0 & \gamma_2 \end{bmatrix},$$

$$\mathbf{u}_0=\begin{bmatrix} u_0 \\ v_0 \end{bmatrix}, \ \mathbf{u}_1=\begin{bmatrix} u_1 \\ v_1 \end{bmatrix}.$$

Next, replacing problem (4) with the first order of accuracy unconditionally stable difference scheme, we get

$$\begin{cases} \tau^{-2}(u_{k+1}-2u_k+u_{k-1})+Au_{k+1}+\gamma_1\sin(\delta_{11}u_k+\delta_{12}v_k)=f_{1k}, \\ f_{1k}=f_1(t_{k+1},u(t_k),v(t_k)), t_{k+1}=(k+1)\tau, \ 1\le k\le N-1, \ N\tau=1, \\ \tau^{-2}(v_{k+1}-2v_k+v_{k-1})+Av_{k+1}+\gamma_2\sin(\delta_{21}u_k+\delta_{22}v_k)=f_{2k}, \\ f_{2k}=f_2(t_{k+1},u(t_k),v(t_k)), t_{k+1}=(k+1)\tau, \ 1\le k\le N-1, \ N\tau=1, \\ u_0=\varphi_1, \ \tau^{-1}(u_1-u_0)=\psi_1, \\ v_0=\varphi_2, \ \tau^{-1}(u_1-u_0)=\psi_2, \end{cases} \tag{9}$$

and the second order of accuracy stable differences scheme, we obtain

$$\begin{cases} \tau^{-2}(u_{k+1}-2u_k+u_{k-1})+\frac{1}{2}Au_k+\frac{1}{4}A(u_{k+1}+u_{k-1}) \\ +\gamma_1\sin(\delta_{11}u_k+\delta_{12}v_k)=f_{1k}, \\ f_{1k}=f_1(t_k,u_k,v_k), \quad t_k=k\tau, \ 1\le k\le N-1, N\tau=1, \\ \tau^{-2}(v_{k+1}-2v_k+v_{k-1})+\frac{1}{2}Av_k+\frac{1}{4}A(v_{k+1}+v_{k-1}) \\ +\gamma_2\sin(\delta_{21}u_k+\delta_{22}v_k)=f_{2k}, \\ f_{2k}=f_2(t_k,u_k,v_k), \quad t_k=k\tau, \ 1\le k\le N-1, N\tau=1, \\ u_0=\varphi_1, v_0=\varphi_2, \\ (I+\tau^2A)\tau^{-1}(u_1-u_0)=\frac{\tau}{2}(f_{10}-Au_0)+\psi_1, \\ (I+\tau^2A)\tau^{-1}(v_1-v_0)=\frac{\tau}{2}(f_{20}-Av_0)+\psi_2. \end{cases} \tag{10}$$

Using the family of grid points (7), we introduce the Hilbert space

$$L_{2h}(\Omega) = L_2(\Omega_h) \tag{11}$$

equipped with the norms

$$\|u_k\|_{L_{2h}(\Omega)} = \left(\sum_{j=1}^{N} \left| u_k^j \right|^2 h \right)^{\frac{1}{2}}, \ \|v_k\|_{L_{2h}(\Omega)} = \left(\sum_{j=1}^{N} \left| v_k^j \right|^2 h \right)^{\frac{1}{2}}.$$

By the results of Theorem 3 in [12], we state the following theorem.

Theorem 1 *Under the assumptions of Theorem 3 in* [12], *there exists a positive constant K, independent of grid parameters τ and h, such that for all $k \in \mathbb{N}$*

$$\left(\left\| \frac{u_{k+1} - u_k}{\tau} \right\|^2_{L_{2h}(\Omega)} + \left\| \frac{u_k - u_{k-1}}{\tau} \right\|^2_{L_{2h}(\Omega)} \right.$$

$$+ \left\| \frac{u_{k+1} - u_{k-1}}{2\tau} \right\|^2_{L_{2h}(\Omega)} + \left\| \frac{v_{k+1} - v_k}{\tau} \right\|^2_{L_{2h}(\Omega)}$$

$$\left. + \left\| \frac{v_k - v_{k-1}}{\tau} \right\|^2_{L_{2h}(\Omega)} + \left\| \frac{v_{k+1} - v_{k-1}}{2\tau} \right\|^2_{L_{2h}(\Omega)} \right) \leq K. \tag{12}$$

Using the family of grid points (7), let us consider problem (9) in the following vector form

$$\begin{cases} \tau^{-2}(\mathbf{u}_{k+1} - 2\mathbf{u}_k + \mathbf{u}_{k-1}) + \mathbf{A}\mathbf{u}_k \\ +\gamma \sin \delta \mathbf{u}_k = \mathbf{f}_k \text{ in } [0, T]_h, \\ \mathbf{u}_0 = \varphi, \tau^{-1}(\mathbf{u}_1 - \mathbf{u}_0) = \psi, \end{cases} \tag{13}$$

where

$$\mathbf{u}_k = \begin{bmatrix} u_k \\ v_k \end{bmatrix}, \ \sin \mathbf{u_k} = \begin{bmatrix} \sin u_k \\ \sin v_k \end{bmatrix},$$

$$\mathbf{f_k} = \begin{bmatrix} f_{1k} \\ f_{2k} \end{bmatrix}, \varphi = \begin{bmatrix} \varphi_1 \\ \varphi_2 \end{bmatrix}, \ \psi = \begin{bmatrix} \psi_1 \\ \psi_2 \end{bmatrix}.$$

Then, by the results of Theorem 2 in [12] we state the following theorem.

Theorem 2 *Suppose that the assumptions of Theorem 2 in* [12] *are satisfied. Then problem* (13) *has a unique solution* $\mathbf{u}_k$ *satisfying*

$$\|\mathbf{u}_{k-1}\|^2_{L_{2h}(\Omega)} + \|\mathbf{u}_k\|^2_{L_{2h}(\Omega)} + \|\mathbf{u}_{k+1}\|^2_{L_{2h}(\Omega)} \leq K \tag{14}$$

where K *is a positive constant independent of the grid parameters* τ *and* h, *for all* $k \in \mathbb{N}$.

3 Numerical Analysis

In the present section the first order of accuracy difference scheme (9) and the second order of accuracy difference scheme (10) are used in combination with fixed point iteration, for the approximate solution of problem (1)–(3). The following initial boundary value problem (IBVP) for one dimensional system of coupled sine-Gordon equation

$$\begin{cases} u_{tt} - u_{xx} = -\sin(u - v) + f_1(t, x), \\ 0 < t < 1, 0 < x < \pi, \\ v_{tt} - v_{xx} = \sin(u - v) + f_2(t, x), \\ 0 < t < 1, 0 < x < \pi, \\ u(0, x) = \varphi_1(x), \ u_t(0, x) = \psi_1(x), \ 0 \leq x \leq \pi, \\ v(0, x) = \varphi_2(x), \ v_t(0, x) = \psi_2(x), \ 0 \leq x \leq \pi, \\ u(t, 0) = u(t, \pi) = 0, \ 0 \leq t \leq 1, \\ v(t, 0) = v(t, \pi) = 0, \ 0 \leq t \leq 1 \end{cases} \tag{15}$$

with some modifications is considered.

3.1 First Order of Accuracy Difference Scheme

In this section the IBVP

$$\begin{cases} u_{tt} - u_{xx} = -\sin(u - v) + \sin(\sin t \sin x - \cos t \sin x), \\ v_{tt} - v_{xx} = \sin(u - v) - \sin(\sin t \sin x - \cos t \sin x), \\ 0 < t < 1, 0 < x < \pi, \\ u(0, x) = 0, \ u_t(0, x) = \sin x, \ 0 \le x \le \pi, \\ v(0, x) = \sin x, \ v_t(0, x) = 0, \ 0 \le x \le \pi, \\ u(t, 0) = u(t, \pi) = 0, \ 0 \le t \le 1, \\ v(t, 0) = v(t, \pi) = 0, \ 0 \le t \le 1 \end{cases} \tag{16}$$

is considered. The exact solution of this problem is

$$u(t, x) = \sin t \sin x,$$
$$v(t, x) = \cos t \sin x.$$

Using the first order of accuracy in t and second order of accuracy in x implicit difference scheme (9) for the approximate solutions of IBVP (16), we get

$$\begin{cases} \frac{{}_m u_n^{k+1} - 2{}_m u_n^k + {}_m u_n^{k-1}}{\tau^2} - \frac{{}_m u_{n+1}^{k+1} - 2{}_m u_n^{k+1} + {}_m u_{n-1}^{k+1}}{h^2} \\ = -\sin({}_{m-1}u_n^k - {}_{m-1} v_n^k) + \sin(\sin t_k \sin x_n - \cos t_k \sin x_n), \\ \frac{{}_m v_n^{k+1} - 2{}_m v_n^k + {}_m v_n^{k-1}}{\tau^2} - \frac{{}_m v_{n+1}^{k+1} - 2{}_m v_n^{k+1} + {}_m v_{n-1}^{k+1}}{h^2} \\ = \sin({}_{m-1}u_n^k - {}_{m-1} v_n^k) - \sin(\sin t_k \sin x_n - \cos t_k \sin x_n), \\ t_{k+1} = (k + 1)\tau, \ 0 \le k \le N, \ N\tau = 1, \\ x_n = nh, \ 1 < n < M - 1, \ Mh = \pi, \\ {}_m u_n^0 = 0, \ \tau^{-1}({}_m u_n^1 - {}_m u_n^0) = \sin x_n, \ x_n = nh, 1 \le n \le M - 1, \\ {}_m v_n^0 = \sin x_n, \ \tau^{-1}({}_m v_n^1 - {}_m v_n^0) = 0, \ x_n = nh, 1 \le n \le M - 1, \\ {}_m u_0^k = {}_m u_M^k = 0, \ {}_m v_0^k = {}_m v_M^k = 0, 0 \le k \le N. \end{cases} \tag{17}$$

Here and in the sequel m is the fixed point iteration number, k and n are the parameters of the grid space (7). In the fixed point iteration, the initials are taken as the identity matrices of the form

Table 1 Errors for the approximate solution of problem (16)

N = M	Error of u	Error of v	m
20	0.0883	0.1564	10
40	0.0439	0.0785	11
80	0.0222	0.0393	12
160	0.0114	0.0196	13

$$
\begin{aligned}
{}_0u_n^k &= I(N+1, 1), \\
{}_0v_n^k &= I(N+1, 1).
\end{aligned} \tag{18}
$$

The algorithm is solved for $m = 1, 2, ..., p$ where p will be obtained with respect to the error tolerance ε such that

$$
\left|{}_pu_n - {}_{p-1}u_n\right| < \varepsilon. \tag{19}
$$

The solution is obtained by using the modified Gauss elimination method.

The iterative computations are terminated when the maximum difference at grid points of two successive iteration results gets less than ε. The errors are computed by the following formulas

$$
\begin{aligned}
&\max_{\substack{1\le k\le N-1 \\ 1\le n\le M-1}} \left|u(t_k, x_n) - u_n^k\right|, \\
&\max_{\substack{1\le k\le N-1 \\ 1\le n\le M-1}} \left|v(t_k, x_n) - v_n^k\right|,
\end{aligned} \tag{20}
$$

where u_n^k is the numerical solution and $u(t_k, x_n)$ is the exact solution of difference scheme (9) at (t_k, x_n). Numerical results are given in the following table.

In Table 1, the errors between the exact and the numerical solution of IBVP (16) are shown for different values of N and M. It can be seen from table that the difference scheme converges for several N, M values. If N are M doubled, the error decreases by a factor of 1/2, and we conclude that the scheme has first order of accuracy.

3.2 Second Order of Accuracy Difference Scheme

In the present section we again consider problem (16). Using the second order of accuracy in t and in x difference scheme (10) for the approximate solutions of IBVP (16), we get

Table 2 Errors for the approximate solution of problem (16)

N = M	Error of u	Error of v	m
20	$0.3619 \cdot 10^{-3}$	$0.9182 \cdot 10^{-3}$	10
40	$0.0897 \cdot 10^{-3}$	$0.2238 \cdot 10^{-3}$	12
80	$0.2231 \cdot 10^{-4}$	$0.5523 \cdot 10^{-4}$	12
160	$0.5561 \cdot 10^{-5}$	$0.1372 \cdot 10^{-4}$	13

$$\begin{cases} \frac{{}_m u_n^{k+1} - 2{}_m u_n^k + {}_m u_n^{k-1}}{\tau^2} - \frac{{}_m u_{n+1}^k - 2{}_m u_n^k + {}_m u_{n-1}^k}{2h^2} - \frac{{}_m u_{n+1}^{k+1} - 2{}_m u_n^{k+1} + {}_m u_{n-1}^{k+1}}{4h^2} - \frac{{}_m u_{n+1}^{k-1} - 2{}_m u_n^{k-1} + {}_m u_{n-1}^{k-1}}{2h^2} \\ = -\sin({}_{m-1}u_n^k - {}_{m-1} v_n^k) + \sin(\sin t_k \sin x_n - \cos t_k \sin x_n), \\ 1 \le k \le N-1,\ N\tau = 1, 1 \le n \le M-1, Mh = \pi, x_n = nh,\ t_k = k\tau \\ \frac{{}_m v_n^{k+1} - 2{}_m v_n^k + {}_m v_n^{k-1}}{\tau^2} - \frac{{}_m v_{n+1}^k - 2{}_m v_n^k + {}_m v_{n-1}^k}{2h^2} - \frac{{}_m v_{n+1}^{k+1} - 2{}_m v_n^{k+1} + {}_m v_{n-1}^{k+1}}{4h^2} - \frac{{}_m v_{n+1}^{k-1} - 2{}_m v_n^{k-1} + {}_m v_{n-1}^{k-1}}{4h^2} \\ = \sin({}_{m-1}u_n^k - {}_{m-1} v_n^k) - \sin(\sin t_k \sin x_n - \cos t_k \sin x_n), \\ {}_m u_n^0 = 0,\ {}_m v_n^0 = \sin x_n,\ x_n = nh, 1 \le n \le M-1, \\ {}_m u_n^0 = 0,\ \tau^{-1}({}_m u_n^1 - {}_m u_n^0) = \sin x_n,\ x_n = nh, 1 \le n \le M-1, \\ {}_m v_n^0 = \sin x_n,\ \tau^{-1}({}_m v_n^1 - {}_m v_n^0) = 0,\ x_n = nh, 1 \le n \le M-1, \\ u_0^k = u_M^k = 0, v_0^k = v_M^k = 0 \le k \le N. \end{cases} \tag{21}$$

Similar to the last section, the modified Gauss elimination method is used with the initial vectors (18) to obtain the approximate solution of problem (16). The iterative computations are terminated when the maximum difference at the grid points of two successive iteration results gets less than ε, in the formula (19). Errors are computed by formula (20).

In Table 2, the errors between the exact and the numerical solution of IBVP (16) are shown for different values of N and M. It can be seen from the table that the difference scheme converges for several N, M values. If N are M doubled, the error decreases by a factor of 1/4, and we conclude that the scheme has second order of accuracy.

The numerical implementations are carried out by MATLAB R2018b. The errors presented in tables indicate the convergence of the difference schemes and the accuracy of the results. It is observed that the second order of accuracy difference scheme converges faster than the first order of accuracy difference scheme.

4 Conclusion

In this paper, the unique solvability of coupled sine Gordon equations is presented and numerical experiments for the simple test problem were provided with error analysis. The first and the second order of accuracy unconditionally stable difference schemes and fixed point theory are used in numerical experiments. The errors presented in the tables indicate the convergence of the difference schemes and the accuracy of the results. We conclude that the second order of accuracy difference scheme increases faster than the first order of accuracy difference scheme.

Acknowledgements The authors would like to thank Professor Allaberen Ashyralyev (Near East University, Department of Mathematics) for his helpful suggestions to the improvement of this paper.

References

1. Zeidler, E.: Nonlinear Functional Analysis and Its Applications II/B: Nonlinear Monotone Operators. Springer, Leipzig, Germany (1990)
2. Ashyralyev, A., Sobolevskii, P.E.: Well Posedness of Parabolic Difference Equations. Birkauser Verlag, Basel (1994)
3. Ashyralyev, A., Sobolevskii, P.E.: New Difference Schemes for Partial Differential Equations, Operator Theory: Advances and Applications, vol. 148. Birkhauser Verlag, Basel (2004)
4. Ashyralyev, A., Sobolevskii, P.E.: A note on the difference schemes for hyperbolic equations. Abstr. Appl. Anal. **6**(2), 63–70 (2001)
5. Ashyralyev, A., Yildirim, O. (2013). On the numerical solution of hyperbolic IBVP with high order stable difference schemes. Bound. Value Probl. **2013**(1), 34 (2013)
6. Ashyralyev, A., Yildirim, O.: On multipoint nonlocal boundary value problems for hyperbolic differential and difference equations. Taiwa. J. Math. **14**(1), 165–194 (2010)
7. Ashyralyev, A., Yildirim, O.: Stable difference schemes for the hyperbolic problems subject to nonlocal boundary conditions with self-adjoint operator. Appl. Math. Comput. **218**(3), 1124–1131 (2011)
8. Yildirim, O., Uzun, M.: On third order stable difference scheme for hyperbolic multipoint nonlocal boundary value problem. Discret. Dyn. Nat. Soc. **2015**(2015), 1–16 (2015)
9. Yildirim, O., Uzun, M.: On the numerical solutions of high order stable difference schemes for the hyperbolic multipoint nonlocal boundary value problems. Appl. Math. Comput. **254**(2015), 210–218 (2015)
10. Yildirim, O., Uzun, M., Altun, O.: On the numerical solution of nonlinear system of coupled sine-Gordon equations. AIP Conf. Proc. **1997**, 020062 (2018). https://doi.org/10.1063/1.5049056
11. Yildirim, O., Uzun, M.: On fourth-order stable difference scheme for hyperbolic multipoint NBVP. Numer. Funct. Anal. Optim. **38**(10), 1305–1324 (2017)
12. Yildirim, O., Uzun, M.: Weak solvability of the unconditionally stable difference scheme for the coupled sine-Gordon system. Nonlinear Anal.: Model. Control **25**, 1–18 (2020)
13. Ashyraliyev, M.: On Gronwall's type integral inequalities with singular kernels. Filomat **31**, 4 (2017)
14. Direk, Z., Ashyraliyev, M.: FDM for the integral-differential equation of the hyperbolic type. Adv. Differ. Equs. **2014**(132), 1–8 (2014)
15. Ashyralyev, A., Ashyralyyev, C.: On the problem of determining the parameter of an elliptic equation in a Banach space. Nonlinear Anal.: Model. Control **19**(3), 350–366 (2014)

16. Ashyralyyev, C.: High order approximation of the inverse elliptic problem with Dirichlet-Neumann conditions. Filomat **28**(5), 947–962 (2014)
17. Ashyralyyev, C.: High order of accuracy difference schemes for the inverse elliptic problem with Dirichlet condition. Bound. Value Probl. **2014**(5), (2014)
18. Macías-Díaz, J.E.: On the simulation of the energy transmission in the forbidden band-gap of a spatially discrete double sine-Gordon system. Comput. Phys. Commun. **181**, 1842–1849 (2010)
19. Ben-Yu, G., Pascual, P.J., Rodriguez, M.J., Vazquez, L.: Numerical solution of the sine-Gordon equation. Appl. Math. Comput. **18**, 1–14 (1986)
20. Ha, J., Nakagiri, S.: Coupled sine-Gordon equations as nonlinear second order evolution equations. Taiwan. J. Math. **5**(2), 297–315 (2001)
21. Hocquet, A., Hofmanová, M.: An energy method for rough partial differential equations. J. Differ. Equs. **265**, 1407–1466 (2018)
22. Wang, Q.F.: Numerical solution for series sine-Gordon equations using variational method and finite element approximation. Appl. Math. Comput. **168**, 567–599 (2005)
23. Bratsos, A.G.: A numerical method for the one-dimensional sine-Gordon equation. Numer. Methods Part. Differ. Equs. **24**(3), 833–844 (2008)
24. Ilati, M., Dehghan, M.: The use of radial basis functions (RBFs) collocation and RBF-QR methods for solving the coupled nonlinear sine-Gordon equations. Eng. Anal. Bound. Elem. **52**, 99–109 (2015)

Comparison of the Rate of Induced Intrinsic Pathway of Apoptosis on COLO-320 and COLO-741

Evren Hincal, Günsu Soykut, Farouk Tijjani Saad, Seda Vatansever, Isa Abdullahi Baba, İhsan Çalış, Bilgen Kaymakamzade, and Eda Becer

Abstract Mathematical modeling is a key tool in understanding complicated behavior of large signal transduction networks. In this paper a mathematical model is developed to understand the complex signaling behavior of induced apoptosis in Colo-320 and Colo-741 cell lines. The model studies the intrinsic pathway of apoptosis induced by Apaf-1, Cyt-c, and Caspase 3. Experimentally these tests were carried out five

E. Hincal (✉) · F. Tijjani Saad · I. Abdullahi Baba · B. Kaymakamzade
Department of Mathematics, Near East University, Mersin 10, Nicosia-Northern Cyprus, Turkey
e-mail: evren.hincal@neu.edu.tr

F. Tijjani Saad
e-mail: farouksaaad@yahoo.co.uk

I. Abdullahi Baba
e-mail: isababa7@yahoo.com

B. Kaymakamzade
e-mail: bilgen.kaymakamzade@neu.edu.tr

G. Soykut
Department of Nutrition and Dietetics, Faculty of Health Sciences, Near East University, Nicosia, Cyprus
e-mail: gunsusoykut@gmail.com

S. Vatansever
Experimental Health Research Center of Health Sciences, Near East University, Nicosia, Cyprus
e-mail: Sedavatansever@yahoo.com

Department of Histology and Embryology, Faculty of Medicine, Celal Bayar University, Manisa, Turkey

İ. Çalış · E. Becer
Faculty of Pharmacy, Near East University, Mersin 10, 99138 Nicosia, Cyprus
e-mail: Ihsan.calis@neu.edu

E. Becer
e-mail: Edabecer@hotmail.com

E. Becer
Department of Biochemistry, Faculty of Pharmacy, Near East University, Mersin 10, 99138 Nicosia, Cyprus

A. Ashyralyev et al. (eds.), *Functional Analysis in Interdisciplinary Applications—II*, Springer Proceedings in Mathematics & Statistics 351,
https://doi.org/10.1007/978-3-030-69292-6_12

times for three groups; namely Control, DCM and Aqueous groups. The main aim is to compare the rate of the occurrence of apoptosis in the two cells Colo-320 and Colo-741 using experimental data. Numerical results reveal that apoptosis takes place at the same time in DCM for Colo-320 and Colo-741, but slightly faster in Aqueous group for Colo-320. It was observed that it took place between the 5–7th h in all the cases.

Keywords HIV · Mathematical model · Basic reproduction number · Stability analysis

1 Introduction

Apoptosis as a form of cell death helps multicellular organisms in removing damaged cells in order to maintain tissue homeostasis. Disregulation of apoptosis leads to many pathological implications, including cancer and neurological problems. Extra- or intracellular stimuli can induce extrinsic and intrinsic apoptotic pathways, respectively. Both pathways lead to activation caspases and other apoptotic phenotypes such as nuclear fragmentation, membrane blebbing and chromatin condensation. These cause the cell to disintegrate [1].

Experimental studies revealed that the qualitative behavior of caspase activation in the intrinsic pathway depends on the cellular context [9, 10, 12, 14, 19]. These qualitative differences in caspase activation suggest that the intrinsic pathway is bi-stable in some cells, but mono-stable in others. Recent mathematical modeling demonstrated that bi-stability can arise from "hidden" or implicit, feedback loops that are usually not explicitly drawn in biochemical reaction schemes [4, 11].

Mathematical modeling is a key tool in understanding complicated behavior of large signal transduction networks. In this paper a mathematical model is developed to understand the complex signaling behavior of induced apoptosis in Colo-320, primary colon adenocarcinoma cells and Colo-741, metastatic colon adenocarcinoma cells. The model studies the intrinsic pathway of apoptosis induced by Apaf-1, Cyt-c, and caspase 3. Experimentally these tests were carried out five times for three groups; namely Control, DCM and Aqueous groups. The groups are extracts of a plant called *Corchorus olitorius* L. and known as medicinal plant which is highly consumed in the Mediterranean area [7]. The main aim is to compare the rate of the occurrence of apoptosis in the two cells Colo-320 and Colo-741 using experimental data. The rest of the paper is organized as follows; in the next section the reactions, and system of ODE derived from the reactions are given. Then results and numerical simulations are presented. Finally the results were discussed.

2 Materials and Method

In this chapter, reaction details for the intrinsic apoptosis for both Colo-320 and Colo-741 are given. From the reactions the model is derived, which consists of a system of 9 ordinary differential equations. Table 1 gives the meaning of the parameters in the model.

2.1 *Extraction and Immunocytochemical Analysis*

The extraction and immunocytochemical analysis were applied as previous study states [10].

2.2 *Reactions*

The following are the chemical reactions involved in the process:

$$Cyt - c + Apaf - 1 \xrightarrow{k_1} C9^*, \tag{1}$$

$$C9^* + C3 \xrightarrow{k_2} C9^* + C3^*, \tag{2}$$

Table 1 Meaning of parameters of model

Parameter	Meaning
k_1	Activation rate of caspase 9
k_2	Activation rate of caspase 3
k_3	Activation rate of proteases and nucleuses
k_4	Activation rate of Nucleic acids and proteins
k_5	Degradation rate of cyt-c
k_6	Degradation rate of Apaf-1
k_7	Degradation rate of caspase 9
k_8	Degradation rate of procaspase 3
k_9	Degradation rate of caspase 3
k_{10}	Degradation rate of proteases
k_{11}	Degradation rate of Nucleases
k_{12}	Degradation rate of proteins
k_{13}	Degradation rate of nucleic acids

$$C3^* \xrightarrow{k_3} Proteases + Nucleases, \tag{3}$$

$$Proteases + Nucleases \xrightarrow{k_4} Proteins + NucleicAcid. \tag{4}$$

When the BAX forces the mitochondrion to release Cyt-c, then the free Cyt-c binds with Apaf-1 and activates procaspse 9. This is given in REACTION 1. REACTION 2 is the activation of C3 by C9. In REACTION 3, C3 cleaves and releases the proteases and nucleases inside the cells, which in turn breakdown the all the proteins and nucleic acids inside the cell respectively, and it is given in REACTION 4.

Note that: C3, C8 refers to procaspse 3, 8 (inactive caspases), while $C3^*$ and $C8^*$ refer to active caspase 3, 8 respectively. The $(*)$ attached to the caspases indicate that they are at their active state.

2.3 The Mathematical Model

The reactions above yields the following systems of ordinary differential equations:

$$\begin{aligned}
\frac{d\,[Cyt-c\,]}{dt} &= -k_1\,[Cyt-c\,]\,[Apaf-1\,] - k_5\,[Cyt-c\,],\\
\frac{d\,[Apaf-1\,]}{dt} &= -k_1\,[Cyt-c\,]\,[Apaf-1\,] - k_6\,[Apaf-1\,],\\
\frac{d\,[C9^*\,]}{dt} &= k_1\,[Cyt-c\,]\,[Apaf-1\,] - k_2\,[C9^*\,]\,[C3\,] - k_7\,[C9^*\,],\\
\frac{d\,[C3\,]}{dt} &= -k_2\,[C9^*\,]\,[C3\,] - k_8\,[C3\,],\\
\frac{d\,[C3^*\,]}{dt} &= k_1\,[Cyt-c\,]\,[Apaf-1\,] - k_2\,[C9^*\,]\,[C3\,] - k_7\,[C9^*\,],\\
\frac{d\,[Proteases\,]}{dt} &= -k_4\,[Proteases\,]\,[Nucleuses\,] + k_3\,[C3^*\,] - k_{10}\,[Proteases\,],\\
\frac{d\,[Nucleuses\,]}{dt} &= -k_4\,[Proteases\,]\,[Nucleuses\,] + k_3\,[C3^*\,] - k_{11}\,[Nucleuses\,],\\
\frac{d\,[Proteins\,]}{dt} &= k_4\,[Proteases\,]\,[Nucleuses\,] - k_{12}\,[Proteins\,],\\
\frac{d\,[NucleicAcids\,]}{dt} &= k_4\,[Proteases\,]\,[Nucleuses\,] - k_{13}\,[NucleicAcids\,].
\end{aligned}$$

3 Results

In this section, we perform the numerical simulations for each of Control group, DCM group, and Aqueous group. We conducted experiments and the results are shown below. With the assumption that, the moment caspase 3 is activated apoptosis will occur; the model is reduced to a system of four ordinary differential equations:

$$\begin{aligned}
\frac{d\,[Cyt-c\,]}{dt} &= -k_1\,[Cyt-c\,]\,[Apaf-1\,] - k_3\,[Cyt-c\,],\\
\frac{d\,[Apaf-1\,]}{dt} &= -k_1\,[Cyt-c\,]\,[Apaf-1\,] - k_4\,[Apaf-1\,],\\
\frac{d\,[C9^*\,]}{dt} &= k_1\,[Cyt-c\,]\,[Apaf-1\,] - k_2\,[C9^*\,]\,[C3\,] - k_5\,[C9^*\,],\\
\frac{d\,[C3\,]}{dt} &= -k_2\,[C9^*\,]\,[C3\,] - k_6\,[C3\,].
\end{aligned} \tag{5}$$

3.1 Numerical Simulations

The numerical simulations results are given in Figs. 1, 2, and 3 in each case for both Colo-320 and Colo-741.

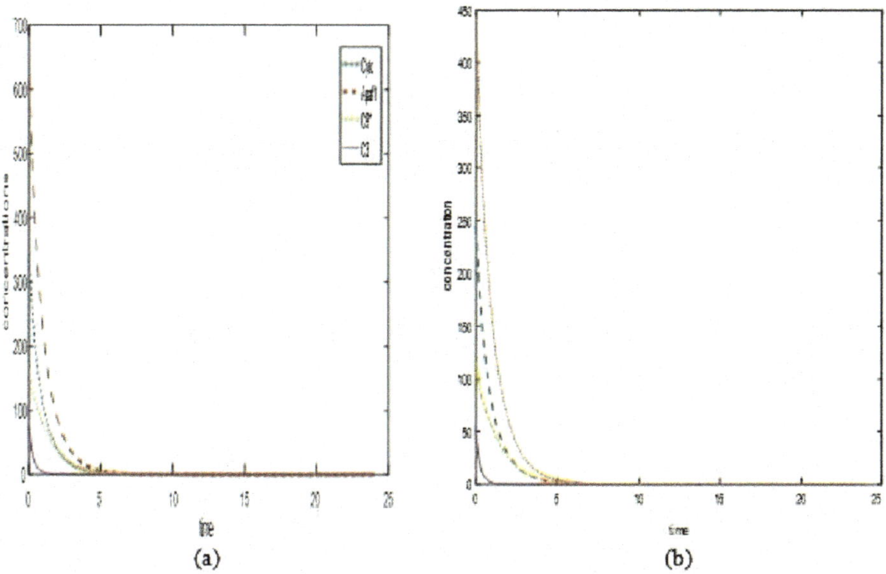

Fig. 1 Numerical results for Control group for Colo-320 (a) and Colo-741 (b)

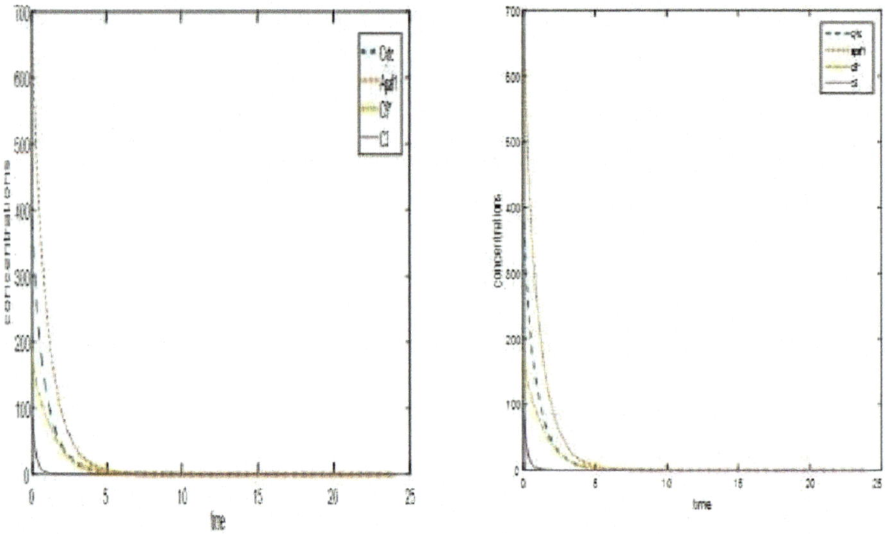

Fig. 2 Numerical results for DCM group for Colo-320 (**a**) and Colo-741 (**b**)

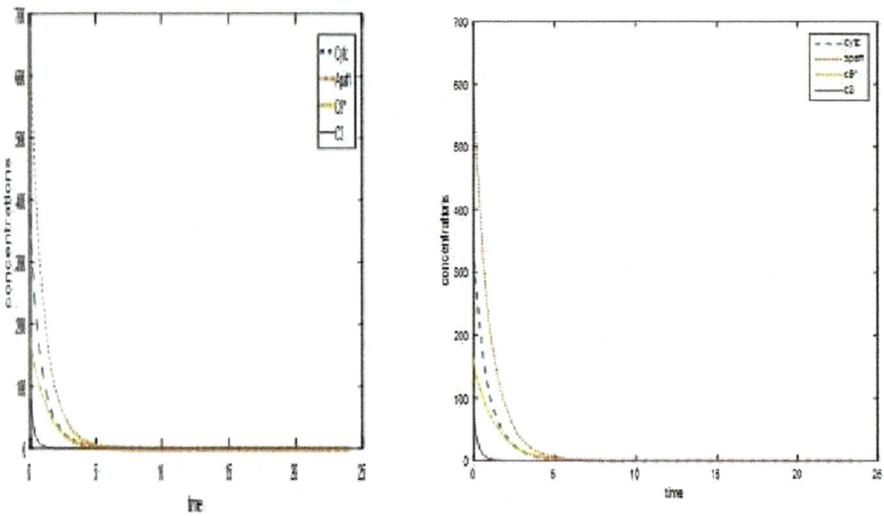

Fig. 3 Numerical results for Aqueous group for Colo-320 (**a**) and Colo-741 (**b**)

4 Discussion

Apoptosis is an important process and necessity for multicellular organisms. It is organism's protection mechanism especially against cancer. In cancer, the malignant, cancerous cells divide uncontrollably and form tumors which might become invasive and metastase into different organs and increase the rate and risk of mortality [6]. Numerous formalisms have been used to model apoptosis to understand behavior of cells [2, 13, 16]. However, mathematical model that is combined with experimental approaches for the specific plant extract triggered apoptosis in colon cancer cells has not been studied.

Based on the numerical results computed, apoptosis takes place at the same time in DCM for both COLO-320 and COLO-741, but slightly faster in aqueous phase in COLO-320 than in COLO-741 cells. We can also observe that in all the cases. Apoptosis takes place between the 5–7th h (Fig. 4).

Figures 4 and 5 compare the apoptotic cells DCM (Colo-320 and Colo-741) and AQUEOUS (Colo-320 and Colo-741) from experimental data for 24 h.

Corchorus olitorius L. plant is known to exert anti-cancer properties due to its rich polyphenolic content [3]. The polyphenols are chemicals found in plants which are usually strong antioxidants, giving most of the plants its cancer protective properties [17]. In vitro cell culture studies are of importance in terms of identifying the effects of plants on certain kind of cancer cells. In the present study, both extracts induced apoptosis while aqueous phase being more effective in terms of induction time of apoptosis. Another experimental study presented parallel results stating aqueous phase extract is more effective in triggering apoptosis especially in metastatic Colo-741 colon cancer cells [15]. These results suggest that combining cell culture immunocytochemical analysis of cancer cells with mathematical modelling give more robust information about the preferred treatment of certain cancer which in this study is colon cancer.

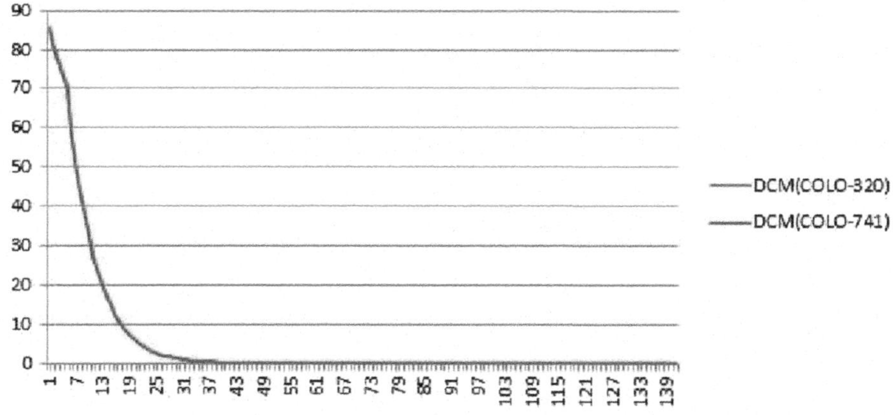

Fig. 4 Behavior of apoptotic cells DCM (Colo-320 and Colo-741) from experimental data for 24 h

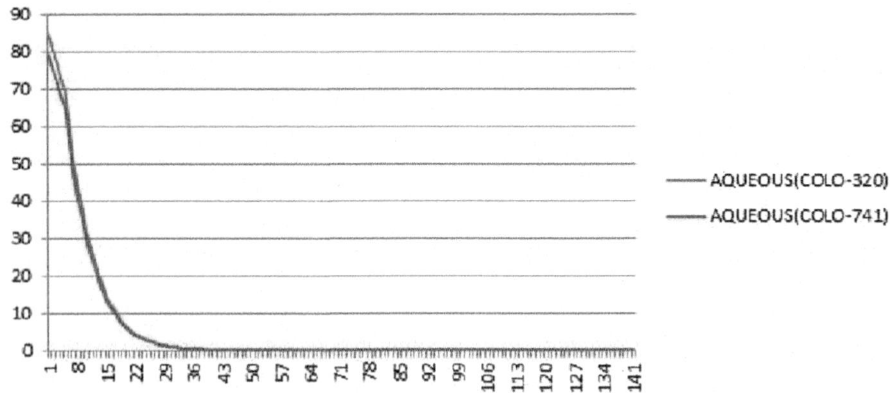

Fig. 5 Behavior of apoptotic cells AQUEOUS (Colo-320 and Colo-741) from experimental data for 24 h

The kinetic mathematical models are used for well understand signal transduction pathways of apoptosis. Kinetic approaches can be used for differential equations that used to generate protein profiles over time in models. These designs give topological networks of protein interaction through apoptotic pathway. The reaction rate depends on substrates, enzymes and products concentrations [5, 18]. On the other hand, experimental approaches of apoptosis include cell death assays, Western Blot, mass spectrometry and single cell analysis. These assays' results may be changed and affected from environmental conditions easily and might not be able to give exact reaction time scales. For this reasons, quantitative mathematical modelling of experimental data provides insights into the correctness of the assumed apoptosis pathway topology and time scales.

By using multidisciplinary sciences of mathematics and molecular biology, might increase chances of having more effective drug treatment in more cost effective way. The mathematical models stating the right dose of drugs and timing of apoptosis might also decrease treatment time and dose which is beneficial to patients whom having long-term cancer treatment and suffering from side effects [8]. When all taken into account, forming a mathematical model to identify and determine the timing of apoptosis of cancer cells has vital role in terms of treatment of cancer.

To conclude, using *Corchorus olitorius* L. extracts and combining the experimental immunocytochemical evaluation results together with mathematical modelling might guide future studies where the plant can be used drug therapy in colon cancer treatment in more cost effective way.

References

1. Chang, H.Y., Yang, X.: Proteases for cell suicide functions and regulation of caspases. Mic. Mol. Biol. Rev. **64**, 821–846 (2000)
2. Fussenegger, M., Bailey, J.E., Varner, J.: A mathematical model of caspase function in apoptosis. Nat. Biotechnol. **18**, 768–774 (2000)
3. Handoussa, H., Hanafi, R., Eddiasty, I., El-Gendy, M., El-Khatip, A., Linscheid, M., et al.: Anti-inflammatory and cytotoxic activities of dietary phenolics isolated from corchorus olitorius and vitis vinifera. J. Funct. Foods **5**, 1204–1216 (2013)
4. Hayer, A., Bhalla, U.S.: Molecular switches at the synapse emerge from receptor and kinase traffic. PLoS Comput. Biol. **1**, 137–154 (2005)
5. Huber, H.F., Bullinger, E., Rehm, M.: Systems biology approaches to the study of apoptosis, essentials of apoptosis. A Guid. Basic Clin. Res. **1**, 283–297 (2009)
6. Huerta, S., Goulet, E.J., Livingston, E.J.: Colon cancer and apoptosis. Am. J. Surg. **191**, 517–526 (2006)
7. Işeri, O.D., Yurtcu, E., Sahin, F.I., Mehmet, H.: Corchorus olitorius (jute) extract induced cytotoxicity and genotoxicity on human multiple myeloma cells (ARH-77). Pharm. Biol. **6**, 766–770 (2013)
8. Jordano, G., Tavares, J.N.: Mathematical models in cancer therapy. BioSystems **162**, 12–23 (2017)
9. Kluck, R.M., Bossy-Wetzel, E., Green, D.R., Newmeyer, D.D.: The release of cytochrome C from mitochondria. A primary site for Bcl-2 regulation of apoptosis. Science **275**, 1132–1136 (1997)
10. Kluck, R.M., Ellerby, L.M., Ellerby, H.M., Naiem, S., Yaffe, M.P., et al.: Determinants of cytochrome c pro-apoptotic activity. The role of lysine 72 trimethylation. J. Biol. Chem. **275**, 16127–16133 (2000)
11. Markevich, N.I., Hoek, J.B., Kholodenko, B.N.: Signaling switches and bistability arising from multisite phosphorylation in protein kinase cascades. J. Cell Biol. **164**, 353–359 (2004)
12. Murphy, B.M., O'Neill, A.J., Adrain, C., Watson, R.W., Martin, S.J.: The apoptosome pathway to caspase activation in primary human neutrophils exhibits dramatically reduced requirements for cytochrome. C. J. Exp. Med. **197**, 625–632 (2003)
13. Schlatter, R., Schmich, K., Avalos Vizcarra, I., Scheurich, P., Sauter, T., Borner, C., Ederer, M., Merfort, I., Sawodny, O.: On/off and beyond–a boolean model of apoptosis. PLoS Comput. Biol. **5**, 1000595 (2009)
14. Slee, E.A., Harte, M.T., Kluck, R.M., Wolf, B.B., Casiano, C.A., et al.: Ordering the cytochrome initiated caspase cascade, hierarchical activation of caspases-2,3,6,7,8, and 10 in a caspase-9-dependent manner. J. Cell. Biol. **144**, 281–292 (1999)
15. Soykut, G., Becer, E., Calis, I., Yucecan, S., Vatansever, S.: Apoptotic effects of corchorus olitorius L. leaf extracts in colon adenocarcinoma cell lines. Prog. Nutr. (2018). https://doi.org/10.23751/pn.v20i4.6892
16. Spencer, S.L., Sorger, P.K.: Measuring and modeling apoptosis in single cells. Cell **144**, 926–939 (2011)
17. Tsao, R., Linscheid, M., et al.: Chemistry and biochemistry of dietary polyphenols. Nutrients **2**, 1231–1246 (2010)
18. Wolkenhauer, O., et al.: The dynamic systems approach to control and regulation of intracellular networks. FEBS. Lett. **8**, 1846–1853 (2005)
19. Yang, J., Liu, X., Bhalla, K., Kim, C.N., Ibrado, A.M., et al.: Prevention of apoptosis by Bcl-2 release of cytochrome c from mitochondria blocked. Science **275**, 1129–1132 (1997)

A Note on Representation Variety of Abelian Groups and Reidemeister Torsion

Fatih Hezenci and Yasar Sozen

Abstract Let S and G denote respectively the $2-$torus and one of the Lie groups $\mathrm{GL}(n, \mathbb{C})$, $\mathrm{SL}(n, \mathbb{C})$, $\mathrm{SO}(n, \mathbb{C})$, $\mathrm{Sp}(2n, \mathbb{C})$. In the present article, we consider the smooth part of the representation variety $\mathrm{Rep}(S, G)$ consisting of conjugacy classes of homomorphisms from fundamental group $\pi_1(S)$ to G. We show the well definiteness of Reidemeister torsion for such representations. In addition, we establish a formula for computing the Reidemeister torsion of such representations in terms of the symplectic structure on $\mathrm{Rep}\,(S, G)$ [51]. This symplectic form is analogous to Atiyah–Bott–Goldman symplectic form of higher genera for the Lie group G.

Keywords Reidemeister torsion · Representation varieties · Abelian varieties · Atiyah-Bott-Goldman Symplectic Form · Chain complex · Symplectic chain complex.

1 Introduction

The role played by character varieties in many branches of mathematics and physics is well known. For instance, if S is a closed orientable surface of genus at leqst 2, then one can inteprete several geometric structures on S as representation of the fundamental group of S to certain Lie groups [1–6]. They have many applications in several branches of mathematics and physics such as in hyperbolic geometry [7–9], in 3-manifold topology theory [10–13], in Casson invariant theory [14–16], in A-polynamial [17–20], in Yang-Mills and Chern-Simons quantum field theories [21–25], in skein theory of quantum invariants of 3-manifolds [26–29], in the moduli spaces of flat connections, holomorphic bundles, and Higgs bundles [30–39].

F. Hezenci (✉)
Department of Mathematics, Faculty of Science and Arts, Duzce University, 81620 Duzce, Turkey
e-mail: fatihhezenci@duzce.edu.tr

Y. Sozen
Department of Mathematics, Faculty of Science, Hacettepe University, 06800 Ankara, Turkey
e-mail: ysozen@hacettepe.edu.tr

A. Ashyralyev et al. (eds.), *Functional Analysis in Interdisciplinary Applications—II*,
Springer Proceedings in Mathematics & Statistics 351,
https://doi.org/10.1007/978-3-030-69292-6_13

Reidemeister torsion is a topological invariantand was introduced by Reidemeister in [40] where he classified 3−dimensional lens spaces. This invariant has many applications in several branches of mathematics and theoretical phsyics such as topology [40–43], representation spaces [25], differential geometry [44, 45], knot theory [46], dynamical systems [47], 3-dimensional Seiberg-Witten theory [48], Chern-Simon theory [24], algebraic K-theory [46], theoretical physics and quantum field theory [24, 25, 49]. See references [7, 50], and the references therein for more information.

Real symplectic chain complex is an algebraic topological instrument and was introduced by Witten in [25]. Using real symplectic chain complex and Reidemeister torsion, he computed the volume of several moduli space $\mathrm{Rep}(\Sigma, G)$ of all conjugacy classes of representations from the fundamental group of a Riemann surface Σ to the compact gauge group $G \in \{\mathrm{SU}(2), \mathrm{SO}(3)\}$.

The present paper considers the smooth part of the representation variety $\mathrm{Rep}(S, G)$ consisting of conjugacy classes of homomorphisms from fundamental group $\pi_1(S)$ of 2−torus S to the Lie group G, where G is one of the groups $\mathrm{GL}(n, \mathbb{C})$, $\mathrm{SL}(n, \mathbb{C})$, $\mathrm{SO}(n, \mathbb{C})$, $\mathrm{Sp}(2n, \mathbb{C})$. It shows that Reidemeister torsion for such representations is well defined (Theorem 3). Furthermore, it establishes a formula for computing the Reidemeister torsion of such representations (Theorem 4) in terms of the symplectic structure on $\mathrm{Rep}\,(S, G)$ [51], which is similar to Atiyah–Bott–Goldman symplectic form of higher genera for the Lie group G.

2 Reidemeister Torsion

In this section, we will provide the basic definitions and facts about Reidemeister torsion and also symplectic chain complex. We refer the reader to [7, 25, 50, 52] for further information.

Let $C_* = \left(0 \to C_n \xrightarrow{\partial_n} C_{n-1} \to \cdots \to C_1 \xrightarrow{\partial_1} C_0 \to 0\right)$ denote a chain complex of finite dimensional vector spaces over the field $\mathbb{C}$ of complex numbers. For $p = 0, \ldots, n$, denote by $Z_p(C_*)$, $B_p(C_*)$, and $H_p(C_*)$, respectively the kernel of ∂_p, the image of ∂_{p+1}, and the pth homology group of the chain complex C_*. From the definition of $Z_p(C_*)$, $B_p(C_*)$, and $H_p(C_*)$ it follows the following short-exact sequences:

$$0 \longrightarrow Z_p(C_*) \hookrightarrow C_p \twoheadrightarrow B_{p-1}(C_*) \longrightarrow 0 \tag{1}$$

and

$$0 \longrightarrow B_p(C_*) \hookrightarrow Z_p(C_*) \twoheadrightarrow H_p(C_*) \longrightarrow 0. \tag{2}$$

For $p = 0, \ldots, n$, let $\mathbf{c}_p$, $\mathbf{b}_p$, and $\mathbf{h}_p$ denote bases of C_p, $B_p(C_*)$, and $H_p(C_*)$, respectively. Assume $\ell_p : H_p(C_*) \to Z_p(C_*)$, $s_p : B_{p-1}(C_*) \to C_p$ are sections of $Z_p(C_*) \to H_p(C_*)$, $C_p \to B_{p-1}(C_*)$, respectively. Using the sequences (1) and (2), we obtain the basis $\mathbf{b}_p \sqcup \ell_p(\mathbf{h}_p) \sqcup s_p(\mathbf{b}_{p-1})$ of C_p. Here, $\sqcup$ is the disjoint union.

Definition 1 Suppose $\mathbf{c}_p$, $\mathbf{b}_p$, $\mathbf{h}_p$, ℓ_p, and s_p are as above. Then, the alternating product $\mathbb{T}\left(C_*, \{\mathbf{c}_p\}_0^n, \{\mathbf{h}_p\}_0^n\right) = \prod_{p=0}^n \left[\mathbf{b}_p \sqcup \ell_p(\mathbf{h}_p) \sqcup s_p(\mathbf{b}_{p-1}), \mathbf{c}_p\right]^{(-1)^{(p+1)}}$ is called *Reidemeister torsion* of the chain complex C_* is with respect to bases $\{\mathbf{c}_p\}_{p=0}^n$, $\{\mathbf{h}_p\}_{p=0}^n$. Here, $\left[\mathbf{e}_p, \mathbf{f}_p\right]$ is determinant of the change-base-matrix from basis $\mathbf{f}_p$ to $\mathbf{e}_p$ of C_p.

As is well known that Reidemeister torsion does not depend on the bases $\mathbf{b}_p$ and sections s_p, ℓ_p. Furthermore, if $\mathbf{c}'_p$, $\mathbf{h}'_p$ are also bases of C_p, $H_p(C_*)$, respectively, then the following change-base-formula holds [46]:

$$\mathbb{T}\left(C_*, \{\mathbf{c}'_p\}_0^n, \{\mathbf{h}'_p\}_0^n\right) = \prod_{p=0}^n \left(\frac{\left[\mathbf{c}'_p, \mathbf{c}_p\right]}{\left[\mathbf{h}'_p, \mathbf{h}_p\right]}\right)^{(-1)^p} \mathbb{T}\left(C_*, \{\mathbf{c}_p\}_0^n, \{\mathbf{h}_p\}_0^n\right). \tag{3}$$

Let

$$0 \longrightarrow A_* \stackrel{\iota}{\longrightarrow} B_* \stackrel{j}{\longrightarrow} D_* \longrightarrow 0 \tag{4}$$

be a short-exact sequence of chain complexes. Assume $\mathbf{c}_p^A$, $\mathbf{c}_p^B$, $\mathbf{c}_p^D$, $\mathbf{h}_p^A$, $\mathbf{h}_p^B$, and $\mathbf{h}_p^D$ are bases of A_p, B_p, D_p, $H_p(A_*)$, $H_p(B_*)$, and $H_p(D_*)$, respectively. For the short-exact sequence (4), one has the Mayer-Vietoris long-exact sequence of vector spaces

$$\mathcal{C}_* : \cdots \longrightarrow H_p(A_*) \stackrel{\iota_p}{\longrightarrow} H_p(B_*) \stackrel{j_p}{\longrightarrow} H_p(D_*) \stackrel{\delta_p}{\longrightarrow} H_{p-1}(A_*) \longrightarrow \cdots.$$

Here, the long-exact sequence $\mathcal{C}_*$ has length $3n+2$, $\mathcal{C}_{3p} = H_p(D_*)$, $\mathcal{C}_{3p+1} = H_p(A_*)$, and $\mathcal{C}_{3p+2} = H_p(B_*)$. Moreover, the bases $\mathbf{h}_p^D$, $\mathbf{h}_p^A$, and $\mathbf{h}_p^B$ are bases of $\mathcal{C}_{3p}$, $\mathcal{C}_{3p+1}$, and $\mathcal{C}_{3p+2}$, respectively.

Theorem 1 ([46]) *Suppose* $\mathbf{c}_p^A$, $\mathbf{c}_p^B$, $\mathbf{c}_p^D$, $\mathbf{h}_p^A$, $\mathbf{h}_p^B$, *and* $\mathbf{h}_p^D$ *are as above. Suppose also* $\left[\mathbf{c}_p^B, \mathbf{c}_p^A \oplus \widetilde{\mathbf{c}_p^D}\right] = \pm 1$, *where* $j\left(\widetilde{\mathbf{c}_p^D}\right) = \mathbf{c}_p^D$. *Then, the following equality holds:*

$$\begin{aligned}\mathbb{T}\left(B_*, \{\mathbf{c}_p^B\}_0^n, \{\mathbf{h}_p^B\}_0^n\right) &= \mathbb{T}\left(A_*, \{\mathbf{c}_p^A\}_0^n, \{\mathbf{h}_p^A\}_0^n\right) \mathbb{T}\left(D_*, \{\mathbf{c}_p^D\}_{p=0}^n, \{\mathbf{h}_p^D\}_0^n\right)\\ &\times \mathbb{T}\left(\mathcal{C}_*, \{\mathbf{c}_{3p}\}_0^{3n+2}, \{0\}_0^{3n+2}\right).\end{aligned}$$

By using Theorem 1, the following the sum-lemma is obtained.

Lemma 1 *Assume* A_*, D_* *are chain complexes of vector spaces and assume also* $\mathbf{c}_p^A$, $\mathbf{c}_p^D$, $\mathbf{h}_p^A$, *and* $\mathbf{h}_p^D$ *are bases of* A_p, D_p, $H_p(A_*)$, *and* $H_p(D_*)$, *respectively. Then, the following formula is valid:*

$$\mathbb{T}(A_* \oplus D_*, \{\mathbf{c}_p^A \sqcup \mathbf{c}_p^D\}_0^n, \{\mathbf{h}_p^A \sqcup \mathbf{h}_p^D\}_0^n) = \mathbb{T}(A_*, \{\mathbf{c}_p^A\}_0^n, \{\mathbf{h}_p^A\}_0^n)\mathbb{T}(D_*, \{\mathbf{c}_p^D\}_0^n, \{\mathbf{h}_p^D\}_0^n).$$

The proof of Lemma 1 can also be found in [53].

Definition 2 A $\mathbb{C}-$symplectic chain complex is a triple $(C_*, \partial_*, \{\omega_{*,q-*}\})$ with the following two conditions:

1. $C_* : 0 \to C_q \xrightarrow{\partial_q} C_{q-1} \to \cdots \to C_{q/2} \to \cdots \to C_1 \xrightarrow{\partial_1} C_0 \to 0$ is a chain complex and $q \equiv 2 \pmod 4$),
2. For $p = 0, \ldots, q$, $\omega_{p,q-p} : C_p \times C_{q-p} \to \mathbb{C}$ is a $\partial-$compatible non-degenerate anti-symmetric bilinear form. To be precise,

$$\begin{cases} \omega_{p,q-p}\left(\partial_{p+1}a, b\right) = (-1)^{p+1}\omega_{p+1,q-(p+1)}\left(a, \partial_{q-p}b\right), \\ \omega_{p,q-p}(a, b) = (-1)^{p(q-p)}\omega_{q-p,p}(b, a). \end{cases}$$

Clearly, we have $\omega_{p,q-p}(a, b) = (-1)^p\omega_{q-p,p}(b, a)$ and the extension of $\omega_{p,q-p}$ to homomologies $[\omega_{p,q-p}] : H_p(C_*) \times H_{q-p}(C_*) \to \mathbb{C}$. In the sequel, for the symplectic chain complex $(C_*, \partial_*, \{\omega_{*,q-*}\})$, we will denote by $\Delta(\mathbf{h}_p, \mathbf{h}_{q-p})$ the determinant of the matrix of the non-degenerate pairing $[\omega_{p,q-p}]$ in the bases $\mathbf{h}_p$, $\mathbf{h}_{q-p}$.

For a symplectic chain complex C_* length q, the bases $\mathbf{c}_p$, $\mathbf{c}_{q-p}$ of C_p, C_{q-p}, are said to be *$\omega-$compatible*, if the matrix of $\omega_{p,q-p}$ in bases $\mathbf{c}_p$, $\mathbf{c}_{q-p}$ is the $k \times k$ identity matrix $\mathrm{Id}_{k\times k}$ when $p \neq q/2$ and $\begin{pmatrix} 0_{l\times l} & \mathrm{Id}_{l\times l} \\ -\mathrm{Id}_{l\times l} & 0_{l\times l} \end{pmatrix}$ when $p = q/2$. Here, k is $\dim C_p = \dim C_{q-p}$ and $2l$ is $\dim C_{q/2}$.

The following theorem gives a formula for computing Reidemeister torsion in connection with intersections pairings.

Theorem 2 ([54]) *Let $(C_*, \partial_*, \{\omega_{*,q-*}\})$ be a $\mathbb{C}-$symplectic chain complex with the $\omega-$compatible bases $\mathbf{c}_p$ and $\mathbf{h}_p$ be a basis of $H_p(C_*)$, $p = 0, \ldots, q$. Then, the following equality holds:*

$$\left|\mathbb{T}\left(C_*, \{\mathbf{c}_p\}_0^q, \{\mathbf{h}_p\}_0^q\right)\right| = \prod_{p=0}^{(q/2)-1} |\Delta(\mathbf{h}_p, \mathbf{h}_{q-p})|^{(-1)^p} \sqrt{\left|\Delta\left(\mathbf{h}_{q/2}, \mathbf{h}_{q/2}\right)\right|}^{(-1)^{q/2}}. \tag{5}$$

In the case $\mathbf{h}_p = \mathbf{h}_{q-p} = 0$, the convention $0 = 1.0$ is used in equation (5) and hence $\Delta(\mathbf{h}_p, \mathbf{h}_{q-p}) = 1$.

Remark 1 Clearly, for a $\mathbb{C}-$symplectic chain complex with the $\omega-$compatible bases $\mathbf{c}_p$, $p = 0, \ldots, q$, there exist bases $\mathbf{h}_p^0$ such that $\mathbb{T}\left(C_*, \{\mathbf{c}_p\}_0^q, \{\mathbf{h}_p^0\}_0^q\right) = 1$. This and Eq. (5) yield the following equality:

$$\mathbb{T}\left(C_*, \{\mathbf{c}_p\}_0^q, \{\mathbf{h}_p\}_0^q\right) = e^{i\theta} \prod_{p=0}^{(q/2)-1} \Delta(\mathbf{h}_p, \mathbf{h}_{q-p})^{(-1)^p} \sqrt{\Delta\left(\mathbf{h}_{q/2}, \mathbf{h}_{q/2}\right)}^{(-1)^{q/2}}.$$

Here, $\theta = \sum_{p=0}^{q} (-1)^p\theta_p$ and $[\mathbf{h}_p^0, \mathbf{h}_p] = \left|[\mathbf{h}_p^0, \mathbf{h}_p]\right| e^{i\theta_p}$.

For more applications of Theorem 2, see [53, 55] and the references therein.

3 Main Theorems

Let S be the closed orientable surface of genus 1. Clearly, the universal cover is $\mathbb{C}$. Let G denote one of the Lie groups $\mathrm{GL}(n, \mathbb{C})$, $\mathrm{SL}(n, \mathbb{C})$, $\mathrm{SO}(n, \mathbb{C})$, $\mathrm{Sp}(2n, \mathbb{C})$, and $\mathcal{G}$ denote the Lie algebra of G with the non-degenerate Killing form B.

For a homomorphism $\varrho : \pi_1(S) \to G$ from the fundamental group of the torus to G, there is the corresponding adjoint bundle $E_\varrho = \mathbb{C} \times \mathcal{G}/\sim$ over S. Here, $(x_1, t_1) \sim (x_2, t_2)$, if $(x_2, t_2) = (\gamma \cdot x_1, \gamma \cdot t_1)$ for some $\gamma \in \pi_1(S)$, γ acts in the first component by deck transformation $(\gamma \cdot x_1 = \gamma(x_1))$ and in the second component by the adjoint action $(\gamma \cdot t_1 = \mathrm{Ad}_{\varrho(\gamma)}(t_1) = \varrho(\gamma) t_1 \varrho(\gamma)^{-1})$.

Let K be a cell-decomposition of S such that the adjoint bundle E_ϱ is trivial over each cell and $\widetilde{K}$ be the lift of K to the universal covering $\mathbb{C}$. If $\mathbb{Z}[\pi_1(S)]$ denotes the integral group ring, then $C_*\left(K; \mathcal{G}_{\mathrm{Ad}_\varrho}\right) = C_*\left(\widetilde{K}; \mathbb{Z}\right) \otimes \mathcal{G}/\sim$. Here, $\sigma \otimes t \sim \gamma \cdot \sigma \otimes \gamma \cdot t$, γ acts in the first component by deck transformation and in the second by adjoint action for all $\gamma \in \pi_1(S)$. Thus, we obtain the following chain complex:

$$0 \longrightarrow C_2\left(K; \mathcal{G}_{\mathrm{Ad}_\varrho}\right) \stackrel{\partial_2 \otimes \mathrm{id}}{\longrightarrow} C_1\left(K; \mathcal{G}_{\mathrm{Ad}_\varrho}\right) \stackrel{\partial_1 \otimes \mathrm{id}}{\longrightarrow} C_0\left(K; \mathcal{G}_{\mathrm{Ad}_\varrho}\right) \longrightarrow 0. \qquad (6)$$

Here, ∂_p is the usual boundary operator.

Clearly, we have homologies $H_*\left(K; \mathcal{G}_{\mathrm{Ad}_\varrho}\right)$ and cohomologies $H^*\left(K; \mathcal{G}_{\mathrm{Ad}_\varrho}\right)$ of the chain complex (6), where $C^*\left(K; \mathcal{G}_{\mathrm{Ad}_\varrho}\right)$ is the set of $\mathbb{Z}[\pi_1(\Sigma)]$-module homomorphisms from $C_*(\widetilde{K}; \mathbb{Z})$ to $\mathcal{G}$. Note that for conjugate homomorphisms $\varrho, \varrho' : \pi_1(\Sigma) \to G$ (that is, $\varrho'(.) = A\varrho(.)A^{-1}$ for some $A \in G$) the chains $C_*\left(K; \mathcal{G}_{\mathrm{Ad}_\varrho}\right)$ and $C_*\left(K; \mathcal{G}_{\mathrm{Ad}_{\varrho'}}\right)$ are isomorphic. In addition, the corresponding cochains $C^*\left(K; \mathcal{G}_{\mathrm{Ad}_\varrho}\right)$ and $C^*\left(K; \mathcal{G}_{\mathrm{Ad}_{\varrho'}}\right)$ are isomorphic. For more information, we refer the reader to [7].

Consider chain complex (6). Let $\left\{e_j^p\right\}_{j=1}^{m_p}$ be a basis of $C_p(K; \mathbb{Z})$. If a lift $\widetilde{e}_j^p$ of e_j^p is fixed for $j = 1, \ldots, m_p$, then we obtain a $\mathbb{Z}[\pi_1(\Sigma)]-$basis $c_p = \left\{\widetilde{e}_j^p\right\}_{j=1}^{m_p}$ of $C_p(\widetilde{K}; \mathbb{Z})$. Suppose $\mathcal{A} = \{a_k\}_{k=1}^{\dim \mathcal{G}}$ is a $B-$*orthonormal basis* of the semi-simple Lie algebra $\mathcal{G}$. More precisely, the matrix of the Killing form B equals to the identity matrix of size $\dim \mathcal{G}$.Thus, we obtain a $\mathbb{C}-$basis $\mathbf{c}_p = c_p \otimes_\varrho \mathcal{A}$ of $C_p(K; \mathcal{G}_{\mathrm{Ad}_\varrho})$. We call such basis of $C_p\left(K; \mathcal{G}_{\mathrm{Ad}_\varrho}\right)$ as a *geometric basis.*

Definition 3 If $\mathbf{c}_p = c_p \otimes_\varrho \mathcal{A}$ and $\mathbf{h_p}$ are respectively the geometric basis of $C_p\left(K; \mathcal{G}_{\mathrm{Ad}_\varrho}\right)$ and a basis of $H_p\left(\Sigma; \mathcal{G}_{\mathrm{Ad}_\varrho}\right)$, then

$$\mathbb{T}\left(C_*\left(K;\mathcal{G}_{\mathrm{Ad}_\varrho}\right),\left\{c_p\otimes_\varrho\mathcal{A}\right\}_{p=0}^2,\left\{\mathbf{h}_p\right\}_{p=0}^2\right)$$

is Reidemeister torsion of the triple K, Ad_ϱ, and $\left\{\mathbf{h}_p\right\}_{p=0}^2$.

Following arguments as in [52, Lemmas 1.4.2 and 2.0.5] conclude that the definition does not depend on basis $\mathcal{A}$, lifts $\widetilde{e}_j^p$, conjugacy class of ϱ, and the cell-decomposition. For sake of completeness, we shall explain the independence of $\mathcal{A}$, lifts $\widetilde{e}_j^p$, and conjugacy class of ϱ. For the independence of the cell-decomposition, the reader is referred to [52, Lemma 2.0.5].

Theorem 3 *Suppose S, K, ϱ, $\mathbf{c}_p = c_p\otimes_\varrho\mathcal{A}$, and $\mathbf{h}_p$, $p = 0, 1, 2$, are as above. Then, $\mathbb{T}\left(C_*\left(K;\mathcal{G}_{\mathrm{Ad}_\varrho}\right),\left\{c_p\otimes_\varrho\mathcal{A}\right\}_{p=0}^2,\left\{\mathbf{h}_p\right\}_{p=0}^2\right)$ is independent of the basis $\mathcal{A}$, lifts $\widetilde{e}_j^p$, conjugacy class of ϱ, and the cell-decomposition K.*

Proof Let $\mathcal{A}'$ be also a $B-$orthonormal basis of $\mathcal{G}$. From the change-base-formula (3) of Reidemeister torsion it follows that

$$\frac{\mathbb{T}\left(C_*\left(K;\mathcal{G}_{\mathrm{Ad}_\varrho}\right),\left\{\mathbf{c}'_p\right\}_{p=0}^2,\left\{\mathbf{h}_p\right\}_{p=0}^2\right)}{\mathbb{T}\left(C_*\left(K;\mathcal{G}_{\mathrm{Ad}_\varrho}\right),\left\{\mathbf{c}_p\right\}_{p=0}^2,\left\{\mathbf{h}_p\right\}_{p=0}^2\right)}$$

equals to determinant of the change-base-matrix from basis $\mathcal{A}'$ to $\mathcal{A}$. Here, $\mathbf{c}_p$ and $\mathbf{c}'_p$ are the geometric bases obtained from K, $\mathcal{A}$, and $\mathcal{A}'$ and χ denotes the Euler characteristic. For $\mathcal{A}$ and $\mathcal{A}'$ being $B-$orthonormal bases of $\mathcal{G}$, we obtain the independence of Reidemeister torsion from the basis $\mathcal{A}$.

Next, let us consider conjugate representations result isomorphic twisted chains and cochains. Hence, Reidemeister torsion does not depend on conjugacy class of ϱ.

Finally, we prove that Reidemeister torsion is independent of the lifts $\widetilde{e}_j^p$. Fix $\gamma\in\pi_1(S)$ and consider the lift $c'_p = \left\{\widetilde{e}_1^p\cdot\gamma, \widetilde{e}_2^p, \ldots, \widetilde{e}_{m_p}^p\right\}$ of $\left\{e_1^p, \ldots, e_{m_p}^p\right\}$.

With the help of the change-base-formula (3), we obtain

$$\frac{\mathbb{T}\left(C_*\left(K;\mathcal{G}_{\mathrm{Ad}_\varrho}\right),\left\{\mathbf{c}'_p\right\}_{p=0}^2,\left\{\mathbf{h}_p\right\}_{p=0}^2\right)}{\mathbb{T}\left(C_*\left(K;\mathcal{G}_{\mathrm{Ad}_\varrho}\right),\left\{\mathbf{c}_p\right\}_{p=0}^2,\left\{\mathbf{h}_p\right\}_{p=0}^2\right)}$$

is equal to determinant of the matrix matrix of the linear map $Ad_{\varrho(\gamma)}:\mathcal{G}\to\mathcal{G}$ with respect to basis $\mathcal{A}$. Here, $\mathbf{c}_p$, $\mathbf{c}'_p$ are the geometric bases obtained from $c_p = \left\{\widetilde{e}_1^p, \widetilde{e}_2^p, \ldots, \widetilde{e}_{m_p}^p\right\}$, $c'_p = \left\{\widetilde{e}_1^p\cdot\gamma, \widetilde{e}_2^p, \ldots, \widetilde{e}_{m_p}^p\right\}$, and $\mathcal{A}$.

We consider the basis $\mathcal{B}$ (see, e.g. [56]) for $\mathcal{G} = \mathfrak{gl}(n,\mathbb{C})$, $\mathfrak{sl}(n,\mathbb{C})$, $\mathfrak{so}(n,\mathbb{C})$, $\mathfrak{sp}(2n,\mathbb{C})$. Let us prove only the general linear group $\mathrm{GL}(n,\mathbb{C})$ case for the computations being similar. Clearly, $\varrho(\gamma)$ is conjugate to $D = \mathrm{Diag}(\lambda_1,\ldots,\lambda_n)$ for some $Q\in\mathrm{GL}(n,\mathbb{C})$. Here, $\lambda_1,\ldots,\lambda_n$ are eigenvalues of $\varrho(\gamma)\in\mathrm{GL}(n,\mathbb{C})$.

We consider the basis $\left\{E_{ij}\right\}_{i,j=1}^n$ for $\mathfrak{gl}(n,\mathbb{C})$. The matrix of $Ad_{\varrho(\gamma)}$ in the basis $\left\{Ad_{Q^{-1}}\left(E_{ij}\right)\right\}_{i,j=1}^n$ of $\mathfrak{gl}(n,\mathbb{C})$ is

$$\text{Diag}\left(\frac{\lambda_k}{\lambda_l},\ k \leq l;\ \frac{\lambda_k}{\lambda_l},\ k > l\right).$$

Since determinant of this diagonal matrix is 1, one concludes the independence of Reidemeister torsion from the lifts $\widetilde{e}_j^p$.

This is the end of the proof of Theorem 3.

We write $\mathbb{T}(S, \{\mathbf{h}_p\}_{p=0}^2)$ instead of $\mathbb{T}\left(C_*\left(K; \mathcal{G}_{\text{Ad}_\varrho}\right), \{c_p \otimes_\varrho \mathcal{A}\}_{p=0}^2, \{\mathbf{h}_p\}_{p=0}^2\right)$. Assume $\varrho : \pi_1(\Sigma) \to G$, where $G = G_1 \times \cdots \times G_d$ and $G_i, i = 1, \ldots, d$, are one of the Lie group from the above list. Reidemeister torsion of such representation is well defined by Theorem 3.

By using $\mathbb{C}-$symplectic chain complex, let us establish a formula for computing Reidemeister torsion of representations in terms of the symplectic structure on $\text{Rep}\,(S, G)$ [51], which is similar to Atiyah–Bott–Goldman symplectic form of higher genera for the Lie group G.

Suppose $S, K, G, \mathcal{G}, \varrho, \mathbf{c}_p{=}c_p \otimes_\varrho \mathcal{A}$ are as above and the dual cell-decomposition K' of S corresponding to the cell-decomposition K. Consider the lifts $\widetilde{K}$ and $\widetilde{K'}$ of K and K', respectively. For $i = 0, 1, 2$, we consider the intersection form

$$(\cdot, \cdot)_{i,2-i} : C_i\left(K; \mathcal{G}_{\text{Ad}_\varrho}\right) \times C_{2-i}\left(K'; \mathcal{G}_{\text{Ad}_\varrho}\right) \longrightarrow \mathbb{C} \tag{7}$$

defined by

$$(\sigma_1 \otimes t_1, \sigma_2 \otimes t_2)_{i,2-i} = \sum_{\gamma \in \pi_1(\Sigma)} \sigma_1 . (\gamma \bullet \sigma_2)\ B\left(t_1, \gamma \bullet t_2\right),$$

where γ acts on σ_2 by deck transformation, on t_2 by the adjoint action and "." is the intersection number pairing.

The intersection form $(\cdot, \cdot)_{i,2-i}$ is $\partial-$compatible and anti-symmetric. It can naturally be extended to twisted homologies and yield the non-degenerate anti-symmetric form

$$[\cdot, \cdot]_{i,2-i} : H_i\left(S; \mathcal{G}_{\text{Ad}_\varrho}\right) \times H_{2-i}\left(S; \mathcal{G}_{\text{Ad}_\varrho}\right) \longrightarrow \mathbb{C}. \tag{8}$$

Consider the chain complex $D_i = C_i\left(K; \mathcal{G}_{\text{Ad}_\varrho}\right) \oplus C_i\left(K'; \mathcal{G}_{\text{Ad}_\varrho}\right)$. By extending the intersection form (7) zero on $C_i\left(K; \mathcal{G}_{\text{Ad}_\varrho}\right) \times C_{2-i}\left(K; \mathcal{G}_{\text{Ad}_\varrho}\right)$ and $C_i\left(K'; \mathcal{G}_{\text{Ad}_\varrho}\right) \times C_{2-i}\left(K'; \mathcal{G}_{\text{Ad}_\varrho}\right)$, we obtain the $\partial-$compatible anti-symmetric bilinear $\omega_{i,2-i} : D_i \times D_{2-i} \to \mathbb{C}$. Thus, D_* is a $\mathbb{C}-$symplectic chain complex. The bases c_i of $C_i\left(\widetilde{K}; \mathbb{Z}\right)$ and c_i' of $C_i\left(\widetilde{K'}; \mathbb{Z}\right)$ corresponding to c_i yield an $\omega-$compatible basis for D_*.

Next, let us recall Kronecker pairing. It is the non-degenerate form $\langle \cdot, \cdot \rangle$: $C^i\left(K; \mathcal{G}_{\text{Ad}_\varrho}\right) \times C_i\left(K; \mathcal{G}_{\text{Ad}_\varrho}\right) \to \mathbb{C}$ defined by $\langle \theta, \sigma \otimes_\varrho t \rangle = B\left(t, \theta\left(\sigma\right)\right)$. It has natural extension to $\langle \cdot, \cdot \rangle : H^i\left(S; \mathcal{G}_{\text{Ad}_\varrho}\right) \times H_i\left(S; \mathcal{G}_{\text{Ad}_\varrho}\right) \to \mathbb{C}$.

Finally, recall the cup product $\cup : C^i\left(K; \mathcal{G}_{\text{Ad}_\varrho}\right) \times C^j\left(K; \mathcal{G}_{\text{Ad}_\varrho}\right) \to C^{i+j}\left(\widetilde{S}; \mathbb{C}\right)$ is defined by $\left(\theta_i \cup \theta_j\right)\left(\sigma_{i+j}\right) = B\left(\theta_i\left(\left(\sigma_{i+j}\right)_{\text{front}}\right), \theta_j\left(\left(\sigma_{i+j}\right)_{\text{back}}\right)\right)$, where $\widetilde{K}$ is the lift of K to the universal covering $\widetilde{S} = \mathbb{C}$ of S, $\theta_i : C_i\left(\widetilde{K};\mathbb{Z}\right) \to \mathcal{G}$, $\theta_j : C_j\left(\widetilde{K};\mathbb{Z}\right) \to \mathcal{G}$ are $\mathbb{Z}[\pi_1(S)]$ -module homomorphisms, and σ_{i+j} is in $C_{i+j}(\widetilde{K};\mathbb{Z})$. We have the cup product

$$\smile_B : C^i\left(K;\mathcal{G}_{\mathrm{Ad}_\varrho}\right) \times C^j\left(K;\mathcal{G}_{\mathrm{Ad}_\varrho}\right) \longrightarrow C^{i+j}\left(K;\mathbb{C}\right).$$

$\smile_B$ has natural extension to twisted cohomologies

$$\smile_B : H^i\left(S;\mathcal{G}_{\mathrm{Ad}_\varrho}\right) \times H^j\left(S;\mathcal{G}_{\mathrm{Ad}_\varrho}\right) \longrightarrow H^{i+j}\left(\Sigma;\mathbb{C}\right),$$

where $[\theta_i] \smile_B [\theta_j] = \left[\theta_i \smile_B \theta_j\right]$.

Kronecker pairing and the isomorphisms induced by intersection form (8) yield the Poincare duality isomorphisms

$$\mathrm{PD} : H_i\left(S;\mathcal{G}_{\mathrm{Ad}_\varrho}\right) \cong H_{2-i}\left(S;\mathcal{G}_{\mathrm{Ad}_\varrho}\right)^* \cong H^{2-i}\left(S;\mathcal{G}_{\mathrm{Ad}_\varrho}\right).$$

From this we obtain the following commutative diagram for $i = 0, 1, 2$

$$\begin{array}{ccccc} H^{2-i}\left(S;\mathcal{G}_{\mathrm{Ad}_\varrho}\right) \times & H^i\left(S;\mathcal{G}_{\mathrm{Ad}_\varrho}\right) & \xrightarrow{\smile_B} & H^2\left(S;\mathbb{C}\right) \\ \uparrow \mathrm{PD} & \uparrow \mathrm{PD} & \circlearrowleft & \uparrow \\ H_i\left(S;\mathcal{G}_{\mathrm{Ad}_\varrho}\right) \times & H_{2-i}\left(S;\mathcal{G}_{\mathrm{Ad}_\varrho}\right) & \xrightarrow{[\cdot,\cdot]_{i,2-i}} & \mathbb{C}. \end{array} \tag{9}$$

Here, the isomorphism $\mathbb{C} \to H^2\left(S;\mathbb{C}\right)$ sends $1 \in \mathbb{C}$ to the fundamental class of $H^2(S;\mathbb{C})$ and the inverse of this isomorphism is integration over S.

Commutative diagram (9) yields the pairing

$$\Omega_{i,2-i} : H^i\left(S;\mathcal{G}_{\mathrm{Ad}_\varrho}\right) \times H^{2-i}\left(S;\mathcal{G}_{\mathrm{Ad}_\varrho}\right) \xrightarrow{\smile_B} H^2\left(S;\mathbb{C}\right) \xrightarrow{\int_S} \mathbb{C}. \tag{10}$$

Moreover, if $\varrho : \pi_1(S) \to G$ is a representation with a Zariski dense image in a maximal torus of G, then $\Omega_{1,1}$ is a symplectic form on $X'_G(\mathbb{Z}^2) = Hom'(\mathbb{Z}^2, G)//G$. Here, $Hom'(\mathbb{Z}^2, G)$ is the space of $G-$valued homomorphisms of $\mathbb{Z}^2$ with a Zariski dense image in a maximal torus of G [51]. Note that all representations in $Hom'(\mathbb{Z}^2, G)//G$ are completely reducible and thus it is the set-theoretic quotient. Symplectic form $\Omega_{1,1}$ is similar to Atiyah–Bott–Goldman symplectic form of higher genera for the Lie group G.

Let us state one of our main results where we establish a formula for computing Reidemeister torsion of representations in terms of $\Omega_{1,1}$.

Theorem 4 *Assume S, K, K', ϱ are as above. Assume also $\mathbf{c}_p$ and $\mathbf{c}'_p$ are the corresponding geometric bases of $C_p\left(K;\mathcal{G}_{\mathrm{Ad}_\varrho}\right)$ and $C_p\left(K';\mathcal{G}_{\mathrm{Ad}_\varrho}\right)$, respectively, $p = 0, 1, 2$. If $\mathbf{h}_p$ is a basis of $H_p\left(S;\mathcal{G}_{\mathrm{Ad}_\varrho}\right)$, $p = 0, 1, 2$, then we have*

$$(i)\ \mathbb{T}\left(S, \{\mathbf{h}_p\}_{p=0}^{2}\right) = e^{\sqrt{-1}\frac{(\pi+\theta)}{2}} \frac{\Delta\left(\mathbf{h}_0, \mathbf{h}_2\right)}{\sqrt{\Delta\left(\mathbf{h}_1, \mathbf{h}_1\right)}},$$

$$(ii)\ \mathbb{T}\left(S, \{\mathbf{h}_p\}_{p=0}^{2}\right) = e^{\sqrt{-1}\frac{(\pi+\theta)}{2}} \frac{\sqrt{\delta\left(\mathbf{h}^1, \mathbf{h}^1\right)}}{\delta\left(\mathbf{h}^2, \mathbf{h}^0\right)},$$

where $\left[\mathbf{h}^0_{p,p}, \mathbf{h}_p \oplus \mathbf{h}_p\right] = \left|\left[\mathbf{h}^0_{p,p}, \mathbf{h}_p \oplus \mathbf{h}_p\right]\right| e^{\sqrt{-1}\theta_p}$, $\theta = \theta_0 - \theta_1 + \theta_2$, *and* $\mathbf{h}^0_{p,p}$ *is a basis of* $H_p\left(S; \mathcal{G}_{\mathrm{Ad}_\varrho}\right) \oplus H_p\left(S; \mathcal{G}_{\mathrm{Ad}_\varrho}\right)$ *such that* $\mathbb{T}\left(D_*, \{\mathbf{c}_p \oplus \mathbf{c}'_p\}_0^2, \{\mathbf{h}^0_{p,p}\}_0^2\right) = 1$. $\Delta\left(\mathbf{h}_p, \mathbf{h}_{2-p}\right)$ *is the determinant of the matrix of the intersection pairing (8) in the bases* $\mathbf{h}_p$ *and* $\mathbf{h}_{2-p}$, $\delta\left(\mathbf{h}^{2-p}, \mathbf{h}^p\right)$ *is the determinant of the matrix of the pairing (10) in the bases* $\mathbf{h}^p$ *and* $\mathbf{h}^{2-p}$, *and* $\mathbf{h}^p$ *is the Poincare dual basis of* $H^p(S; \mathcal{G}_{\mathrm{Ad}_\varrho})$ *corresponding to the basis* $\mathbf{h}_p$ *of* $H_p(S; \mathcal{G}_{\mathrm{Ad}_\varrho})$, $p = 0, 1, 2$.

Proof Clearly, $D_* = C_*\left(K; \mathcal{G}_{\mathrm{Ad}_\varrho}\right) \oplus C_*\left(K'; \mathcal{G}_{\mathrm{Ad}_\varrho}\right)$ is a $\mathbb{C}-$symplectic chain complex with $\omega-$compatible basis obtained by the geometric bases $\mathbf{c}_p$, $\mathbf{c}'_p$. Combining Theorem 2 and Remark 1, we obtain basis $\mathbf{h}^0_{p,p}$ of $H_p\left(\Sigma; \mathcal{G}_{\mathrm{Ad}_\varrho}\right) \oplus H_p\left(\Sigma; \mathcal{G}_{\mathrm{Ad}_\varrho}\right)$, $p = 0, 1, 2$ such that

$$\mathbb{T}\left(D_*, \{\mathbf{c}_p \oplus \mathbf{c}'_p\}_{p=0}^2, \{\mathbf{h}^0_{p,p}\}_{p=0}^2\right) = 1$$

. We also have

$$\mathbb{T}\left(D_*, \{\mathbf{c}_p \oplus \mathbf{c}'_p\}_{p=0}^2, \{\mathbf{h}_p \oplus \mathbf{h}_p\}_{p=0}^2\right) = e^{\sqrt{-1}(\pi+\theta)} \frac{\Delta\left(\mathbf{h}_0, \mathbf{h}_2\right)^2}{\Delta\left(\mathbf{h}_1, \mathbf{h}_1\right)}, \tag{11}$$

where $\left[\mathbf{h}^0_{p,p}, \mathbf{h}_p \oplus \mathbf{h}_p\right] = \left|\left[\mathbf{h}^0_{p,p}, \mathbf{h}_p \oplus \mathbf{h}_p\right]\right| e^{\sqrt{-1}\theta_p}$ and $\theta = \theta_0 - \theta_1 + \theta_2$.

From Eq. (11) and Lemma 1 it follows

$$\mathbb{T}\left(S, \{\mathbf{h}_p\}_{p=0}^{2}\right) = e^{\sqrt{-1}\frac{(\pi+\theta)}{2}} \frac{\Delta\left(\mathbf{h}_0, \mathbf{h}_2\right)}{\sqrt{\Delta\left(\mathbf{h}_1, \mathbf{h}_1\right)}}. \tag{12}$$

Using commutative diagram (9), it follows $\delta\left(\mathbf{h}^{2-p}, \mathbf{h}^p\right) \cdot \Delta\left(\mathbf{h}_p, \mathbf{h}_{2-p}\right) = 1$. This finishes the proof of Theorem 4.

References

1. Goldman, W.M.: Representations of fundamental groups of surfaces. In: Geometry and Topology. Lecture Notes in Mathematics, vol. 1167, pp. 95–117. Springer, Berlin (1985)
2. Goldman, W.M.: Geometric structures on manifolds and varieties of representations. In: The Geometry of Group Representations. American Mathematical Society Contemporary Mathe-

matics, pp. 169–198 (1988) 3. Geddes, K.O., Czapor, S.R., Labahn, G.: Algorithms for Computer Algebra. Kluwer, Boston (1992) 4. Hamburger, C.: Quasimonotonicity, regularity and duality for nonlinear systems of partial differential equations. Ann. Mat. Pura. Appl. 169, 321–354 (1995)

3. Choi, S., Goldman, W.M.: Convex real projective structures on closed surfaces are closed. Proc. Am. Math. Soc. **118**(2), 657–661 (1993)
4. Guichard, O., Wienhard, A.: Topological invariants of Anosov representations. J. Topol. **3**(3), 578–642 (2010)
5. Hitchin, N.: Lie groups and Teichmuller spaces. Topology **31**(3), 449–473 (1992)
6. Labourie, F.: Anasov flows, surface groups and curves in projective space. Invent. Math. **165**(1), 51–114 (2006)
7. Porti, J.: Torsion de Reidemeister pour les Varietes Hyperboliques. Mem. Am. Math. Soc. **128**(612), 139 (1997)
8. Heusener, M., Porti, J.: The variety of characters in PSL2(C). Bol. Soc. Mat. Mex. **10**(3), 221–237 (2004)
9. Heusener, M., Porti, J.: Deformations of reducible representations of 3-manifold groups into PSL(2, C). Alg. Geom. Top **5**, 965–997 (2005)
10. Motegi, K.: Haken manifolds and representations of their fundamental groups in SL(2, C). Topol. Appl. **29**(3), 207–212 (1988)
11. Cooper, D., Culler, M., Gillet, H., Long, D.D., Shalen, P.B.: Plane curves associated to character varieties of 3-manifolds. Invent. Math. **118**, 47–84 (1994)
12. Shalen, P.: Representations of 3-manifold groups. In: Daverman, R., Sher, R. (eds.) Handbook of Geometric Topology, pp. 955–1044. North-Holland, Amsterdam (2002)
13. Boileau, M., Boyer, S.: On character varieties, sets of discrete characters, and nonzero degree maps. Am. J. Math. **134**(2), 285–347 (2012)
14. Boden, H., Herald, C.: The SU(3) Casson invariant for integral homology 3-spheres. J. Differ. Geom. **50**, 147–206 (1998)
15. Boden, H.U., Curtis, C.L.: Splicing and the SL2(C) Casson invariant. Proc. Am. Math. Soc. **136**(7), 2615–2623 (2008)
16. Curtis, C.L.: An intersection theory count of the SL2(C)-representations of the fundamental group of a 3-manifold. Topology **40**, 773–787 (2001)
17. Boden, H.U., Friedl, S.: Metabelian SL(n, C) representations of knot groups. Pacific J. Math. **238**(1), 7–25 (2008)
18. Cooper, D., Long, D.: Representation theory and the A-polynomial of a knot. Chaos Solitons Fractals **9**, 749–763 (1998)
19. Gelca, R.: On the relation between the A-polynomial and the Jones polynomial. Proc. Am. Math. Soc. **130**(4), 1235–1241 (2002)
20. Le, T.Q.: The colored Jones polynomial and the A-polynomial of knots. Adv. Math. **207**(2), 782–804 (2006)
21. Gukov, S.: Three-dimensional quantum gravity, Chern-Simons theory, and the A-polynomial. Commun. Math. Phys. **255**(3), 577–627 (2005)
22. Kirk, P., Klassen, E.: Chern-Simons invariants of 3-manifolds decomposed along tori and the circle bundle over the representation space of T2. Commun. Math. Phys. **153**(3), 521–557 (1993)
23. Weitsman, J.: Quantization via real polarization of the moduli space of flat connections and Chern-Simons gauge theory in genus 1. Commun. Math. Phys. **137**, 175–190 (1991)
24. Witten, E.: Quantum field theory and the Jones polynomial. Commun. Math. Phys. **121**(3), 351–399 (1989)
25. Witten, E.: On quantum gauge theories in two dimensions. Commun. Math. Phys. **141**, 153–209 (1991)
26. Bullock, D.: Rings of SL2(C)-characters and the Kauffman bracket skein module. Comment. Math. Helv. **72**, 521–542 (1997)
27. Frohman, C., Gelca, R.: Skein modules and the noncommutative torus. Trans. Am. Math. Soc. **352**, 4877–4888 (2000)

28. Przytycki, J.H., Sikora, A.S.: On Skein algebras and Sl2(C)-character varieties. Topology **39**(1), 115–148 (2000)
29. Sikora, A.S.: Skein theory for SU(n)-quantum invariants. Algebr. Geom. Topol. **5**, 865–897 (2005)
30. Atiyah, M., Bott, R.: The Yang–Mills equations over Riemann surfaces. Philos. Trans. R. Soc. London Ser. A. **308**, 523–615 (1983)
31. Alekseev, A., Meinrenken, E., Woodward, C.: Duistermaat-Heckman measures and moduli spaces of flat bundles over surfaces. Geom. Funct. Anal. **12**(1), 1–31 (2002)
32. Florentino, C., Lawton, S.: The topology of moduli spaces of free group representations. Math. Ann. **345**(2), 453–489 (2009)
33. Garcia-Prada, O., Gothen, P.B., Mundet i Riera, I.: Higgs bundles and surface group representations in the real symplectic group. J. Topol. **6**, 64–118 (2013)
34. Hausel, T., Thaddeus, M.: Relations in the cohomology ring of the moduli space of rank 2 Higgs bundles. J. Am. Math. Soc. **16**(2), 303–327 (2003)
35. Hitchin, N.J.: The self-duality equations on a Riemann surface. Proc. Lond. Math. Soc. **55**, 59–126 (1987)
36. Jeffrey, L.C.: Flat connections on oriented 2-manifolds. Bull. Lond. Math. Soc. **37**, 1–14 (2005)
37. Meinrenken, E.: Witten's formulas for intersection pairings on moduli spaces of flat G-bundles. Adv. Math. **197**, 140–197 (2005)
38. Simpson, C.: Moduli of representations of the fundamental group of a smooth projective variety I. Publ. Math. Inst. Hautes Etud. Sci. **79**, 47–129 (1994)
39. Simpson, C.: Moduli of representations of the fundamental group of a smooth projective variety II. Publ. Math. Inst. Hautes Etud. Sci. **80**, 5–79 (1994)
40. Reidemeister, K.: Homotopieringe und linsenraume. Abh. Math. Semin. Univ. Hambg. **11**, 102–109 (1935)
41. Franz, W.: Uber die Torsion einer Uberdeckung. J. Reine Angew. Math. **173**, 245–254 (1935)
42. Milnor, J.: Two complexes which are homeomorphic but combinatorially distinct. Ann. Math. **74**(3), 575–590 (1961)
43. Milnor, J.: A duality theorem for Reidemeister torsion. Ann. Math. **76**(1), 137–147 (1962)
44. Cheeger, J.: Analytic torsion and the heat equation. Ann. Math. **109**(2), 259–321 (1979)
45. Muller, W.: Analytic torsion and R-torsion of Riemannian manifolds. Adv. Math. **28**(3), 233–305 (1978)
46. Milnor, J.: Whitehead torsion. Bull. Am. Math. Soc. **72**, 358–426 (1966)
47. Hutchings, M., Lee, Y.: Circle valued Morse theory and Reidemeister torsion. Geom. Topol. **3**, 369–396 (1999)
48. Meng, G., Taubes, C.H.: SW = Milnor torsion. Math. Res. Lett. **3**(5), 661–674 (1996)
49. Schwarz, A.: The partition function of degenerate quadratic functional and Ray-Singer torsion. Lett. Math. Phys. **2**(3), 247–252 (1978)
50. Turaev, V.: Torsions of 3-Dimensional Manifolds. Progress in Mathematics, Birkhauser, Basel (2002)
51. Sikora, A.S.: Character varieties of abelian groups. Math. Z. **277**, 241–256 (2014)
52. Sozen, Y.: On Reidemeister torsion of a symplectic complex. Osaka J. Math. **45**(1), 1–39 (2008)
53. Sozen, Y.: Symplectic chain complex and Reidemeister torsion of compact manifolds. Math. Scand. **111**(1), 65–91 (2012)
54. Sozen, Y.: A note on Reidemeister torsion and pleated surfaces. J. Knot Theory Ramif. **23**(3), 1–32 (2014)
55. Hezenci, F., Sozen, Y.: A note on exceptional groups and Reidemeister torsion. J. Math. Phys. **59**, 081704 (2018)
56. Humphreys, J.E.: Introduction to Lie algebras and representation theory. In: Graduate Texts in Mathematics, vol. 9. Springer, New York, Berlin (1978)

A Space-Dependent Source Identification Problem for Hyperbolic-Parabolic Equations

Maksat Ashyraliyev, Allaberen Ashyralyev, and Victor Zvyagin

Abstract In the present paper, a space-dependent source identification problem for the hyperbolic-parabolic equation with unknown parameter p

$$\begin{cases} u''(t) + Au(t) = p + f(t), \ 0 < t < 1, \\ u'(t) + Au(t) = p + g(t), \ \ -1 < t < 0, \\ u(0^+) = u(0^-), \ u'(0^+) = u'(0^-), \\ u(-1) = \varphi, \ \int\limits_0^1 u(z)dz = \psi \end{cases}$$

in a Hilbert space H with self-adjoint positive definite operator A is investigated. The stability estimates for the solution of this identification problem are established. In applications, the stability estimates for the solutions of four space-dependent source identification hyperbolic-parabolic problems are obtained.

Keywords Hyperbolic-parabolic equation · Source identification problem · Stability

M. Ashyraliyev (✉)
Department of Software Engineering, Bahcesehir University, 34353 Istanbul, Turkey
e-mail: maksat.ashyralyyev@eng.bau.edu.tr

A. Ashyralyev
Department of Mathematics, Near East University, Mersin 10, Nicosia, TRNC, Turkey
e-mail: allaberen.ashyralyev@neu.edu.tr

Peoples' Friendship University of Russia (RUDN University), 6 Miklukho-Maklaya St, Moscow 117198, Russian Federation

Institute of Mathematics and Mathematical Modeling, Almaty 050010, Kazakhstan

V. Zvyagin
Voronezh State University, Universitetskaya 1, Voronezh 394018, Russia
e-mail: zvg_vsu@mail.ru

A. Ashyralyev et al. (eds.), *Functional Analysis in Interdisciplinary Applications—II*,
Springer Proceedings in Mathematics & Statistics 351,
https://doi.org/10.1007/978-3-030-69292-6_14

1 Introduction

Local and nonlocal boundary value problems for mixed type partial differential equations have applications in many different fields of science and engineering. The theory of these type of problems has been extensively investigated by many researchers (see, e.g., [19, 33, 38] and the references given therein). Specifically, the theory of nonlocal boundary value problems for hyperbolic-parabolic differential equations and numerical methods for their solutions have been the subject of recent studies (see, [8, 9, 11, 18] and the references therein).

The differential equations with unknown parameters are used to model the behavior of real-life systems in many different areas of science and technology. They have been studied extensively by many authors (see, [1–3, 5–7, 12, 13, 20–24, 26–31, 34–37, 40] and the references therein). However, the identification problems have not been sufficiently investigated in general for mixed type partial differential equations.

The main goal of this study is to investigate a space-dependent source identification problem for the hyperbolic-parabolic equations. It is known that various space-dependent source identification problems for hyperbolic-parabolic equations can be reduced to the space-dependent source identification problem for the hyperbolic-parabolic differential equation

$$\begin{cases} u''(t) + Au(t) = p + f(t), \ 0 < t < 1, \\ u'(t) + Au(t) = p + g(t), \ -1 < t < 0, \\ u(0^+) = u(0^-), \ u'(0^+) = u'(0^-), \\ u(-1) = \varphi, \ \displaystyle\int_0^1 u(z)dz = \psi \end{cases} \tag{1}$$

in a Hilbert space H with self-adjoint positive definite operator A satisfying $A \geq \delta I$, where $\delta \geq 2$. Here, $u(t)$ and p denote

$$u(t) = u\big(t; f(t), g(t), \varphi, \psi\big), \quad p = p\big(f(t), g(t), \varphi, \psi\big).$$

By a solution of inverse problem (1) we mean a pair $\big(u(t), p\big)$ satisfying the following conditions:

1. $u(t) \in D(A)$ for all $t \in [-1, 1]$, $p \in H$, and the function $Au(t)$ is continuous on $[-1, 1]$. Here, $D(A)$ is the domain of an operator A.
2. $u(t)$ has continuous second derivative on the interval $[0, 1]$ and continuous first derivative on the interval $[-1, 0]$. The derivatives at the endpoints of the intervals are understood as the appropriate one-sided derivatives.
3. $\big(u(t), p\big)$ satisfies the equations and boundary conditions (1).

A solution of problem (1) defined in this way will from now on be referred to as a solution of problem (1) in the space $C(H) \times H$. Here $C(H) = C([-1, 1], H)$ is the space of continuous H-valued functions $u(t)$ defined on $[-1, 1]$, equipped with the norm

$$\|u\|_{C(H)} = \max_{-1 \leq t \leq 1} \|u(t)\|_H .$$

We note that the corresponding inverse problem when the integral condition in (1) is replaced by $u(\lambda) = \psi$, $-1 \leq \lambda \leq 1$ was previously studied by authors. The well-posedness of that identification problem and a few different applications were provided in [4]. The first and second order of accuracy difference schemes for its approximate solution were studied in [14, 17]. These schemes were implemented for simple test problems in [15, 16].

In the present paper, we will prove the main theorem on the stability of inverse problem (1) in the space $C(H) \times H$. In applications, the stability estimates for the solutions of four space-dependent source identification hyperbolic-parabolic problems will be obtained.

To formulate the main results, we make use of the theory of cosine and sine operator-functions (see [25, 32]). Strongly continuous cosine and sine operator-functions are defined respectively by the following formulas

$$c(t) = \frac{e^{itA^{1/2}} + e^{-itA^{1/2}}}{2}, \quad s(t) = A^{-1/2} \frac{e^{itA^{1/2}} - e^{-itA^{1/2}}}{2i}, \quad t \geq 0.$$

Lemma 1 *The following estimates hold:*

$$\|c(t)\|_{H \to H} \leq 1, \quad \|s(t)\|_{H \to H} \leq t, \quad \left\|A^{1/2} s(t)\right\|_{H \to H} \leq 1, \quad t \geq 0, \tag{2}$$

$$\left\|A^{-1/2}\right\|_{H \to H} \leq \frac{1}{\sqrt{\delta}}, \tag{3}$$

$$\left\|e^{-tA}\right\|_{H \to H} \leq e^{-\delta t}, \quad t \geq 0, \quad \left\|A^{1/2} e^{-tA}\right\|_{H \to H} \leq \frac{1}{\sqrt{2et}}, \quad t > 0. \tag{4}$$

Proofs of these estimates are based on the spectral representation of the self-adjoint positive definite operator in a Hilbert space.

Lemma 2 *The operator*

$$Q = I - e^{-A} \left(s(1) - A \int_0^1 s(\tau) d\tau \right)$$

has an inverse and the following estimate holds

$$\|Q^{-1}\|_{H\to H}\le \frac{1}{1-e^{-\delta}\left(2+\frac{1}{\sqrt{\delta}}\right)}. \tag{5}$$

Proof Applying the definitions of $c(t)$ and $s(t)$, the positivity and self-adjointness property of A, we obtain

$$\left\| e^{-A}\left(s(1)-A\int_0^1 s(\tau)d\tau\right)\right\|_{H\to H}$$
$$\le \sup_{\delta\le\rho<\infty} e^{-\rho}\left|\frac{e^{i\sqrt{\rho}}-e^{-i\sqrt{\rho}}}{2i\sqrt{\rho}}-1+\frac{e^{i\sqrt{\rho}}+e^{-i\sqrt{\rho}}}{2}\right|\le e^{-\delta}\left(2+\frac{1}{\sqrt{\delta}}\right).$$

Using this result, we obtain the estimate (5). □

2 The Main Theorem on Stability

In this section, we will prove the following main theorem on continuous dependence of the solution of source identification problem (1) on the given data.

Theorem 1 *Assume that $\varphi\in D(A)$, $\psi\in D(A)$, $f(t)$ has continuous first derivative on $[0,1]$ and $g(t)$ has continuous first derivative on $[-1,0]$. Then, for the solution $(u(t),p)$ of problem (1) in $C(H)\times H$ the following stability inequalities*

$$\|u\|_{C(H)}+\left\|A^{-1}p\right\|_H$$
$$\le M(\delta)\left[\|\varphi\|_H+\|\psi\|_H+\max_{0\le t\le 1}\|f(t)\|_H+\max_{-1\le t\le 0}\|g(t)\|_H\right], \tag{6}$$

$$\max_{0\le t\le 1}\left\|u''(t)\right\|_H+\max_{-1\le t\le 0}\left\|u'(t)\right\|_H+\|Au\|_{C(H)}+\|p\|_H\le M(\delta)\Big[\|A\varphi\|_H$$
$$+\|A\psi\|_H+\max_{0\le t\le 1}\left\|f'(t)\right\|_H+\max_{-1\le t\le 0}\left\|g'(t)\right\|_H+\|g(0)\|_H+\|f(0)\|_H\Big] \tag{7}$$

hold, where $M(\delta)$ is independent of φ, ψ, $f(t)$, and $g(t)$.

Proof Let us denote

$$u(t)=v(t)+A^{-1}p,\quad -1\le t\le 1, \tag{8}$$

where $v(t)$ is the solution of the following nonlocal boundary value problem

$$\begin{cases} v''(t) + Av(t) = f(t), \ 0 < t < 1, \\ v'(t) + Av(t) = g(t), \ \ -1 < t < 0, \\ v(0^+) = v(0^-), \ v'(0^+) = v'(0^-), \\ \int\limits_0^1 v(z)dz - v(-1) = \psi - \varphi \end{cases} \tag{9}$$

for the differential equation in a Hilbert space H with self-adjoint positive definite operator A. We will first obtain the formulas for the solution of nonlocal boundary value problem (9). It is known that for the smooth data the following boundary value problems

$$\begin{cases} v'(t) + Av(t) = g(t), \ \ -1 < t < 0, \\ v(-1) = v_{-1} \end{cases}$$

and

$$\begin{cases} v''(t) + Av(t) = f(t), \ 0 < t < 1, \\ v(0) = v_0, \ v'(0) = v_0' \end{cases}$$

have unique solutions, given by formulas

$$v(t) = e^{-(1+t)A}v_{-1} + \int\limits_{-1}^{t} e^{-(t-\tau)A}g(\tau)d\tau, \ \ -1 \leq t \leq 0 \tag{10}$$

and

$$v(t) = c(t)v_0 + s(t)v_0' + \int\limits_0^t s(t-\tau)f(\tau)d\tau, \ 0 \leq t \leq 1, \tag{11}$$

respectively. With expressions for $v(0)$ and $v'(0)$, obtained from (10), we can rewrite (11) as following

$$\begin{aligned} v(t) &= c(t)\Big(e^{-A}v_{-1} + \int\limits_{-1}^{0} e^{\tau A}g(\tau)d\tau\Big) \\ &+s(t)\Big(- Ae^{-A}v_{-1} + g(0) - A\int\limits_{-1}^{0} e^{\tau A}g(\tau)d\tau\Big) \\ &+ \int\limits_0^t s(t-\tau)f(\tau)d\tau, \ 0 \leq t \leq 1. \end{aligned} \tag{12}$$

Applying formula (12) and condition $\int\limits_0^1 v(z)dz - v(-1) = \psi - \varphi$, we get

$$v(-1) = s(1)\Big(e^{-A}v_{-1} + \int_{-1}^{0} e^{\tau A}g(\tau)d\tau\Big)$$
$$+\int_{0}^{1} s(z)dz\Big(- Ae^{-A}v_{-1} + g(0) - A\int_{-1}^{0} e^{\tau A}g(\tau)d\tau\Big)$$
$$+\int_{0}^{1}\int_{0}^{z} s(z-\tau)f(\tau)d\tau dz + \varphi - \psi.$$

By Lemma 1, there exists the operator $\Big(I - e^{-A}\Big[s(1) - A\int_{0}^{1} s(z)dz\Big]\Big)^{-1}$. Therefore, the following formula holds

$$v_{-1} = \Big(I - e^{-A}\Big[s(1) - A\int_{0}^{1} s(z)dz\Big]\Big)^{-1}\Big\{g(0)\int_{0}^{1} s(z)dz + \varphi - \psi$$
$$+\Big[s(1) - A\int_{0}^{1} s(z)dz\Big]\int_{-1}^{0} e^{\tau A}g(\tau)d\tau + A^{-1}\int_{0}^{1}\big(I - c(1-\tau)\big)f(\tau)d\tau\Big\}. \quad (13)$$

To conclude, we have obtained formulas (10) and (12) for the solution of nonlocal boundary value problem (9), with v_{-1} being found by formula (13). Then, by using (8) and

$$p = A\psi - A\int_{0}^{1} v(z)dz, \quad (14)$$

we obtain the solution of problem (1).

Now, let us obtain the estimate (6). Using formula (13) and estimates (2), (4), and (5), we get

$$\|v_{-1}\|_H \le M_1(\delta)\Big[\ \|\varphi\|_H + \|\psi\|_H + \max_{0\le t\le 1}\|f(t)\|_H + \max_{-1\le t\le 0}\|g(t)\|_H\Big]. \quad (15)$$

Next, using formula (10) and estimates (4), we obtain the following estimate

$$\|v(t)\|_H \le \|v_{-1}\|_H + \max_{-1\le t\le 0}\|g(t)\|_H, \quad -1 \le t \le 0. \quad (16)$$

Finally, by using formula (12) and estimates (2) and (4), we have

$$\|v(t)\|_H \le 2\,\|v_{-1}\|_H + \frac{1}{\sqrt{\delta}}\max_{0\le t\le 1}\|f(t)\|_H + \Big(2 + \frac{1}{\sqrt{\delta}}\Big)\max_{-1\le t\le 0}\|g(t)\|_H \quad (17)$$

for every $0 \leq t \leq 1$. Estimates (15), (16), and (17) allow us to obtain the following stability inequality

$$\|v\|_{C(H)} \leq M_2(\delta) \left[\|\varphi\|_H + \|\psi\|_H + \max_{0\leq t\leq 1} \|f(t)\|_H + \max_{-1\leq t\leq 0} \|g(t)\|_H \right].$$

Using this result and formulas (8) and (14), we finish the proof of estimate (6).

Now, let us obtain the estimate (7). Applying integration by parts in the formula (13), we obtain

$$Av_{-1} = \left(I - e^{-A}\left[s(1) - A\int_0^1 s(z)dz \right] \right)^{-1} \left\{ \left(I - c(1)\right)g(0) + A\varphi - A\psi \right.$$
$$+ \left[s(1) - A\int_0^1 s(z)dz \right]\left(g(0) - e^{-A}g(-1) - \int_{-1}^0 e^{\tau A} g'(\tau)d\tau \right)$$
$$\left. + \int_0^1 \left(I - c(1-\tau)\right) f(\tau)d\tau \right\}.$$

Using this formula and estimates (2), (3), (4), and (5), we have

$$\|Av_{-1}\|_H \leq M_3(\delta)\left[\|A\varphi\|_H + \|A\psi\|_H + \|g(0)\|_H + \|f(0)\|_H \right.$$
$$\left. + \max_{0\leq t\leq 1} \left\|f'(t)\right\|_H + \max_{-1\leq t\leq 0} \left\|g'(t)\right\| \right]. \tag{18}$$

Applying integration by parts in the formula (10), we obtain

$$Av(t) = e^{-(t+1)A}Av_{-1} + g(t) - e^{-(t+1)A}g(-1) - \int_{-1}^{t} e^{-(t-\tau)A}g'(\tau)d\tau, \quad -1 \leq t \leq 0.$$

Using this formula and estimates (4), we have

$$\|Av(t)\|_H \leq \|Av_{-1}\|_H + 3 \max_{-1\leq t\leq 0} \left\|g'(t)\right\|_H + 2\,\|g(0)\|_H\,, \quad -1 \leq t \leq 0. \tag{19}$$

Similarly, applying integration by parts in the formula (12), we obtain

$$Av(t) = c(t)\Big(e^{-A}Av_{-1} + g(0) - e^{-A}g(-1) - \int_{-1}^{0} e^{\tau A}g'(\tau)d\tau\Big)$$
$$+As(t)\Big(- Ae^{-A}v_{-1} + e^{-A}g(-1) + \int_{-1}^{0} e^{\tau A}g'(\tau)d\tau\Big)$$
$$+f(t) - c(t)f(0) - \int_{0}^{t} c(t-\tau)f'(\tau)d\tau,\ 0 \le t \le 1.$$

Using this formula and estimates (2), (3), and (4), we get

$$\|Av(t)\|_H \le M_4(\delta)\Big[\ \|Av_{-1}\|_H + \|g(0)\|_H + \|f(0)\|_H + \max_{0\le t\le 1}\left\|f'(t)\right\|_H + \max_{-1\le t\le 0}\left\|g'(t)\right\|_H\Big],\ 0 \le t \le 1. \tag{20}$$

Taking into account the estimates (18), (19), and (20), we prove the following result

$$\|Av\|_{C(H)} \le M_5(\delta)\Big[\ \|A\varphi\|_H + \|A\psi\|_H + \|g(0)\|_H + \|f(0)\|_H + \max_{0\le t\le 1}\left\|f'(t)\right\|_H + \max_{-1\le t\le 0}\left\|g'(t)\right\|_H\Big]. \tag{21}$$

Finally, by using the estimate (21), formulas (8) and (14), and triangle inequality in (1), we complete the proof of estimate (7). □

3 Applications

In this section, we illustrate a few applications of the main result. We consider four space-dependent source identification hyperbolic-parabolic problems and use the result of Theorem 1 to obtain the stability estimates for the solutions of these problems.

First, we consider the space-dependent source identification problem for the hyperbolic-parabolic equation

$$\begin{cases} u_{tt} - \left(a(x)u_x\right)_x + \delta u = p(x) + f(t,x),\ 0 < t < 1,\ 0 < x < 1, \\ u_t - \left(a(x)u_x\right)_x + \delta u = p(x) + g(t,x),\ \ -1 < t < 0,\ 0 < x < 1, \\ u(0^+,x) = u(0^-,x),\ u_t(0^+,x) = u_t(0^-,x),\ 0 \le x \le 1, \\ u(t,0) = u(t,1),\ u_x(t,0) = u_x(t,1),\ \ -1 \le t \le 1, \\ u(-1,x) = \varphi(x),\ \displaystyle\int_0^1 u(z,x)dz = \psi(x),\ 0 \le x \le 1. \end{cases} \tag{22}$$

Under compatibility conditions problem (22) has a unique smooth solution $\big(u(t,x), p(x)\big)$ for the given smooth functions $a(x) \geq a > 0$, $x \in (0,1)$, $a(1) = a(0)$, $\varphi(x)$, $\psi(x)$, $x \in [0,1]$, $f(t,x)$, $t \in [0,1]$, $x \in [0,1]$, $g(t,x)$, $t \in [-1,0]$, $x \in [0,1]$ and constant $\delta \geq 2$. This allows us to reduce the space-dependent source identification problem (22) to the abstract space-dependent source identification problem (1) in a Hilbert space $H = L_2[0,1]$ with a self-adjoint positive definite operator A^x defined by formula

$$A^x u(x) = -\big(a(x)u_x\big)_x + \delta u(x) \tag{23}$$

with domain

$$D(A^x) = \Big\{u(x) \;\Big|\; u(x), u_x(x), \big(a(x)u_x\big)_x \in L_2[0,1],\ u(1) = u(0),\ u_x(1) = u_x(0)\Big\}.$$

Let the Sobolev space $W_2^2[0,1]$ be defined as the set of all functions f defined on $[0,1]$ such that f and second order derivative function f'' are both locally integrable in $L_2[0,1]$, equipped with the norm

$$\|f\|_{W_2^2[0,1]} = \left(\int_0^1 \Big[\big(f(x)\big)^2 + \big(f''(x)\big)^2\Big]dx\right)^{1/2}.$$

Theorem 2 *For the solution of problem (22), we have the following stability inequalities*

$$\|u\|_{C(L_2[0,1])} + \left\|(A^x)^{-1} p\right\|_{L_2[0,1]}$$
$$\leq M(\delta)\Big[\|\varphi\|_{L_2[0,1]} + \|\psi\|_{L_2[0,1]} + \max_{0\leq t\leq 1}\|f(t)\|_{L_2[0,1]} + \max_{-1\leq t\leq 0}\|g(t)\|_{L_2[0,1]}\Big],$$

$$\max_{0\leq t\leq 1}\|u_{tt}\|_{L_2[0,1]} + \max_{-1\leq t\leq 0}\|u_t\|_{L_2[0,1]} + \|u\|_{C(W_2^2[0,1])} + \|p\|_{L_2[0,1]}$$
$$\leq M(\delta)\Big[\|\varphi\|_{W_2^2[0,1]} + \|\psi\|_{W_2^2[0,1]} + \max_{0\leq t\leq 1}\left\|f'(t)\right\|_{L_2[0,1]} + \max_{-1\leq t\leq 0}\left\|g'(t)\right\|_{L_2[0,1]}$$
$$+ \|g(0)\|_{L_2[0,1]} + \|f(0)\|_{L_2[0,1]}\Big],$$

where $M(\delta)$ is independent of $\varphi(x)$, $\psi(x)$, $f(t,x)$, and $g(t,x)$.

The proof of Theorem 2 is based on Theorem 1 and the symmetry properties of the space operator A^x defined by formula (23).

Second, we consider the space-dependent source identification problem for one-dimensional differential equations of hyperbolic-parabolic type with involution

$$
\begin{cases}
u_{tt}(t,x) - \big(a(x)u_x(t,x)\big)_x - \beta\big(a(-x)u_x(t,-x)\big)_x + \delta u(t,x) \\
\quad = p(x) + f(t,x),\ 0 < t < 1,\ -\ell < x < \ell, \\
u_t(t,x) - \big(a(x)u_x(t,x)\big)_x - \beta\big(a(-x)u_x(t,-x)\big)_x + \delta u(t,x) \\
\quad = p(x) + g(t,x),\ -1 < t < 0,\ -\ell < x < \ell, \\
u(0^+,x) = u(0^-,x),\ u_t(0^+,x) = u_t(0^-,x),\ -\ell \le x \le \ell, \\
u(t,-\ell) = u(t,\ell) = 0,\ -1 \le t \le 1, \\
u(-1,x) = \varphi(x),\ \displaystyle\int_0^1 u(z,x)dz = \psi(x),\ -\ell \le x \le \ell.
\end{cases}
\tag{24}
$$

We will assume that $0 < \underline{a} \le a(-x) = a(x) \le \overline{a},\ x \in (-\ell,\ell)$ and $\underline{a} - \overline{a}|\beta| \ge 0$. Under compatibility conditions problem (24) has a unique smooth solution $\big(u(t,x), p(x)\big)$ for the smooth functions $a(x), x \in (-\ell,\ell), \varphi(x), \psi(x), x \in [-\ell,\ell]$, $f(t,x),\ t \in [0,1],\ x \in [-\ell,\ell],\ g(t,x),\ t \in [-1,0],\ x \in [-\ell,\ell]$ and constant $\delta \ge 2$. This allows us to reduce space-dependent source identification problem (24) to abstract space-dependent source identification problem (1) in a Hilbert space $H = L_2[-\ell,\ell]$ with a self-adjoint positive definite operator A^x defined by formula

$$
A^x u(x) = -\big(a(x)u_x(x)\big)_x - \beta\big(a(-x)u_x(-x)\big)_x + \delta u(x) \tag{25}
$$

with the domain $D(A^x) = \Big\{u \in W_2^2[-\ell,\ell] \ \Big|\ u(-\ell) = u(\ell) = 0\Big\}$.

Let the Sobolev space $W_2^2[-\ell,\ell]$ be defined as the set of all functions f defined on $[-\ell,\ell]$ such that f and second order derivative function f'' are both locally integrable in $L_2[-\ell,\ell]$, equipped with the norm

$$
\|f\|_{W_2^2[-\ell,\ell]} = \left(\int_{-\ell}^{\ell} \Big[\big(f(x)\big)^2 + \big(f''(x)\big)^2\Big]dx\right)^{1/2}.
$$

Theorem 3 *For the solution of problem (24), we have the following stability inequalities*

$$
\begin{aligned}
&\|u\|_{C(L_2[-\ell,\ell])} + \left\|\left(A^x\right)^{-1} p\right\|_{L_2[-\ell,\ell]} \\
&\le M(\delta)\Big[\|\varphi\|_{L_2[-\ell,\ell]} + \|\psi\|_{L_2[-\ell,\ell]} + \max_{0\le t\le 1}\|f(t)\|_{L_2[-\ell,\ell]} + \max_{-1\le t\le 0}\|g(t)\|_{L_2[-\ell,\ell]}\Big],
\end{aligned}
$$

$$
\begin{aligned}
&\max_{0\le t\le 1}\|u_{tt}\|_{L_2[-\ell,\ell]} + \max_{-1\le t\le 0}\|u_t\|_{L_2[-\ell,\ell]} + \|u\|_{C(W_2^2[-\ell,\ell])} + \|p\|_{L_2[-\ell,\ell]} \\
&\le M(\delta)\Big[\|\varphi\|_{W_2^2[-\ell,\ell]} + \|\psi\|_{W_2^2[-\ell,\ell]} + \max_{0\le t\le 1}\left\|f'(t)\right\|_{L_2[-\ell,\ell]} + \|f(0)\|_{L_2[-\ell,\ell]} \\
&\quad + \max_{-1\le t\le 0}\left\|g'(t)\right\|_{L_2[-\ell,\ell]} + \|g(0)\|_{L_2[-\ell,\ell]}\Big],
\end{aligned}
$$

where $M(\delta)$ is independent of $\varphi(x)$, $\psi(x)$, $f(t,x)$, and $g(t,x)$.

The proof of Theorem 3 is based on Theorem 1 and the symmetry properties of the space operator A^x defined by formula (25).

Third, let Ω be the unit open cube with boundary S in the n-dimensional Euclidean space $\mathbb{R}^n$, so that $\Omega = \left\{ \mathrm{x} = (x_1, \ldots, x_n) \,\middle|\, 0 < x_k < 1,\ k = 1, \ldots, n \right\}$. Denote $\overline{\Omega} = \Omega \cup S$. In $[-1, 1] \times \Omega$, the space-dependent source identification problem for the multidimensional hyperbolic-parabolic equation

$$\begin{cases} u_{tt} - \sum\limits_{r=1}^{n} \left(a_r(\mathrm{x}) u_{x_r}\right)_{x_r} + \delta u = p(\mathrm{x}) + f(t, \mathrm{x}),\ 0 < t < 1,\ \mathrm{x} \in \Omega, \\ u_t - \sum\limits_{r=1}^{n} \left(a_r(\mathrm{x}) u_{x_r}\right)_{x_r} + \delta u = p(\mathrm{x}) + g(t, \mathrm{x}),\ \ -1 < t < 0,\ \mathrm{x} \in \Omega, \\ u(0^+, \mathrm{x}) = u(0^-, \mathrm{x}),\ u_t(0^+, \mathrm{x}) = u_t(0^-, \mathrm{x}),\ \mathrm{x} \in \overline{\Omega}, \\ u(t, \mathrm{x}) = 0,\ \mathrm{x} \in S,\ \ -1 \le t \le 1, \\ u(-1, \mathrm{x}) = \varphi(\mathrm{x}),\ \ \int\limits_0^1 u(z, x) dz = \psi(\mathrm{x}),\ \mathrm{x} \in \overline{\Omega} \end{cases} \tag{26}$$

with the Dirichlet condition is considered. We introduce the Hilbert space $L_2(\overline{\Omega})$ of all square integrable functions defined on $\overline{\Omega}$, equipped with the norm

$$\| f \|_{L_2(\overline{\Omega})} = \left(\int \cdots \int\limits_{\mathrm{x} \in \overline{\Omega}} \left(f(\mathrm{x})\right)^2 dx_1 \cdots dx_n \right)^{1/2}.$$

Under compatibility conditions problem (26) has a unique smooth solution $\left(u(t, \mathrm{x}), p(\mathrm{x})\right)$ for the smooth functions $\varphi(\mathrm{x})$, $\psi(\mathrm{x})$, $a_r(\mathrm{x}) \ge a > 0$, $f(t, \mathrm{x})$, $g(t, \mathrm{x})$ and constant $\delta > 2$. This allows us to reduce the space-dependent source identification problem (26) to the abstract space-dependent source identification problem (1) in the Hilbert space $H = L_2(\overline{\Omega})$ with a self-adjoint positive definite operator A^x defined by formula

$$A^x u(\mathrm{x}) = -\sum_{r=1}^{n} \left(a_r(\mathrm{x}) u_{x_r}\right)_{x_r} + \delta u(\mathrm{x}) \tag{27}$$

with domain

$$D(A^x) = \left\{ u(\mathrm{x}) \,\middle|\, u(\mathrm{x}), u_{x_r}(\mathrm{x}), \left(a_r(\mathrm{x}) u_{x_r}\right)_{x_r} \in L_2(\overline{\Omega}),\ 1 \le r \le n,\ u(\mathrm{x}) = 0,\ \mathrm{x} \in S \right\}.$$

Theorem 4 *For the solution of problem (26), the following stability inequalities*

$$\|u\|_{C(L_2(\overline{\Omega}))} + \left\|(A^x)^{-1} p\right\|_{L_2(\overline{\Omega})}$$
$$\leq M(\delta)\left[\|\varphi\|_{L_2(\overline{\Omega})} + \|\psi\|_{L_2(\overline{\Omega})} + \max_{0\leq t\leq 1} \|f(t)\|_{L_2(\overline{\Omega})} + \max_{-1\leq t\leq 0} \|g(t)\|_{L_2(\overline{\Omega})} \right],$$

$$\max_{0\leq t\leq 1} \|u_{tt}\|_{L_2(\overline{\Omega})} + \max_{-1\leq t\leq 0} \|u_t\|_{L_2(\overline{\Omega})} + \|u\|_{C(W_2^2(\overline{\Omega}))} + \|p\|_{L_2(\overline{\Omega})}$$
$$\leq M(\delta)\left[\|\varphi\|_{W_2^2(\overline{\Omega})} + \|\psi\|_{W_2^2(\overline{\Omega})} + \max_{0\leq t\leq 1} \left\|f'(t)\right\|_{L_2(\overline{\Omega})} + \max_{-1\leq t\leq 0} \left\|g'(t)\right\|_{L_2(\overline{\Omega})} \right.$$
$$\left. + \|g(0)\|_{L_2(\overline{\Omega})} + \|f(0)\|_{L_2(\overline{\Omega})} \right]$$

hold, where $M(\delta)$ does not depend on $\varphi(\mathrm{x})$, $\psi(\mathrm{x})$, $f(t,\mathrm{x})$, and $g(t,\mathrm{x})$.

Here and in future, the Sobolev space $W_2^2(\overline{\Omega})$ is defined as the set of all functions f defined on $\bar{\Omega}$ such that f and all second order partial derivative functions $f_{x_r x_r}$, $r = 1, \ldots, n$ are locally integrable in $L_2(\overline{\Omega})$, equipped with the norm

$$\|f\|_{W_2^2(\overline{\Omega})} = \left(\int \cdots \int_{x\in\overline{\Omega}} \left[(f(\mathrm{x}))^2 + \sum_{r=1}^{n} \left(f_{x_r x_r}\right)^2 \right] dx_1 \cdots dx_n \right)^{1/2}.$$

The proof of Theorem 4 is based on Theorem 1 and the symmetry properties of the operator A^x defined by formula (27) and the following theorem on the coercivity inequality for the solution of the elliptic differential problem in $L_2(\overline{\Omega})$.

Theorem 5 *For the solution of the elliptic differential problem*

$$\begin{cases} A^x u(\mathrm{x}) = \omega(\mathrm{x}), \ \mathrm{x} \in \Omega, \\ u(\mathrm{x}) = 0, \ \mathrm{x} \in S \end{cases}$$

the following coercivity inequality holds [39]

$$\sum_{r=1}^{n} \left\|u_{x_r x_r}\right\|_{L_2(\overline{\Omega})} \leq M_1 \|\omega\|_{L_2(\overline{\Omega})},$$

where M_1 does not depend on $\omega(\mathrm{x})$.

Fourth, in $[-1, 1] \times \Omega$, the space-dependent source identification problem for the multidimensional hyperbolic-parabolic equation

$$\begin{cases} u_{tt} - \sum_{r=1}^{n} \left(a_r(\mathrm{x})u_{x_r}\right)_{x_r} + \delta u = p(\mathrm{x}) + f(t,\mathrm{x}),\ 0 < t < 1,\ \mathrm{x} \in \Omega, \\ u_t - \sum_{r=1}^{n} \left(a_r(\mathrm{x})u_{x_r}\right)_{x_r} + \delta u = p(\mathrm{x}) + g(t,\mathrm{x}),\ \ -1 < t < 0,\ \mathrm{x} \in \Omega, \\ u(0^+,\mathrm{x}) = u(0^-,\mathrm{x}),\ u_t(0^+,\mathrm{x}) = u_t(0^-,\mathrm{x}),\ \mathrm{x} \in \overline{\Omega}, \\ \frac{\partial u(t,\mathrm{x})}{\partial \mathbf{n}} = 0,\ \mathrm{x} \in S,\ \ -1 \le t \le 1, \\ u(-1,\mathrm{x}) = \varphi(\mathrm{x}),\ \int_0^1 u(z,x)dz = \psi(\mathrm{x}),\ \mathrm{x} \in \overline{\Omega} \end{cases} \tag{28}$$

with the Neumann condition is considered. Here, $\mathbf{n}$ is the normal vector to S. Under compatibility conditions problem (28) has a unique smooth solution $\left(u(t,\mathrm{x}),\ p(\mathrm{x})\right)$ for the given smooth functions $a_r(\mathrm{x}) \ge a > 0$, $\mathrm{x} \in \Omega$, $\varphi(\mathrm{x})$, $\psi(\mathrm{x})$, $x \in \overline{\Omega}$, $f(t,\mathrm{x})$, $t \in [0,1]$, $\mathrm{x} \in \Omega$, $g(t,\mathrm{x})$, $t \in [-1,0]$, $\mathrm{x} \in \Omega$ and constant $\delta > 2$. This allows us to reduce the space-dependent source identification problem (28) to the abstract space-dependent source identification problem (1) in the Hilbert space $H = L_2(\overline{\Omega})$ with a self-adjoint positive definite operator A^x defined by formula (27) with domain

$$D(A^x) = \left\{u(\mathrm{x}) \;\middle|\; u(\mathrm{x}), u_{x_r}(\mathrm{x}), \left(a_r(\mathrm{x})u_{x_r}\right)_{x_r} \in L_2(\overline{\Omega}),\ 1 \le r \le n,\ \frac{\partial u(\mathrm{x})}{\partial \mathbf{n}} = 0,\ \mathrm{x} \in S\right\}.$$

Theorem 6 *For the solution of problem (28), the following stability inequalities*

$$\|u\|_{C(L_2(\overline{\Omega}))} + \left\|\left(A^x\right)^{-1} p\right\|_{L_2(\overline{\Omega})}$$
$$\le M(\delta)\left[\ \|\varphi\|_{L_2(\overline{\Omega})} + \|\psi\|_{L_2(\overline{\Omega})} + \max_{0\le t\le 1} \|f(t)\|_{L_2(\overline{\Omega})} + \max_{-1\le t\le 0} \|g(t)\|_{L_2(\overline{\Omega})}\ \right],$$

$$\max_{0\le t\le 1} \|u_{tt}\|_{L_2(\overline{\Omega})} + \max_{-1\le t\le 0} \|u_t\|_{L_2(\overline{\Omega})} + \|u\|_{C(W_2^2(\overline{\Omega}))} + \|p\|_{L_2(\overline{\Omega})}$$
$$\le M(\delta)\Bigg[\ \|\varphi\|_{W_2^2(\overline{\Omega})} + \|\psi\|_{W_2^2(\overline{\Omega})} + \max_{0\le t\le 1} \left\|f'(t)\right\|_{L_2(\overline{\Omega})} + \max_{-1\le t\le 0} \left\|g'(t)\right\|_{L_2(\overline{\Omega})}$$
$$+ \|g(0)\|_{L_2(\overline{\Omega})} + \|f(0)\|_{L_2(\overline{\Omega})}\ \Bigg],$$

hold, where $M(\delta)$ does not depend on $\varphi(\mathrm{x})$, $\psi(\mathrm{x})$, $f(t,\mathrm{x})$, and $g(t,\mathrm{x})$.

The proof of Theorem 6 is based on Theorem 1 and the symmetry properties of the operator A^x defined by formula (27) and the following theorem on the coercivity inequality for the solution of the elliptic differential problem in $L_2(\overline{\Omega})$.

Theorem 7 *For the solution of the elliptic differential problem*

$$\begin{cases} A^x u(\mathrm{x}) = \omega(\mathrm{x}), \ \mathrm{x} \in \Omega, \\ \dfrac{\partial u(\mathrm{x})}{\partial \mathbf{n}} = 0, \ \mathrm{x} \in S \end{cases}$$

the following coercivity inequality holds [39]

$$\sum_{r=1}^{n} \left\| u_{x_r x_r} \right\|_{L_2(\overline{\Omega})} \leq M_1(\delta) \left\| \omega \right\|_{L_2(\overline{\Omega})},$$

where $M_1(\delta)$ is independent of $\omega(\mathrm{x})$.

4 Conclusion

In the present paper, the well-posedness of the space-dependent source identification problem (1) is established. In applications, the stability inequalities for the solutions of four space-dependent source identification problems for hyperbolic-parabolic equations are obtained.

Finally, we note that using the techniques introduced in [10], stable difference schemes can be constructed for the approximate solution of inverse problem (1).

Acknowledgements The publication has been prepared with the support of the "RUDN University Program 5-100".

References

1. Ashyralyev, A.: On a problem of determining the parameter of a parabolic equation. Ukrainian Math. J. **62**, 1200–1210 (2010)
2. Ashyralyev, A., Agirseven, D.: On source identification problem for a delay parabolic equation. Nonlinear Anal. Model. Control **19**, 335–349 (2014)
3. Ashyralyev, A., Ashyralyyev, C.: On the problem of determining the parameter of an elliptic equation in a Banach space. Nonlinear Anal. Model. Control **19**, 350–366 (2014)
4. Ashyralyev, A., Ashyralyyeva, M.A.: On source identification problem for a hyperbolic-parabolic equation. Contemp. Anal. Appl. Math. **3**, 88–103 (2015)
5. Ashyralyev, A., Emharab, F.: Source identification problems for hyperbolic differential and difference equations. J. Inverse Ill-Posed Probl. **27**, 301–315 (2018)
6. Ashyralyev, A., Erdogan, A.S.: Well-posedness of the right-hand side identification problem for a parabolic equation. Ukrainian Math. J. **66**, 165–177 (2014)
7. Ashyralyev, A., Erdogan, A.S., Demirdag, O.: On the determination of the right-hand side in a parabolic equation. Appl. Numer. Math. **62**, 1672–1683 (2012)
8. Ashyralyev, A., Ozdemir, Y.: Stability of difference schemes for hyperbolic-parabolic equations. Comput. Math. Appl. **50**, 1443–1476 (2005)
9. Ashyralyev, A., Ozdemir, Y.: On nonlocal boundary value problems for hyperbolic-parabolic equations. Taiwanese J. Math. **11**, 1075–1089 (2007)
10. Ashyralyev, A., Sobolevskii, P.E.: New Difference Schemes for Partial Differential Equations. Operator Theory Advances and Applications. Birkhauser Verlag, Basel, Boston, Berlin (2004)

11. Ashyralyev, A., Yurtsever, A.: On a nonlocal boundary value problem for semilinear hyperbolic-parabolic equations. Nonlinear Anal. **47**, 3585–3592 (2001)
12. Ashyralyyev, C.: High order of accuracy difference schemes for the inverse elliptic problem with Dirichlet condition. Bound. Value Probl. **2014**, 5 (2014)
13. Ashyralyyev, C.: Well-posedness of boundary value problems for reverse parabolic equation with integral condition. e-J. Anal. Appl. Math. **2018**, 11–21 (2018)
14. Ashyralyyeva, M., Ashyraliyev, M.: On a second order of accuracy stable difference scheme for the solution of a source identification problem for hyperbolic-parabolic equations. AIP Conf. Proc. **1759**, 020023 (2016)
15. Ashyralyyeva, M., Ashyraliyev, M.: Numerical solutions of source identification problem for hyperbolic-parabolic equations. AIP Conf. Proc. **1997**, 020048 (2018)
16. Ashyralyyeva, M.A., Ashyraliyev, M.: On the numerical solution of identification hyperbolic-parabolic problems with the Neumann boundary condition. Bull. Karaganda Univ.-Math. **91**, 69–74 (2018)
17. Ashyralyyeva, M.A., Ashyralyyev, A.: Stable difference scheme for the solution of the source identification problem for hyperbolic-parabolic equations. AIP Conf. Proc. **1676**, 020024 (2015)
18. Berdyshev, A.S., Cabada, A., Karimov, E.T., Akhtaeva, N.S.: On the Volterra property of a boundary problem with integral gluing condition for a mixed parabolic-hyperbolic equation. Bound. Value Probl. **2013**, 94 (2013)
19. Bitsadze, A.V.: Equations of Mixed Type. Pergamon Press, Oxford (1964)
20. Blasio, G.D., Lorenzi, A.: Identification problems for parabolic delay differential equations with measurement on the boundary. J. Inverse Ill-Posed Probl. **15**, 709–734 (2007)
21. Dehghan, M.: Determination of a control parameter in the two-dimensional diffusion equation. Appl. Numer. Math. **37**, 489–502 (2001)
22. Eidelman, Y.S.: Boundary value problems for differential equations with parameters. Ph.D. thesis, Voronezh State University, Russia (1984)
23. Emharab, F.: Source identification problems for hyperbolic differential and difference equations. Ph.D. thesis, Near East University, Lefkoşa (2019)
24. Erdogan, A.S.: A note on the right-hand side identification problem arising in biofluid mechanics. Abstr. Appl. Anal. **2012**, 548508 (2012)
25. Fattorini, H.O.: Second Order Linear Differential Equations in Banach Spaces. Notas de Matematica, North-Holland (1985)
26. Gryazin, Y.A., Klibanov, M.V., Lucas, T.R.: Imaging the diffusion coefficient in a parabolic inverse problem in optical tomography. Inverse Probl. **15**, 373–397 (1999)
27. Ivanchov, N.I.: On the determination of unknown source in the heat equation with nonlocal boundary conditions. Ukrainian Math. J. **47**, 1647–1652 (1995)
28. Kabanikhin, S.I., Krivorotko, O.I.: Identification of biological models described by systems of nonlinear differential equations. J. Inverse Ill-Posed Probl. **23**, 519–527 (2015)
29. Kimura, T., Suzuki, T.: A parabolic inverse problem arising in a mathematical model for chromatography. SIAM J. Appl. Math. **53**, 1747–1761 (1993)
30. Orlovsky, D., Piskarev, S.: On approximation of inverse problems for abstract elliptic problems. J. Inverse Ill-Posed Probl. **17**, 765–782 (2009)
31. Ozbilge, E., Demir, A.: Semigroup approach for identification of the unknown diffusion coefficient in a linear parabolic equation with mixed output data. Bound. Value Probl. **2013**, 43 (2013)
32. Piskarev, S., Shaw, S.Y.: On certain operator families related to cosine operator function. Taiwanese J. Math. **1**, 3585–3592 (1997)
33. Rassias, J.M.: Lecture Notes on Mixed Type Partial Differential Equations. World Scientific, Singapore (1990)
34. Sadybekov, M.A.: Stable difference scheme for a nonlocal boundary value heat conduction problem. e-J. Anal. Appl. Math. **2018**, 1–10 (2018)
35. Safari, A.R., Mekhtiyev, M.F., Sharifov, Y.A.: Maximum principle in the optimal control problems for systems with integral boundary conditions and its extension. Abstr. Appl. Anal. **2013**, 946910 (2013)

36. Sazaklioglu, A.U., Erdogan, A.S., Ashyralyev, A.: Existence and uniqueness results for an inverse problem for a semilinear equation with final overdetermination. Filomat **32**, 847–858 (2018)
37. Shakhmurov, V.B., Sahmurova, A.: Abstract parabolic problems with parameter and application. Appl. Math. Comput. **219**, 9561–9571 (2013)
38. Smirnov, M.M.: Equations of Mixed Type. American Mathematical Society, Providence (1978)
39. Sobolevskii, P.E.: Difference Methods for the Approximate Solution of Differential Equations. Voronezh State University Press, Voronezh, Russia (1975)
40. Wu, B., Wu, S.: Existence and uniqueness of an inverse source problem for a fractional integrodifferential equation. Comput. Math. Appl. **68**, 1123–1136 (2014)

Numerical Solution of a Parabolic Source Identification Problem with Involution and Neumann Condition

On the Stability of the Time Delay Telegraph Equation with Neumann Condition

Allaberen Ashyralyev, Koray Turk, and Deniz Agirseven

Abstract In this paper, we consider the initial-boundary value problem for time delay telegraph equation with Neumann condition. Stability estimates for the solution of this problem are obtained. A first order of accuracy difference scheme is constructed. Stability estimates for the solution of this difference scheme are established. As a test problem, one-dimensional time delay telegraph equation with Neumann condition is considered. Numerical results are given.

Keywords Stability · Delay telegraph equations

1 Introduction

Telegraph equations occur in many areas of science and engineering, especially in physics, e.g., hydrodynamics [1], electromagnetic [2], random walk theory and random motions [3, 4], statistical physics [5]. Several questions (stability, obtaining numerical and exact solutions, asymptotic formulas) for telegraph equations have been investigated by many scientists (see, [6–9] and the references given therein). In most studies related to the telegraph equation, the delay term is not taken into account. However, in order to find more realistic solutions, the delay term should be taken into

A. Ashyralyev
Department of Mathematics, Near East University, TRNC, Mersin 10, Nicosia, Turkey
e-mail: allaberen.ashyralyev@neu.edu.tr

Peoples' Friendship University of Russia (RUDN University), 6 Miklukho-Maklaya St, Moscow 117198, Russian Federation

Institute of Mathematics and Mathematical Modeling, 050010 Almaty, Kazakhstan

K. Turk (✉) · D. Agirseven
Department of Mathematics, Trakya University, Edirne, Turkey
e-mail: korayturk@trakya.edu.tr

D. Agirseven
e-mail: denizagirseven@trakya.edu.tr

A. Ashyralyev et al. (eds.), *Functional Analysis in Interdisciplinary Applications—II*, Springer Proceedings in Mathematics & Statistics 351,
https://doi.org/10.1007/978-3-030-69292-6_15

account in the modeling of many problems. It is known that analysis of differential and difference equations with delay term is more difficult than equations without delay term. Nevertheless, many problems containing the delay term have been studied (see, [10–13] and references given therein). In this paper, we consider differential and difference problem for the multidimensional telegraph equation with time delay. The paper is organized as follows. Section 1 is an introduction. In Sect. 2, stability estimates for the solution of initial-boundary value problem for the multidimensional delay telegraph equation with Neumann condition are obtained. In Sect. 3, a first order of accuracy difference scheme for this differential problem is constructed. Stability estimates for the solution of this difference problem are established. In Sect. 4, as a test problem, one dimensional time delay telegraph equation with Neumann condition is considered. Numerical results are given in a table of errors of the difference scheme.

2 Stability of Differential Problem

In this section, let $\Omega \subset R^n$ be an open bounded domain with smooth boundary S, $\overline{\Omega} = \Omega \cup S$. In $[0, \infty) \times \Omega$, the initial-boundary value problem for the multidimensional delay telegraph equation with Neumann condition

$$\begin{cases} u_{tt}(t,x) + \alpha u_t(t,x) - \sum\limits_{r=1}^{n} (a_r(x) u_{x_r}(t,x))_{x_r} + \delta u(t,x) \\ = b\left(- \sum\limits_{r=1}^{n} (a_r(x) u_{x_r}([t],x))_{x_r} + \delta u([t],x) \right), \\ x = (x_1, \dots, x_n) \in \Omega, \ 0 < t < \infty, \\ u(0,x) = \varphi(x), \ \frac{\partial u(0,x)}{\partial t} = \psi(x), \ x \in \overline{\Omega}, \\ \frac{\partial u(t,x)}{\partial \vec{m}} = 0, \ x \in S, \ 0 \le t < \infty \end{cases} \tag{1}$$

is considered. Here, $\vec{m}$ is the normal vector to S. Under compatibility conditions, the problem (1) has a unique smooth solution $u(t,x)$ for smooth functions $a_r(x) > a_0 > 0, \ (x \in \Omega), \ \delta > 0, \ \varphi(x), \ \psi(x), \ (x \in \overline{\Omega})$ and $b \in R^1$. We introduce the Hilbert space $H = L_2(\overline{\Omega})$, the Sobolev spaces $W_2^1(\overline{\Omega})$ and $W_2^1(\overline{\Omega})$, these spaces of all integrable functions defined on $\overline{\Omega}$, equipped with the norms

$$\|f\|_{L_2(\overline{\Omega})} = \left(\int \cdots \int_{x \in \overline{\Omega}} |f(x)|^2 \, dx_1 \dots dx_n \right)^{1/2},$$

$$\|f\|_{W_2^1(\overline{\Omega})} = \left(\int \cdots \int_{x\in\overline{\Omega}} \left[|f(x)|^2 + \sum_{r=1}^{n} \left| f_{x_r}(x) \right|^2 \right] dx_1 \ldots dx_n \right)^{1/2},$$

$$\|f\|_{W_2^2(\overline{\Omega})} = \left(\int \cdots \int_{x\in\overline{\Omega}} \left[|f(x)|^2 + \sum_{r=1}^{n} \left| f_{x_r x_r}(x) \right|^2 \right] dx_1 \ldots dx_n \right)^{1/2},$$

respectively. The following stability results are obtained.

Theorem 1 *For solution of the problem (1), the following stability inequalities hold:*

$$\max_{0\le t\le 1} \|u(t,.)\|_{L_2(\overline{\Omega})} \le M_1 \left\{ \|\varphi\|_{L_2(\overline{\Omega})} + \|\psi\|_{L_2(\overline{\Omega})} \right\},$$

$$\max_{0\le t\le 1} \|u(t,.)\|_{W_2^1(\overline{\Omega})} + \max_{0\le t\le 1} \|u_t(t,.)\|_{L_2(\overline{\Omega})} \le M_2 \left\{ \|\varphi\|_{W_2^1(\overline{\Omega})} + \|\psi\|_{L_2(\overline{\Omega})} \right\},$$

$$\max_{n\le t\le n+1} \|u(t,.)\|_{L_2(\overline{\Omega})}$$

$$\le M_3 \left\{ \max_{n-1\le t\le n} \|u(t,.)\|_{L_2(\overline{\Omega})} + \max_{n-1\le t\le n} \|u_t(t,.)\|_{L_2(\overline{\Omega})} \right\}, \ n = 1, 2, \ldots,$$

$$\max_{n\le t\le n+1} \|u(t,.)\|_{W_2^1(\overline{\Omega})} + \max_{n\le t\le n+1} \|u_t(t,.)\|_{L_2(\overline{\Omega})}$$

$$\le M_4 \left\{ \max_{n-1\le t\le n} \|u(t,.)\|_{W_2^1(\overline{\Omega})} + \max_{n-1\le t\le n} \|u_t(t,.)\|_{L_2(\overline{\Omega})} \right\}, \ n = 1, 2, \ldots,$$

where M_1, M_2, M_3 and M_4 do not depend on $\varphi(x)$ or $\psi(x)$.

Proof The problem (1) can be written in abstract form

$$\begin{cases} \frac{d^2u(t)}{dt^2} + \alpha \frac{du(t)}{dt} + A^x u(t) = bA^x u([t]), \ t > 0, \\ u(0) = \varphi, \quad u'(0) = \psi \end{cases}$$

in a Hilbert space $H = L_2(\overline{\Omega})$ with self-adjoint positive definite operator A^x defined by the formula

$$A^x u(x) = -\sum_{r=1}^{n} (a_r(x) u_{x_r}(x))_{x_r} + \delta u(x) \tag{2}$$

with domain

$$D(A^x) = \left\{ u(x) : u(x), u_{x_r}(x), \left(a_r(x)u_{x_r}(x)\right)_{x_r} \in L_2(\overline{\Omega}), 1 \leq r \leq n, \frac{\partial u(t,x)}{\partial \overrightarrow{m}} = 0, x \in S \right\}.$$

Here, $u(t) = u(t, x)$ is unknown abstract function defined on $\overline{\Omega}$ with values in $H = L_2(\overline{\Omega})$. So, the estimates in Theorem 1 follow from the estimates of the main theorem of [13] and on the following theorem on the coercivity inequality for the solution of the elliptic differential problem in $L_2(\overline{\Omega})$, (see [15]). Therefore, the proof of Theorem 1 is completed. □

Theorem 2 *For the solutions of the elliptic differential problem*

$$\begin{cases} A^x u(x) = \omega(x), \ x \in \Omega, \\ \frac{\partial u(t,x)}{\partial \overrightarrow{m}} = 0, \ x \in S, \end{cases}$$

the coercivity inequality

$$\sum_{r=1}^{n} \|u_{x_r x_r}\|_{L_2(\overline{\Omega})} \leq M_5(\delta) \|\omega\|_{L_2(\overline{\Omega})}$$

is satisfied. Here $M_5(\delta)$ is independent of $\omega(x)$.

3 Stability of Difference Problem

In this section, we present a first order of accuracy difference scheme for the differential problem (1). Stability estimates for the solution of this difference scheme are obtained. The discretization of problem (1) is carried out in two steps. In the first step, we define the grid space

$$\begin{aligned} \overline{\Omega}_h = \{x = x_r = (h_1 j_1, \ldots, h_n j_n)\,, \ j = (j_1, \ldots, j_n)\,, \\ 0 \leq j_r \leq N_r, N_r h_r = 1, \ r = 1, \ldots, n\}\,, \\ \Omega_h = \overline{\Omega}_h \cap \Omega, \ S_h = \overline{\Omega}_h \cap S. \end{aligned}$$

We introduce the Hilbert spaces L_{2h}=$L_2(\overline{\Omega}_h)$, $W_{2h}^1 = W_2^1(\overline{\Omega}_h)$ and $W_{2h}^2 = W_2^2(\overline{\Omega}_h)$ of the grid functions $\varphi^h(x) = \{\varphi(h_1 r_1, \ldots, h_n r_n)\}$ defined on $\overline{\Omega}_h$, equipped with the norms

$$\|\varphi^h\|_{L_{2h}} = \left(\sum_{x \in \overline{\Omega}_h} \left|\varphi^h(x)\right|^2 h_1 \cdots h_n \right)^{1/2},$$

$$\left\|\varphi^h\right\|_{W_{2h}^1}=\left\|\varphi^h\right\|_{L_{2h}}+\left(\sum_{x\in\overline{\Omega}_h}\sum_{r=1}^{m}\left|\varphi^h_{x_r,j_r}(x)\right|^2 h_1\cdots h_n\right)^{1/2},$$

$$\left\|\varphi^h\right\|_{W_{2h}^2}=\left\|\varphi^h\right\|_{L_{2h}}+\left(\sum_{x\in\overline{\Omega}_h}\sum_{r=1}^{m}\left|\varphi^h_{x_r x_r,j_r}(x)\right|^2 h_1\cdots h_n\right)^{1/2},$$

respectively. To the differential operator A^x defined by (2), we assign the difference operator A_h^x by the formula

$$A_h^x u^h(x)=-\sum_{r=1}^{n}\left(a_r(x)u^h_{x_r}(x)\right)_{x_r,j_r}+\delta u^h(x), \tag{3}$$

acting in the space of grid functions $u^h(x)$ satisfying the conditions

$$D^h u^h(x)=0,\ x\in S_h.$$

Here D^h is approximation of operator $\frac{\partial}{\partial \vec{m}}$. With the help of A_h^x, we reach the initial value problem

$$\begin{cases}\frac{d^2u^h(t,x)}{dt^2}+\alpha\frac{du^h(t,x)}{dt}+A_h^x u^h(t,x)=bA_h^x u^h([t],x),\\ 0<t<\infty,\ \ x\in\Omega_h,\\ u^h(0,x)=\varphi^h(x),\ \ u_t^h(0,x)=\psi^h(x),\ \ x\in\overline{\Omega}_h\end{cases} \tag{4}$$

for an infinite system of ordinary differential equations. In the second step, we replace (4) with the first order of accuracy difference scheme and we get

$$\begin{cases}\frac{u^h_{k+1}(x)-2u^h_k(x)+u^h_{k-1}(x)}{\tau^2}+\alpha\frac{u^h_{k+1}(x)-u^h_k(x)}{\tau}+A_h^x u^h_{k+1}(x)=bA_h^x u^h_{\left[\frac{k-mN}{N+1}\right]N+mN}(x),\\ t_k=k\tau,\ \ x\in\Omega_h,\ N\tau=1,\ (m-1)N+1\le k\le mN-1,\ \ m=1,2,\ldots,\\ u_0^h(x)=\varphi^h(x),\ ((1+\alpha\tau)\,I_h+\tau^2A_h^x)\frac{u_1^h(x)-u_0^h(x)}{\tau}=\psi^h(x),\ x\in\overline{\Omega}_h,\\ ((1+\alpha\tau)\,I_h+\tau^2A_h^x)\frac{u^h_{mN+1}(x)-u^h_{mN}(x)}{\tau}=\frac{u^h_{mN}(x)-u^h_{mN-1}(x)}{\tau},\ m=1,2,\ldots,\ x\in\overline{\Omega}_h.\end{cases} \tag{5}$$

Theorem 3 *Suppose that $\delta>\frac{\alpha^2}{4}$. Then, for the solution $\left\{u_k^h(x)\right\}_0^N$ of problem (5) the following stability estimates hold:*

$$\max_{1\le k\le N}\|u_k^h\|_{L_{2h}} \le M_6\left\{\|\varphi^h\|_{L_{2h}}+\|\psi^h\|_{L_{2h}}\right\},$$

$$\max_{1\le k\le N}\|u_k^h\|_{W_{2h}^1}+\max_{1\le k\le N}\left\|\frac{u_k^h-u_{k-1}^h}{\tau}\right\|_{L_{2h}} \le M_7\left\{\|\varphi^h\|_{W_{2h}^1}+\|\psi^h\|_{L_{2h}}\right\},$$

$$\max_{mN+1\le k\le (m+1)N}\|u_k^h\|_{L_{2h}}$$

$$\le M_8\left\{\max_{(m-1)N\le k\le mN}\|u_k^h\|_{L_{2h}}+\max_{(m-1)N+1\le k\le mN}\left\|\frac{u_k^h-u_{k-1}^h}{\tau}\right\|_{L_{2h}}\right\},\ m=1,2,\ldots,$$

$$\max_{mN+1\le k\le (m+1)N}\|u_k^h\|_{W_{2h}^1}+\max_{mN+1\le k\le (m+1)N}\left\|\frac{u_k^h-u_{k-1}^h}{\tau}\right\|_{L_{2h}}$$

$$\le M_9\left\{\max_{(m-1)N\le k\le mN}\|u_k^h\|_{W_{2h}^1}+\max_{(m-1)N+1\le k\le mN}\left\|\frac{u_k^h-u_{k-1}^h}{\tau}\right\|_{L_{2h}}\right\},\ m=1,2,\ldots,$$

where M_6, M_7, M_8 and M_9 do not depend on $\varphi^h(x)$ or $\psi^h(x)$.

Proof Difference scheme (5) can be written in the abstract form

$$\begin{cases}\frac{u_{k+1}^h-2u_k^h+u_{k-1}^h}{\tau^2}+\alpha\frac{u_{k+1}^h-u_k^h}{\tau}+A_h^x u_{k+1}^h=aA_h^x u^h_{\left[\frac{k-mN}{N+1}\right]N+mN},\\ t_k=k\tau,\ \ N\tau=1,\ (m-1)N+1\le k\le mN-1,\ \ m=1,2,\ldots,\\ u_0^h=\varphi^h,\ ((1+\alpha\tau)I_h+\tau^2A_h^x)\frac{u_1^h-u_0^h}{\tau}=\psi^h,\\ ((1+\alpha\tau)I_h+\tau^2A_h^x)\frac{u_{mN+1}^h-u_{mN}^h}{\tau}=\frac{u_{mN}^h-u_{mN-1}^h}{\tau},\ m=1,2,\ldots\end{cases}$$

in a Hilbert space L_{2h} with self-adjoint positive definite operator A_h^x by formula (3). Here, $u_k^h=u_k^h(x)$ is unknown abstract mesh function defined on Ω_h with the values in $H=L_{2h}$. Therefore, estimates of Theorem 3 follow the theorem of the stability of difference scheme in paper [14] and the following theorem on the coercivity inequality for the solution of the elliptic difference problem in [15]. Therefore, the proof of Theorem 3 is completed. □

Theorem 4 *For the solutions of the elliptic difference problem*

$$\begin{cases}A_h^x u^h(x)=\omega^h(x),\ x\in\Omega_h,\\ D^h u^h(x)=0,\ x\in S_h,\end{cases}$$

the following coercivity inequality holds:

$$\sum_{r=1}^{n} \left\| u^h{}_{x_r \bar{x_r}} \right\|_{L_{2h}} \leq M_{10} ||\omega^h||_{L_{2h}},$$

where M_{10} does not depend on h and ω^h.

4 Numerical Results

We consider the initial-boundary value problem

$$\begin{cases} u_{tt}(t,x) + 2u_t(t,x) - u_{xx}(t,x) + u(t,x) = 0.001\left(-u_{xx}([t],x) + u([t],x)\right), \\ 0 < t < \infty,\ 0 < x < \pi, \\ u(t,x) = e^{-t}\cos(x),\ -1 \leq t \leq 0,\ 0 \leq x \leq \pi, \\ u_x(t,0) = u_x(t,\pi) = 0,\ 0 \leq t < \infty \end{cases} \tag{6}$$

for the delay telegraph differential equation with Neumann condition. By using step by step method and Fourier series method, it can be shown that the exact solution of the problem (6) is

$$u(t,x) = T_n(t)\cos(x),\ n-1 \leq t \leq n,\ n = 1, 2, \ldots,$$

where

$$T_1(t) = \frac{999}{1000}e^{-t}\cos(t) - \frac{1}{1000}e^{-t}\sin(t) + \frac{1}{1000},$$

$$T_{n+1}(t) = T_n(n)e^{-t}\cos(t) + \left(T_n(n) + T_n'(n)\right)e^{-t}\sin(t)$$
$$+\frac{T_n(n)}{1000}\left(1 - e^{-(t-n)}\cos(t-n) - e^{-(t-n)}\sin(t-n)\right),\ n = 1, 2 \ldots.$$

Using first order of accuracy difference scheme for the approximate solution of problem (6), we get the following system of equations

$$\begin{cases} \frac{u_n^{k+1}-2u_n^k+u_n^{k-1}}{\tau^2} + 2\frac{u_n^{k+1}-u_n^k}{\tau} - \frac{u_{n+1}^{k+1}-2u_n^{k+1}+u_{n-1}^{k+1}}{h^2} + u_n^{k+1} \\ = 0.001\left(-\frac{u_{n+1}^{\left[\frac{k-mN}{N+1}\right]N+mN}-2u_n^{\left[\frac{k-mN}{N+1}\right]N+mN}+u_{n-1}^{\left[\frac{k-mN}{N+1}\right]N+mN}}{h^2} + u_n^{\left[\frac{k-mN}{N+1}\right]N+mN}\right), \\ t_k = k\tau,\ N\tau = 1,\ mN \le k \le (m+1)N-1,\ m = 0,1,2,\ldots, \\ x_n = nh,\ Mh = \pi,\ 1 \le n \le M-1, \\ u_n^0 = \cos(nh),\ (1+2\tau)\frac{u_n^1-u_n^0}{\tau} + \tau^2\left(-\frac{u_{n+1}^1-2u_n^1+u_{n-1}^1}{h^2} + u_n^1\right) \\ +\tau^2\left(\frac{u_{n+1}^0-2u_n^0+u_{n-1}^0}{h^2} - u_n^0\right) = -\tau\cos(nh),\ 0 \le n \le M, \\ (1+2\tau)\frac{u_n^{mN+1}-u_n^{mN}}{\tau} + \tau^2\left(-\frac{u_{n+1}^{mN+1}-2u_n^{mN+1}+u_{n-1}^{mN+1}}{h^2} + u_n^{mN+1}\right) \\ +\tau^2\left(\frac{u_{n+1}^{mN}-2u_n^{mN}+u_{n-1}^{mN}}{h^2} - u_n^{mN}\right) = u_n^{mN} - u_n^{mN-1},\ 0 \le n \le M,\ \ m = 1,2,\ldots, \\ u_1^k - u_0^k = u_M^k - u_{M-1}^k = 0,\ mN \le k \le (m+1)N,\ m = 0,1,2,\ldots. \end{cases} \tag{7}$$

We can rewrite system (7) in the matrix form

$$\begin{cases} CU^{k+1} + DU^k + EU^{k-1} = \varphi\left(U^{\left[\frac{k-mN}{mN+1}\right]N+mN}\right),\ k = 1,2,3,\ldots, \\ U^0 = \begin{bmatrix} 1 \\ \cos(h) \\ \cdot \\ \cdot \\ \cdot \\ \cos((M-1)h) \\ -1 \end{bmatrix},\ U^1 = U^0 - \tau F^{-1}\begin{bmatrix} 0 \\ \cos(h) \\ \cdot \\ \cdot \\ \cdot \\ \cos((M-1)h) \\ 0 \end{bmatrix}, \\ U^{mN+1} = U^{mN} - F^{-1}\begin{bmatrix} 0 \\ U_1^{mN} \\ \cdot \\ \cdot \\ \cdot \\ U_{M-1}^{mN} \\ 0 \end{bmatrix} - F^{-1}\begin{bmatrix} 0 \\ U_1^{mN-1} \\ \cdot \\ \cdot \\ \cdot \\ U_{M-1}^{mN-1} \\ 0 \end{bmatrix},\ m = 1,2,\ldots, \end{cases} \tag{8}$$

where C, D, E, and F are $(M+1)\times(M+1)$ matrices, $\varphi\left(U^{\left[\frac{k-mN}{mN+1}\right]N+mN}\right)$ and U^r, $r=k,k\pm 1$ are $(M+1)\times 1$ column vectors defined by

$$C=\begin{bmatrix} -1 & 1 & 0 & 0 & . & 0 & 0 & 0 & 0 \\ a & b & a & 0 & . & 0 & 0 & 0 & 0 \\ 0 & a & b & a & . & 0 & 0 & 0 & 0 \\ . & . & . & . & . & . & . & . & . \\ 0 & 0 & 0 & 0 & . & a & b & a & 0 \\ 0 & 0 & 0 & 0 & . & 0 & a & b & a \\ 0 & 0 & 0 & 0 & . & 0 & 0 & -1 & 1 \end{bmatrix}_{(M+1)\times(M+1)},$$

$$D=\begin{bmatrix} 0 & 0 & 0 & . & 0 & 0 & 0 & 0 \\ 0 & c & 0 & . & 0 & 0 & 0 & 0 \\ 0 & 0 & c & . & 0 & 0 & 0 & 0 \\ . & . & . & . & . & . & . & . \\ 0 & 0 & 0 & . & 0 & c & 0 & 0 \\ 0 & 0 & 0 & . & 0 & 0 & c & 0 \\ 0 & 0 & 0 & . & 0 & 0 & 0 & 0 \end{bmatrix}_{(M+1)\times(M+1)},$$

$$E=\begin{bmatrix} 0 & 0 & 0 & . & 0 & 0 & 0 & 0 \\ 0 & d & 0 & . & 0 & 0 & 0 & 0 \\ 0 & 0 & d & . & 0 & 0 & 0 & 0 \\ . & . & . & . & . & . & . & . \\ 0 & 0 & 0 & . & 0 & d & 0 & 0 \\ 0 & 0 & 0 & . & 0 & 0 & d & 0 \\ 0 & 0 & 0 & . & 0 & 0 & 0 & 0 \end{bmatrix}_{(M+1)\times(M+1)},$$

$$F=\begin{bmatrix} -1 & 1 & 0 & 0 & . & 0 & 0 & 0 & 0 \\ e & p & e & 0 & . & 0 & 0 & 0 & 0 \\ 0 & e & p & e & . & 0 & 0 & 0 & 0 \\ . & . & . & . & . & . & . & . & . \\ 0 & 0 & 0 & 0 & . & e & p & e & 0 \\ 0 & 0 & 0 & 0 & . & 0 & e & p & e \\ 0 & 0 & 0 & 0 & . & 0 & 0 & -1 & 1 \end{bmatrix}_{(M+1)\times(M+1)},$$

$$\varphi\left(U^{\left[\frac{k-mN}{mN+1}\right]N+mN}\right)=\begin{bmatrix} 0 \\ \varphi_1^k \\ . \\ . \\ . \\ \varphi_{M-1}^k \\ 0 \end{bmatrix}_{(M+1)\times 1},\quad U^r=\begin{bmatrix} U_0^r \\ U_1^r \\ . \\ . \\ . \\ U_{M-1}^r \\ U_M^r \end{bmatrix}_{(M+1)\times 1}\qquad \text{for } r=k,k\mp 1,$$

Table 1 Errors of difference scheme (7)

	N = M = 40	N = M = 80	N = M = 160
$t \in [0, 1]$	0.015476	0.007927	0.004026
$t \in [1, 2]$	0.013674	0.006878	0.003453
$t \in [2, 3]$	0.010620	0.005326	0.002685

where $\varphi_n^k = -0.001 \frac{u_{n+1}^{\left[\frac{k-mN}{N+1}\right]N+mN} - 2u_n^{\left[\frac{k-mN}{N+1}\right]N+mN} + u_{n-1}^{\left[\frac{k-mN}{N+1}\right]N+mN}}{h^2} + 0.001 u_n^{\left[\frac{k-mN}{N+1}\right]N+mN}$ for $k = 1, 2, \ldots, 1 \leq n \leq M - 1$. Here, we denote $a = -1/h^2$, $b = 1/\tau^2 + 2/\tau + 2/h^2 + 1$, $c = -2/\tau^2 - 2/\tau$, $d = 1/\tau^2$, $e = -\tau^2/h^2$ and $p = 1 + 2\tau + \tau^2 + 2\tau^2/h^2$. Hence, we have a second order of difference equation with matrix coefficients. We find the numerical solutions for different values of N and M and here, u_n^k represents the numerical solutions of the difference scheme at (t_k, x_n).

For $N = M = 40$, $N = M = 80$ and $N = M = 160$ in $t \in [0, 1]$, $t \in [1, 2]$ and $t \in [2, 3]$, the errors computed by the following formula are given in Table 1.

$$E_M^N = \max_{\substack{mN+1 \leq k \leq (m+1)N,\ m = 0, 1, \ldots \\ 0 \leq n \leq M}} \left| u(t_k, x_n) - u_n^k \right|$$

As it is seen in Table 1, the errors in the first order of accuracy difference scheme decrease approximately by a factor of 1/2 when the values of M and N are doubled.

5 Conclusion

In this study, we study the initial-boundary value problem for multidimensional delay telegraph equation with Neumann condition. Theorem on stability estimates for the solution of this problem is proved. We construct the difference scheme for this problem and establish stability estimates for solution of this difference scheme. Initial-boundary value problem for one dimensional delay telegraph equation with Neumann condition is considered as numerical experiment.

Acknowledgements We are grateful to the TUBITAK Graduate Scholarship Programme for supporting Koray Turk. The publication has been prepared with the support of the "RUDN University Program 5–100".

References

1. Lamb, H.: Hydrodynamics, 6th edn. Cambridge University Press, Cambridge (1993)
2. Heaviside, O.: Electromagnetic Theory. Cambridge University Press, Cambridge (2011)
3. Banasiak, J., Mika, J.R.: Singularly perturbed telegraph equations with applications in the random walk theory. J. Appl. Math. Stoch. Anal. **11**, 9–28 (1997)
4. Pogorui, A.A., Rodriquez-Dagnino, R.M.: Goldstein-Kac telegraph equations and random flights in higher dimensions. Appl. Math. Comput. **361**, 617–629 (2019)
5. Leonenko, N., Vaz Jr., J.: Spectral analysis of fractional hyperbolic diffusion equations with random data. J. Stat. Phys. **179**, 155–175 (2020)
6. Hassani, H., Avazzadeh, Z., Machado, J.A.T.: Numerical approach for solving variable-order space-time fractional telegraph equation using transcendental Bernstein series. Eng. Comput. **36**, 867–878 (2020)
7. Zhou, Y., Qu, W., Gu, Y., Gao, H.: Numerical approach for solving variable-order space-time fractional telegraph equation using transcendental Bernstein series. Eng. Anal. Boundary Elem. **115**, 21–27 (2020)
8. Singh, S., Devi, V., Tohidi, E., Singh, V.K.: An efficient matrix approach for two-dimensional diffusion and telegraph equations with Dirichlet boundary conditions. Phys. A-Stat. Mech. Its Appl. **545**, Article Number 123784 (2020)
9. Ashyralyev, A., Modanli, M.: An operator method for telegraph partial differential and difference equations. Bound. Value Probl. **2015**, Article Number 41 (2015)
10. Lu, X.: Iterative methods for numerical solutions of parabolic problems with time delays. Appl. Math. Comput. **89**, 213–224 (1998)
11. Ashyralyev, A., Agirseven, D.: Bounded solutions of semilinear time delay hyperbolic differential and difference equations. Mathematics **7**(12), Article Number 1163 (2019)
12. Shang, Y.: On the delayed scaled consesus problems. Appl. Sci. **7**(7), Article Number 713 (2017)
13. Ashyralyev, A., Agirseven, D., Turk, K.: On the stability of the telegraph equation with time delay. AIP Conf. Proc. **1759**, Article Number 020022 (2016)
14. Ashyralyev, A., Turk, K., Agirseven, D.: On the stable difference scheme for the time delay telegraph equation. Bull. Karaganda Univ.-Math. **3** (2020). (In press)
15. Sobolevskii, P.E.: Difference Methods for The Approximate Solution of Differential Equations. Izdat Voronezh Gosud University, Voronezh, Russia (1975)

A Note on a Hyperbolic-Parabolic Problem with Involution

Maksat Ashyraliyev and Maral A. Ashyralyyeva

Abstract In the present paper, a boundary value problem for a one-dimensional hyperbolic-parabolic equation with involution and the Dirichlet condition is studied. The stability estimates for the solution of the hyperbolic-parabolic problem are established. The first order of accuracy stable difference scheme for the approximate solution of the problem under consideration is constructed. Numerical algorithm for implementation of this scheme is presented. Numerical results are provided for a simple test problem.

Keywords Hyperbolic-parabolic equation · Involution · Difference scheme · Stability

1 Introduction

Local and nonlocal boundary value problems for mixed type partial differential equations are used in mathematical modeling of various real-life systems. Theory of mixed type differential equations have been extensively studied by researchers (see, e.g., [10, 12, 15] and the references given therein). In particular, the theory of boundary value problems for hyperbolic-parabolic differential equations as well as the numerical methods for their approximate solutions have been the subject of recent studies (see, [2, 3, 8, 9, 13] and the references therein). Partial differential equations with the involution have been recently investigated in [1, 4–6, 11]. Nonetheless, hyperbolic-parabolic problems with involution have not been investigated yet.

M. Ashyraliyev (✉)
Department of Software Engineering, Bahcesehir University, Istanbul 34353, Turkey
e-mail: maksat.ashyralyyev@eng.bau.edu.tr

M. A. Ashyralyyeva
Department of Applied Mathematics and Informatics, Magtymguly Turkmen State University, Ashgabat, Turkmenistan
e-mail: ashyrmaral2010@mail.ru

A. Ashyralyev et al. (eds.), *Functional Analysis in Interdisciplinary Applications—II*, Springer Proceedings in Mathematics & Statistics 351,
https://doi.org/10.1007/978-3-030-69292-6_16

The present paper is devoted to the study of boundary value problems for hyperbolic-parabolic differential and difference equations with involution. The stability of these boundary value problems is established. Numerical results are presented for a simple test problem.

2 Stability of Differential Equation

We consider the following problem

$$\begin{cases} u_{tt}(t,x) - \big(a(x)u_x(t,x)\big)_x - \beta\big(a(-x)u_x(t,-x)\big)_x + \delta u(t,x) = f(t,x), \\ \quad -\ell < x < \ell,\ 0 < t < 1, \\ u_t(t,x) - \big(a(x)u_x(t,x)\big)_x - \beta\big(a(-x)u_x(t,-x)\big)_x + \delta u(t,x) = g(t,x), \\ \quad -\ell < x < \ell,\ \ -1 < t < 0, \\ u(0^+,x) = u(0^-,x),\ u_t(0^+,x) = u_t(0^-,x),\ \ -\ell \le x \le \ell, \\ u(t,-\ell) = u(t,\ell) = 0,\ \ -1 \le t \le 1, \\ u(-1,x) = \varphi(x),\ \ -\ell \le x \le \ell \end{cases} \tag{1}$$

for one-dimensional hyperbolic-parabolic differential equation with involution and Dirichlet boundary condition. Throughout this paper, we assume that

$$\overline{a} \ge a(x) = a(-x) \ge \underline{a} > 0,\ x \in (-\ell,\ell),\quad \underline{a} - \overline{a}|\beta| \ge 0.$$

Under compatibility conditions, problem (1) has a unique smooth solution $u(t,x)$ for the given smooth functions $a(x)$, $\varphi(x)$, $f(t,x)$, $g(t,x)$ and constants $\delta > 0$ and β.

Let the Sobolev space $W_2^2[-\ell,\ell]$ be defined as the set of all functions $v(x)$ defined on $[-\ell,\ell]$ such that $v(x)$ and the second order derivative function $v''(x)$ are both locally integrable in $L_2[-\ell,\ell]$, equipped with the norm

$$\|v(x)\|_{W_2^2[-\ell,\ell]} = \left(\int_{-\ell}^{\ell} |v(x)|^2 dx\right)^{1/2} + \left(\int_{-\ell}^{\ell} \left|v''(x)\right|^2 dx\right)^{1/2}.$$

Theorem 1 *Suppose that $\varphi \in W_2^2[-\ell,\ell]$. Let $f(t,x)$ and $g(t,x)$ be continuously differentiable functions in t on $[0,1] \times [-\ell,\ell]$ and $[-1,0] \times [-\ell,\ell]$, respectively. Then the solution of the problem (1) satisfies the following stability estimates*

$$\|u\|_{C([-1,1],L_2[-\ell,\ell])} \le M_1\Big[\|\varphi\|_{L_2[-\ell,\ell]} + \|f\|_{C([0,1],L_2[-\ell,\ell])} + \|g\|_{C([-1,0],L_2[-\ell,\ell])}\Big],$$

$$\begin{aligned} &\|u\|_{C^{(2)}([0,1],L_2[-\ell,\ell])} + \|u\|_{C^{(1)}([-1,0],L_2[-\ell,\ell])} + \|u\|_{C([-1,1],W_2^2[-\ell,\ell])} \\ &\le M_2\Big[\|\varphi\|_{W_2^2[-\ell,\ell]} + \|f\|_{C^{(1)}([0,1],L_2[-\ell,\ell])} + \|g\|_{C^{(1)}([-1,0],L_2[-\ell,\ell])}\Big], \end{aligned}$$

where M_1 and M_2 do not depend on $\varphi(x)$, $f(t,x)$ and $g(t,x)$.

Proof Problem (1) can be written in the following abstract form

$$\begin{cases} u''(t) + Au(t) = f(t),\ 0 < t < 1, \\ u'(t) + Au(t) = g(t),\ \ -1 < t < 0, \\ u(0^+) = u(0^-),\ u'(0^+) = u'(0^-), \\ u(-1) = \varphi \end{cases}$$

in a Hilbert space $L_2[-\ell,\ell]$ with self-adjoint positive definite operator $A = A^x$ defined by the formula

$$A^x u(x) = -\big(a(x)u_x(x)\big)_x - \beta\big(a(-x)u_x(-x)\big)_x + \delta u(x) \tag{2}$$

with the domain $D(A^x) = \left\{u \in W_2^2[-\ell,\ell] \,\middle|\, u(-\ell) = u(\ell) = 0\right\}$. Here, $f(t) = f(t,x)$ and $g(t) = g(t,x)$ are given abstract functions and $u(t) = u(t,x)$ is an unknown function. Then, the proof of Theorem 1 is based on the self-adjointness and positive definiteness of the space operator A^x (see [4]).

3 Stability of Difference Scheme

We construct the first order of accuracy stable difference scheme for the approximate solution of problem (1).

The discretization of problem (1) is carried out in two steps. In the first step, the spatial discretization of problem (1) is obtained. We define the grid space

$$[-\ell,\ell]_h = \left\{x = x_n \,\middle|\, x_n = nh,\ \ -M \le n \le M,\ Mh = \ell\right\}.$$

We introduce the Hilbert space $L_{2h} = L_2([-\ell,\ell]_h)$ of the grid functions $\phi^h(x) = \{\phi^n\}_{-M}^{M}$ defined on $[-\ell,\ell]_h$, equipped with the norm

$$\|\phi^h\|_{L_{2h}} = \left(\sum_{x\in[-\ell,\ell]_h} |\phi^h(x)|^2 h\right)^{1/2}.$$

To the differential operator A^x defined by the formula (2), we assign the difference operator A_h^x by the formula

$$A_h^x\phi^h(x) = \left\{-\big(a(x)\phi_{\bar{x}}^n\big)_x - \beta\big(a(-x)\phi_{\bar{x}}^{-n}\big)_x + \delta\phi^n\right\}_{-M+1}^{M-1}, \tag{3}$$

acting in the space of grid functions $\phi^h(x) = \{\phi^n\}_{-M}^{M}$ and satisfying the conditions $\phi_{-M} = \phi_M = 0$. Here,

$$\phi_{\bar{x}}^n = \frac{\phi^n - \phi^{n-1}}{h}, \quad -M+1 \le n \le M, \quad \phi_x^n = \frac{\phi^{n+1} - \phi^n}{h}, \quad -M \le n \le M-1.$$

It is known that A_h^x, defined by (3), is a self-adjoint positive definite operator in L_{2h}. With the help of A_h^x, the first discretization step results in the following problem

$$\begin{cases} u_{tt}^h(t,x) + A_h^x u^h(t,x) = f^h(t,x), \ x \in [-\ell,\ell]_h, \ 0 < t < 1, \\ u_t^h(t,x) + A_h^x u^h(t,x) = g^h(t,x), \ x \in [-\ell,\ell]_h, \ -1 < t < 0, \\ u^h(0^+,x) = u^h(0^-,x), \ u_t^h(0^+,x) = u_t^h(0^-,x), \ x \in [-\ell,\ell]_h, \\ u^h(-1,x) = \varphi^h(x), \ x \in [-\ell,\ell]_h. \end{cases} \tag{4}$$

In the second step, we replace the problem (4) with the first order of accuracy difference scheme in t

$$\begin{cases} \dfrac{u_{k+1}^h(x) - 2u_k^h(x) + u_{k-1}^h(x)}{\tau^2} + A_h^x u_{k+1}^h(x) = f_k^h(x), \ 1 \le k \le N-1, \\ \dfrac{u_k^h(x) - u_{k-1}^h(x)}{\tau} + A_h^x u_k^h(x) = g_k^h(x), \ -N+1 \le k \le 0, \\ u_1^h(x) - u_0^h(x) = u_0^h(x) - u_{-1}^h(x), \ u_{-N}^h(x) = \varphi^h(x), \\ f_k^h(x) = f^h(t_k,x), \ 1 \le k \le N-1, \ g_k^h(x) = g(t_k,x), \ -N+1 \le k \le 0, \end{cases} \tag{5}$$

where $x \in [-\ell,\ell]_h$, $\tau = \dfrac{1}{N}$ and $t_k = k\tau$, $-N \le k \le N$.

Theorem 2 *Let τ and h be sufficiently small numbers. For the solution $\{u_k^h(x)\}_{-N}^{N}$ of problem (5) the following stability estimates*

$$\max_{-N \le k \le N} \|u_k\|_{L_{2h}} \le \tilde{M}_1 \left[\|\varphi^h\|_{L_{2h}} + \max_{-N+1 \le k \le 0} \|g_k^h\|_{L_{2h}} + \max_{1 \le k \le N-1} \|f_k^h\|_{L_{2h}} \right],$$

$$\max_{1 \le k \le N-1} \left\| \frac{u_{k+1}^h - 2u_k^h + u_{k-1}^h}{\tau^2} \right\|_{L_{2h}} + \max_{-N+1 \le k \le 0} \left\| \frac{u_k^h - u_{k-1}^h}{\tau} \right\|_{L_{2h}} + \max_{-N \le k \le N} \|u_k^h\|_{W_{2h}^2}$$
$$\le \tilde{M}_2 \left[\|\varphi^h\|_{W_{2h}^2} + \|g_0^h\|_{L_{2h}} + \max_{-N+1 \le k \le -1} \left\| \frac{g_k^h - g_{k-1}^h}{\tau} \right\|_{L_{2h}} + \|f_1^h\|_{L_{2h}} \right.$$
$$\left. + \max_{2 \le k \le N-1} \left\| \frac{f_k^h - f_{k-1}^h}{\tau} \right\|_{L_{2h}} \right]$$

hold, where $\tilde{M}_1$ and $\tilde{M}_2$ do not depend on τ, h, $f_k^h(x)$, $g_k^h(x)$ and $\varphi^h(x)$.

Proof Difference scheme (5) can be written in the following abstract form

$$\begin{cases} \dfrac{u_{k+1}^h - 2u_k^h + u_{k-1}^h}{\tau^2} + A_h u_{k+1}^h = f_k^h, \ 1 \le k \le N-1, \\ \dfrac{u_k^h - u_{k-1}^h}{\tau} + A_h u_k^h = g_k^h, \ \ -N+1 \le k \le 0, \\ u_1^h - u_0^h = u_0^h - u_{-1}^h, \ u_{-N}^h = \varphi^h \end{cases}$$

in a Hilbert space L_{2h} with operator $A_h = A_h^x$ defined by formula (3). Here, $f_k^h = f_k^h(x)$ and $g_k^h = g_k^h(x)$ are given abstract functions and $u_k^h = u_k^h(x)$ is an unknown mesh function. Then, the proof of Theorem 2 is based on the self-adjointness and positive definiteness of the space operator A_h in L_{2h} [7].

4 Numerical Experiments

The numerical methods for obtaining the approximate solutions of partial differential equations play an important role in applied mathematics, especially when the analytical solutions cannot be found. In this section, we apply the above described first order of accuracy difference scheme to approximate the solution of a simple test problem. We use a procedure of modified Gauss elimination method to solve the resulting problem.

We consider the following problem

$$\begin{cases} u_{tt}(t,x) - u_{xx}(t,x) - \dfrac{1}{2}\left(u_x(t,-x)\right)_x + u(t,x) = f(t,x), \\ \quad x \in (-\pi,\pi), \ t \in (0,1), \\ u_t(t,x) - u_{xx}(t,x) - \dfrac{1}{2}\left(u_x(t,-x)\right)_x + u(t,x) = g(t,x), \\ \quad x \in (-\pi,\pi), \ t \in (-1,0), \\ u(-1,x) = \varphi(x), \ x \in [-\pi,\pi], \\ u(t,-\pi) = u(t,\pi) = 0, \ t \in [-1,1] \end{cases} \tag{6}$$

for one-dimensional hyperbolic-parabolic equation with involution and the Dirichlet condition, where

$$f(t,x) = \frac{1}{2}\cos t \sin x, \ x \in (-\pi,\pi), \ t \in (0,1),$$
$$g(t,x) = \left(\frac{3}{2}\cos t - \sin t\right)\sin x, \ x \in (-\pi,\pi), \ t \in (-1,0),$$
$$\varphi(x) = \cos 1 \sin x, \ x \in [-\pi,\pi].$$

The exact solution of problem (6) is given by

$$u(t,x) = \cos t \sin x, \quad -\pi \le x \le \pi, \quad -1 \le t \le 1.$$

We define the set $[-1,1]_\tau \times [-\pi,\pi]_h$ of all grid points as following:

$$[-1,1]_\tau \times [-\pi,\pi]_h$$
$$= \left\{(t_k, x_n) \mid t_k = k\tau, -N \le k \le N, \ N\tau = 1, \ x_n = nh, -M \le n \le M, \ Mh = \pi\right\}.$$

For the numerical solution of problem (6), we construct the above described first order of accuracy difference scheme in t

$$\begin{cases} \dfrac{u_n^{k+1} - 2u_n^k + u_n^{k-1}}{\tau^2} - \dfrac{u_{n+1}^{k+1} - 2u_n^{k+1} + u_{n-1}^{k+1}}{h^2} - \dfrac{u_{-n+1}^{k+1} - 2u_{-n}^{k+1} + u_{-n-1}^{k+1}}{2h^2} \\ +u_n^{k+1} = f(t_k, x_n), \ 1 \le k \le N-1, \ -M+1 \le n \le M-1, \\ \dfrac{u_n^k - u_n^{k-1}}{\tau} - \dfrac{u_{n+1}^k - 2u_n^k + u_{n-1}^k}{h^2} - \dfrac{u_{-n+1}^k - 2u_{-n}^k + u_{-n-1}^k}{2h^2} + u_n^k \\ = g(t_k, x_n), \ -N+1 \le k \le 0, \ -M+1 \le n \le M-1, \\ u_n^1 - u_n^0 = u_n^0 - u_n^{-1}, \ u_n^{-N} = \varphi(x_n), \ -M \le n \le M, \\ u_{-M}^k = u_M^k = 0, \ -N \le k \le N, \end{cases} \tag{7}$$

where u_n^k denotes the numerical approximation of $u(t,x)$ at $(t,x) = (t_k, x_n)$. To obtain the solution of difference scheme (7), we first rewrite it in the matrix form

$$\begin{cases} AU_{n+1} + BU_n + AU_{n-1} + CU_{-n+1} + DU_{-n} + CU_{-n-1} = F_n, \\ \quad -M+1 \le n \le M-1, \\ U_{-M} = U_M = \tilde{0}, \end{cases} \tag{8}$$

where $\tilde{0}$ is $(2N+1) \times 1$ zero vector and

$$A = \begin{bmatrix} 0 & 0 & \cdots & 0 & 0 & 0 & \cdots & 0 \\ 0 & a & \cdots & 0 & 0 & 0 & \cdots & 0 \\ \vdots & \vdots & \ddots & \vdots & \vdots & \vdots & & \vdots \\ 0 & 0 & \cdots & a & 0 & 0 & \cdots & 0 \\ 0 & 0 & \cdots & 0 & 0 & 0 & \cdots & 0 \\ 0 & 0 & \cdots & 0 & 0 & b & \cdots & 0 \\ \vdots & \vdots & & \vdots & \vdots & \vdots & \ddots & \vdots \\ 0 & 0 & \cdots & 0 & 0 & 0 & \cdots & b \end{bmatrix}_{(2N+1)\times(2N+1)} \qquad F_n = \begin{bmatrix} \varphi(x_n) \\ \tau g\,(t_{-N+1}, x_n) \\ \vdots \\ \tau g\,(t_0, x_n) \\ 0 \\ \tau^2 f\,(t_1, x_n) \\ \vdots \\ \tau^2 f\,(t_{N-1}, x_n) \end{bmatrix}_{(2N+1)\times 1}$$

$$C = \begin{bmatrix} 0 & 0 & \cdots & 0 & 0 & 0 & \cdots & 0 \\ 0 & q & \cdots & 0 & 0 & 0 & \cdots & 0 \\ \vdots & \vdots & \ddots & \vdots & \vdots & \vdots & & \vdots \\ 0 & 0 & \cdots & q & 0 & 0 & \cdots & 0 \\ 0 & 0 & \cdots & 0 & 0 & 0 & \cdots & 0 \\ 0 & 0 & \cdots & 0 & 0 & r & \cdots & 0 \\ \vdots & \vdots & & \vdots & \vdots & \vdots & \ddots & \vdots \\ 0 & 0 & \cdots & 0 & 0 & 0 & \cdots & r \end{bmatrix}_{(2N+1)\times(2N+1)} \qquad U_n = \begin{bmatrix} u_n^{-N} \\ u_n^{-N+1} \\ \vdots \\ u_n^0 \\ u_n^1 \\ u_n^2 \\ \vdots \\ u_n^N \end{bmatrix}_{(2N+1)\times 1}$$

$$B = \begin{bmatrix} 1 & 0 & \cdots & 0 & 0 & 0 & 0 & \cdots & 0 & 0 & 0 \\ -1 & c & \cdots & 0 & 0 & 0 & 0 & \cdots & 0 & 0 & 0 \\ \vdots & \vdots & \ddots & \ddots & \vdots & \vdots & \vdots & & \vdots & \vdots & \vdots \\ 0 & 0 & \cdots & -1 & c & 0 & 0 & \cdots & 0 & 0 & 0 \\ 0 & 0 & \cdots & 1 & -2 & 1 & 0 & \cdots & 0 & 0 & 0 \\ 0 & 0 & \cdots & 0 & 1 & -2 & d & \cdots & 0 & 0 & 0 \\ \vdots & \vdots & & \vdots & \vdots & \vdots & \vdots & \ddots & \ddots & \ddots & \vdots \\ 0 & 0 & \cdots & 0 & 0 & 0 & 0 & \cdots & 1 & -2 & d \end{bmatrix}_{(2N+1)\times(2N+1)}$$

$$D = \begin{bmatrix} 0 & 0 & \cdots & 0 & 0 & 0 & \cdots & 0 \\ 0 & s & \cdots & 0 & 0 & 0 & \cdots & 0 \\ \vdots & \vdots & \ddots & \vdots & \vdots & \vdots & & \vdots \\ 0 & 0 & \cdots & s & 0 & 0 & \cdots & 0 \\ 0 & 0 & \cdots & 0 & 0 & 0 & \cdots & 0 \\ 0 & 0 & \cdots & 0 & 0 & \sigma & \cdots & 0 \\ \vdots & \vdots & & \vdots & \vdots & \vdots & \ddots & \vdots \\ 0 & 0 & \cdots & 0 & 0 & 0 & \cdots & \sigma \end{bmatrix}_{(2N+1)\times(2N+1)} \qquad U_{-n} = \begin{bmatrix} u_{-n}^{-N} \\ u_{-n}^{-N+1} \\ \vdots \\ u_{-n}^{0} \\ u_{-n}^{1} \\ u_{-n}^{2} \\ \vdots \\ u_{-n}^{N} \end{bmatrix}_{(2N+1)\times 1}$$

with $a = -\frac{\tau}{h^2}$, $b = -\frac{\tau^2}{h^2}$, $c = 1 + \frac{2\tau}{h^2} + \tau$, $d = 1 + \frac{2\tau^2}{h^2} + \tau^2$, $q = -\frac{\tau}{2h^2}$, $r = -\frac{\tau^2}{2h^2}$, $s = \frac{\tau}{h^2}$, $\sigma = \frac{\tau^2}{h^2}$. Next, we rewrite the system (8) as following

$$\begin{cases} \tilde{A} Z_{n+1} + \tilde{B} Z_n + \tilde{A} Z_{n-1} = \psi_n, \ 1 \le n \le M-1, \\ \tilde{C} Z_1 + \tilde{B} Z_0 = \psi_0, \\ Z_M = \hat{0}, \end{cases} \tag{9}$$

where $\tilde{A} = \begin{bmatrix} A & C \\ C & A \end{bmatrix}$, $\tilde{B} = \begin{bmatrix} B & D \\ D & B \end{bmatrix}$ and $\tilde{C} = \tilde{A} + \begin{bmatrix} C & A \\ A & C \end{bmatrix}$ are $(4N+2) \times (4N+2)$ matrices, $Z_n = \begin{bmatrix} U_n \\ U_{-n} \end{bmatrix}$ and $\psi_n = \begin{bmatrix} F_n \\ F_{-n} \end{bmatrix}$ are $(4N+2) \times 1$ column vectors, $\hat{0}$ is $(4N+2) \times 1$ zero vector. Now, the matrix equation (9) can be solved by using the modified Gauss elimination method [14]. We seek a solution of the matrix equation (9) in the following form:

$$\begin{cases} Z_n = \alpha_{n+1} Z_{n+1} + \beta_{n+1}, \ n = M-1, \ldots, 2, 1, \\ Z_M = \hat{0}, \end{cases}$$

where α_n are $(4N+2) \times (4N+2)$ square matrices and β_n are $(4N+2) \times 1$ column vectors, calculated by

Table 1 The errors between the exact solution of the problem (6) and the numerical solutions computed by using the first order of accuracy difference scheme (7) for different values of $h = \frac{\pi}{M}$ and $\tau = \frac{1}{N}$

	$\|E_u\|_\infty$	Order
$N = M = 20$	3.6282×10^{-2}	–
$N = M = 40$	1.8961×10^{-2}	0.936
$N = M = 80$	9.6920×10^{-3}	0.968
$N = M = 160$	4.8997×10^{-3}	0.984
$N = M = 320$	2.4634×10^{-3}	0.992

$$\begin{cases} \alpha_{n+1} = -\left(\tilde{B} + \tilde{A}\alpha_n\right)^{-1}\tilde{A}, \\ \beta_{n+1} = \left(\tilde{B} + \tilde{A}\alpha_n\right)^{-1}(\psi_n - \tilde{A}\beta_n) \end{cases}$$

for $n = 1, 2, \ldots, M-1$. Here, $\alpha_1 = -\tilde{B}^{-1}\tilde{C}$ and $\beta_1 = \tilde{B}^{-1}\psi_0$.

The numerical solutions of the first order of accuracy difference scheme (7) are computed for different values of M and N by using the above described algorithm. We measure the error between the exact and numerical solutions by

$$\|E_u\|_\infty = \max_{\substack{-N+1 \le k \le N-1 \\ -M+1 \le n \le M-1}} \left|u(t_k, x_n) - u_n^k\right|,$$

where $u(t_k, x_n)$ is the exact value of $u(t, x)$ at (t_k, x_n) and u_n^k represents the corresponding numerical solution. Table 1 shows the errors between the exact solution of the problem (6) and the numerical solutions of the first order of accuracy difference scheme (7). We observe that the scheme (7) has the first order convergence as it is expected to be.

5 Conclusion

In the present paper, the stability of boundary value problems for hyperbolic-parabolic differential and difference equations with involution and the Dirichlet condition is established. Numerical illustration for a simple test problem is provided to support our findings.

We note that by using the same method one can easily obtain the similar results for boundary value problem for one-dimensional hyperbolic-parabolic equation with involution and the Neumann condition.

Finally, we would like to emphasize that the scheme (7) could be solved, in principle, without using the procedure of modified Gauss elimination method. However, the numerical algorithm which we have introduced here has an important advantage compared to other methods if instead of last condition in (1) we had a nonlocal condition. So, this method can be readily used for nonlocal hyperbolic-parabolic problems with involution.

References

1. Ashyralyev, A., Karabaeva, B., Sarsenbi, A.M.: Stable difference scheme for the solution of an elliptic equation with involution. AIP Conf. Proc. **1759**, 020111 (2016)
2. Ashyralyev, A., Ozdemir, Y.: Stability of difference schemes for hyperbolic-parabolic equations. Comput. Math. Appl. **50**, 1443–1476 (2005)
3. Ashyralyev, A., Ozdemir, Y.: On nonlocal boundary value problems for hyperbolic-parabolic equations. Taiwanese J. Math. **11**, 1075–1089 (2007)
4. Ashyralyev, A., Sarsenbi, A.M.: Well-posedness of an elliptic equation with involution. Electron. J. Differ. Equ. **2015**(284), 1–8 (2015)
5. Ashyralyev, A., Sarsenbi, A.M.: Stability of a hyperbolic equation with the involution. Springer Proc. Math. Stat. **216**, 204–212 (2016)
6. Ashyralyev, A., Sarsenbi, A.: Well-posedness of a parabolic equation with involution. Numer. Funct. Anal. Optim. **38**, 1295–1304 (2017)
7. Ashyralyev, A., Sobolevskii, P.E.: New Difference Schemes for Partial Differential Equations. Operator Theory Advances and Applications. Birkhauser, Basel (2004)
8. Ashyralyev, A., Yurtsever, A.: On a nonlocal boundary value problem for semilinear hyperbolic-parabolic equations. Nonlinear Anal. **47**, 3585–3592 (2001)
9. Berdyshev, A.S., Cabada, A., Karimov, E.T., Akhtaeva, N.S.: On the Volterra property of a boundary problem with integral gluing condition for a mixed parabolic-hyperbolic equation. Bound. Value Probl. **2013**, 94 (2013)
10. Bitsadze, A.V.: Equations of Mixed Type. Pergamon Press, Oxford (1964)
11. Cabada, A., Tojo, F., Adrian, F.: Differential Equations with Involutions. Atlantis Press, Amsterdam (2015)
12. Rassias, J.M.: Lecture Notes on Mixed Type Partial Differential Equations. World Scientific, Singapore (1990)
13. Sadybekov, M.A.: Stable difference scheme for a nonlocal boundary value heat conduction problem. e-J. Anal. Appl. Math. **2018**, 1–10 (2018)
14. Samarskii, A.A., Nikolaev, E.S.: Numerical Methods for Grid Equations: Iterative Methods. Birkhauser, Basel (1989)
15. Smirnov, M.M.: Equations of Mixed Type. American Mathematical Society, Providence (1978)

Numerical Solution of a Parabolic Source Identification Problem with Involution and Neumann Condition

Allaberen Ashyralyev and Abdullah S. Erdogan

Abstract In this paper, a space source identification problem for parabolic equation with involution and Neumann condition is studied. The well-posedness theorem on the differential equation of the source identification parabolic problem is established. For the approximate solution of the problem, a stable difference scheme and its stability estimates are presented. The theoretical results are supported by the numerical results of a test problem.

Keywords Source identification problem · Involution · Stability estimates · Finite difference method

1 Introduction

Numerical solutions and theory of various source identification problems for partial differential equations have been studied by several authors (for instance, see [1–9]). Also, the theory and applications of several source identification problems for hyperbolic-parabolic equations have been investigated in [10–13] and the references therein. Differential equations with involution, well-posedness of a parabolic equation with involution, and stability of a hyperbolic equation with the involution were investigated in [14–18]. However, source identification problems for parabolic equations with involution need to be investigated.

A. Ashyralyev
Department of Mathematics, Near East University, Nicosia TRNC Mersin 10, Turkey
e-mail: allaberen.ashyralyev@neu.edu.tr

Friendship' University of Russia (RUDN University), Ul Miklukho Maklaya 6, Moscow 117198, Russia

Institute of Mathematics and Mathematical Modeling, 050010 Almaty, Kazakhstan

A. S. Erdogan (✉)
Palm Beach State College, Palm Beach Gardens, FL 33410, USA
e-mail: aserdogan@gmail.com

A. Ashyralyev et al. (eds.), *Functional Analysis in Interdisciplinary Applications—II*,
Springer Proceedings in Mathematics & Statistics 351,
https://doi.org/10.1007/978-3-030-69292-6_17

This paper is devoted to the study of a space source identification problem for parabolic equation with involution and Neumann condition. The well-posedness theorem on the differential equation of the source identification parabolic problem is proved. For the approximate solution of this problem, a stable difference scheme is constructed. Moreover, stability estimates for the difference scheme of the source identification parabolic problem are established. Theoretical results are supported by a numerical experiment.

2 The Differential Problem

In this paper, for the one dimensional parabolic differential equation we consider the space source identification problem with involution and Neumann condition

$$\begin{cases} u_t(t,x) - (a(x)u_x(t,x))_x - \beta\left(a(-x)u_x(t,-x)\right)_x + \sigma u(t,x) \\ = p(x) + f(t,x), \quad -l < x < l, \ 0 < t < T, \\ u(0,x) = \xi(x), u(T,x) = \psi(x), \quad -l \le x \le l, \\ u_x(t,-l) = u_x(t,l) = 0, \ 0 \le t \le T. \end{cases} \tag{1}$$

This problem has a unique solution $(u(t,x), p(x))$ for the smooth functions $f(t,x)$ $((t,x) \in (0,T) \times (-l,l))$, $a \ge a(x) = a(-x) \ge \delta > 0$, $\delta - a|\beta| \ge 0$ $(x \in (-l,l))$, $\xi(x)$, $\psi(x)$, $x \in [-l,l]$, and $\sigma > 0$.

Throughout the paper, we use the following spaces:

- $C_0^\alpha([0,T], H)$ $(0 < \alpha < 1)$ is the Banach space of all abstract continuous functions $\varphi(t)$ defined on $[0,T]$ with values in H satisfying a Hölder condition with weight t^α for which the following norm is finite

$$\|\varphi\|_{C_0^\alpha([0,T],H)} = \|\varphi\|_{C([0,T],H)} + \sup_{0 \le t < t+\tau \le T} \frac{(t+\tau)^\alpha \|\varphi(t+\tau) - \varphi(t)\|_H}{\tau^\alpha}.$$

- $C([0,T], H)$ is the Banach space of all abstract continuous functions $\varphi(t)$ defined on $[0,T]$ with values in H equipped with the norm

$$\|\varphi\|_{C([0,T],H)} = \max_{0 \le t \le T} \|\varphi(t)\|_H.$$

- $W_2^2[-l,l]$ is the Sobolev space defined as the set of all functions $u(x)$ defined on $[0,l]$ such that $u(x)$ and the second order derivative function $u''(x)$ are all locally integrable in $L_2[-l,l]$, equipped with the norm

$$\|u\|_{W_2^2[-l,l]} = \left(\int_{-l}^{l} |u(x)|^2 dx\right)^{\frac{1}{2}} + \left(\int_{-l}^{l} |u''(x)|^2 dx\right)^{\frac{1}{2}}.$$

In a Hilbert space $H = L_2[-l, l]$, with self-adjoint positive definite operator $A = A^x$ defined by the formula

$$A^x u(x) = -(a(x)u_x(x))_x - \beta\,(a(-x)u_x(-x))_x + \sigma u(x) \tag{2}$$

with the domain $D(A^x) = \left\{u \in W_2^2[-l, l] : u'(-l) = u'(l) = 0\right\}$, we can write problem (1) in the abstract form as follows [14]:

$$\begin{cases} \frac{du(t)}{dt} + Au(t) = p + f(t), 0 < t < T, \\ u(0) = \xi, u(T) = \psi. \end{cases} \tag{3}$$

This abstract form allows us to give the following stability estimates.

Theorem 1 *Let $f(t, x)$ be continuously differentiable function in t on $[0, T] \times [-l, l]$ and suppose that $\xi, \psi \in W_2^2[-l, l]$. Then, for the solution of the identification problem (1), the stability estimates*

$$\|u\|_{C([0,T],L_2[-l,l])} + \left\|(A^x)^{-1}p\right\|_{L_2[-l,l]}$$

$$\leq M_1(\delta, \sigma, \beta, l)\left[\|\xi\|_{L_2[-l,l]} + \|\psi\|_{L_2[-l,l]} + \|f\|_{C([0,T],L_2[-l,l])}\right],$$

$$\|u\|_{C^{(1)}([0,T],L_2[-l,l])} + \|u\|_{C([0,T],W_2^2[-l,l])} + \|p\|_{L_2[-l,l]}$$

$$\leq M_2(\delta, \sigma, \beta, l)\left[\|\xi\|_{W_2^2[-l,l]} + \|\psi\|_{W_2^2[-l,l]} + \|f\|_{C^{(1)}([0,T],L_2[0,l])}\right]$$

are satisfied. Here, the coefficients $M_1(\delta, \sigma, \beta, l)$ and $M_2(\delta, \sigma, \beta, l)$ only depend on the given variables and do not depend on $\xi(x)$, $\psi(x)$ and $f(t, x)$.

The proof of Theorem 1 is based on the stability results given in the following theorem and also on the symmetry properties of the space operator A.

Theorem 2 ([5]) *Assume that $\xi, \psi \in D(A)$ and let $f(t)$ be continuously differentiable in t on $[0, T]$ function. Then, for the solution $\{u(t), p\}$ of the source identification problem (3), stability inequalities*

$$\|u\|_{C([0,T],H)} + \left\|A^{-1}p\right\|_H \leq M\left[\|\xi\|_H + \|\psi\|_H + \|f\|_{C([0,T],H)}\right]$$

and

$$\begin{aligned} &\|u\|_{C^{(1)}([0,T],H)} + \|Au\|_{C([0,T],H)} + \|p\|_H \\ &\leq M\left[\|A\xi\|_H + \|A\psi\|_H + \|f\|_{C^{(1)}([0,T],H)}\right] \end{aligned}$$

hold. Here, M is independent of ξ, ψ and $f(t)$.

Moreover, the solution of the identification problem (1) satisfies coercive stability estimate.

Theorem 3 *Suppose that* $\xi, \psi \in W_2^2[-l,l]$ *and* $f(t,x) \in C_0^\alpha([0,T], L_2[-l,l])$. *Then, the coercive stability estimate*

$$\|u_t\|_{C_0^\alpha([0,T],L_2[-l,l])} + \|u\|_{C_0^\alpha([0,T],W_2^2[-l,l])} + \|p\|_{L_2[-l,l]}$$

$$\leq M(\delta,\sigma,\alpha,\beta,l)\left[\|\xi\|_{W_2^2[-l,l]} + \|\psi\|_{W_2^2[-l,l]} + \|f\|_{C_0^\alpha([0,T],L_2[-l,l])}\right]$$

is valid for the solution of the identification problem (1). Here, $M(\delta,\sigma,\alpha,\beta,l)$ *is independent of* $\xi(x)$, $\psi(x)$ *and* $f(t,x)$.

The proof of Theorem 3 is based on the following abstract Theorem 4 which is on coercive stability of the identification problem (3) in $C_0^\alpha([0,T],H)$ spaces. It is also based on self-adjointness and positive definiteness of the unbounded operator A given by (2) in $L_2[-l,l]$ space.

Theorem 4 *Assume that* $\xi, \psi \in D(A)$ *and* $f \in C_0^\alpha([0,T],H)$ $(0 < \alpha < 1)$. *Then, for the solution* $\{u(t), p\}$ *of the source identification problem (3) the following coercive stability inequalities hold:*

$$\|u'\|_{C_0^\alpha([0,T],H)} + \|Au\|_{C_0^\alpha([0,T],H)} + \|p\|_H$$

$$\leq M\left[\|A\xi\|_H + \|A\psi\|_H + \frac{1}{\alpha(1-\alpha)}\|f\|_{C([0,T],H)}\right],$$

where M *is independent of* ξ, ψ, *and* $f(t)$.

3 The Difference Scheme and Its Stability

In this section, for the approximate solution of identification problem (1) the stable difference scheme is considered. Before defining the discretization and giving the theoretical results, let's introduce the following spaces:

- The Hilbert spaces $L_{2h} = L_2([-l,l]_h)$ of the grid functions $\varphi^h(x) = \{\varphi^r\}_{-M}^M$ defined on $[-l,l]_h$ is equipped with the norm

$$\left\|\varphi^h\right\|_{L_{2h}} = \left(\sum_{x\in[-l,l]_h} \left|\varphi^h(x)\right|^2 h\right)^{1/2}$$

- The Sobolev space $W_{2h}^2 = W_2^2([-l,l]_h)$ of the grid functions $\varphi^h(x) = \{\varphi^r\}_{-M}^M$ defined on $[-l,l]_h$ is equipped with the norm

$$\left\| \varphi^h \right\|_{W_{2h}^2} = \left\| \varphi^h \right\|_{L_{2h}} + \left(\sum_{x \in [-l,l]_h} \left| \left(\varphi^h\right)_{x\overline{x},j} \right|^2 h \right)^{1/2}.$$

- For a given number $\alpha \in (0, 1)$, $C_\tau(H)$ and $C_\tau^\alpha(H)$ be Banach spaces of H-valued grid functions $w_\tau = \{w_k\}_{k=0}^N$ with norms

$$\|w_\tau\|_{C_\tau(H)} = \max_{0 \le k \le N} \|w_k\|_H ,$$

$$\|w_\tau\|_{C_\tau^\alpha(H)} = \sup_{1 \le k < k+n \le N} (n\tau)^{-\alpha} (k\tau)^\alpha \|w_{k+n} - w_k\|_H + \|w_\tau\|_{C_\tau(H)} ,$$

respectively.

Now, we define the discretization of source identification problem (1). In the first step, the grid space

$$[-l, l]_h = \{x = x_n : x_n = nh, \ -M \le n \le M, \ Mh = l\}$$

is defined. To the differential operator A generated by problem (2), we assign the difference operator A_h^x by the formula

$$A_h^x \varphi^h(x) = \{-(a(x)\varphi_{\overline{x}}(x))_{x,r} - \beta\, (a(-x)\varphi_{\overline{x}}(-x))_{x,r} + \sigma \varphi_r\}_{-M+1}^{M-1}, \tag{4}$$

acting in the space of grid functions $\varphi^h(x) = \{\varphi_r\}_{-M}^M$ satisfying the conditions $\varphi_{-M} - \varphi_{-M+1} = \varphi_M - \varphi_{M-1} = 0$.

It is well-known that A_h^x is a self-adjoint positive definite operator in L_{2h}. Then, using A_h^x we reach to the following identification problem

$$\begin{cases} u_t^h(t,x) + A_h^x u^h(t,x) = p^h(x) + f^h(t,x), \ x \in [-l,l]_h, \ 0 < t < T, \\ u^h(0,x) = \xi^h(x), u^h(T,x) = \psi^h(x), \ x \in [-l,l]_h . \end{cases} \tag{5}$$

In the next step, we replace source identification problem (5) with a first order of accuracy difference scheme

$$\begin{cases} \frac{u_k^h(x) - u_{k-1}^h(x)}{\tau} + A_h^x u_k^h(x) = p^h(x) + f_k^h, \ f_k^h(x) - f(t_k, x), \\ t_k = k\tau, 1 \le k \le N, N\tau = T, x \in [-l,l]_h , \\ u_0^h(x) = \xi^h(x), u_N^h(x) = \psi^h(x), \ x \in [-l,l]_h . \end{cases} \tag{6}$$

For the solution $\left\{ \left\{ u_k^h(x) \right\}_0^N, p^h(x) \right\}$ of problem (6), Theorems 5–8 hold.

Theorem 5 *The following stability estimates*

$$\left\| \left\{ u_k^h \right\}_0^N \right\|_{C_\tau(L_{2h})} + \left\| \left(A_h^x\right)^{-1} p^h \right\|_{L_{2h}}$$

$$\leq M_3(\delta,\sigma,\beta,l)\left[\left\|\xi^h\right\|_{L_{2h}}+\left\|\psi^h\right\|_{L_{2h}}+\left\|\left\{f_k^h\right\}_1^N\right\|_{C_\tau(L_{2h})}\right],$$

$$\left\|\left\{\frac{u_k^h-u_{k-1}^h}{\tau}\right\}_1^N\right\|_{C_\tau(L_{2h})}+\left\|\left\{u_k^h\right\}_0^N\right\|_{C_\tau(W_{2h}^2)}+\left\|p^h\right\|_{L_{2h}}\leq M_4(\delta,\sigma,\beta,l)$$

$$\times\left[\left\|\xi^h\right\|_{W_{2h}^2}+\left\|\psi^h\right\|_{W_{2h}^2}+\left\|f_1^h\right\|_{L_{2h}}+\max_{2\leq k\leq N}\left\|\left\{\frac{1}{\tau}\left(f_k^h-f_{k-1}^h\right)\right\}_2^N\right\|_{L_{2h}}\right]$$

hold, where $M_3(\delta,\sigma,\beta,l)$ *and* $M_4(\delta,\sigma,\beta,l)$ *do not depend on* $\tau, h, f_k^h,\ 1\leq k\leq N,\ \xi^h(x)$ *and* $\psi^h(x)$.

Proof We can write difference scheme (6) in the abstract form as

$$\begin{cases}\frac{u_k-u_{k-1}}{\tau}+Au_k=p+f_k,\ 1\leq k\leq N,\\ u_0=\xi,\ u_N=\psi\end{cases}\tag{7}$$

in a Hilbert space $H=L_{2h}$ with operator $A=A_h^x$ by formula (4). Here, $u_k=u_k^h(x)$ is the unknown mesh function and $p=p^h(x)$ is the unknown mesh element of L_{2h}. The mesh function $f_k=f_k^h(x)$ is a given abstract function.

The proof of Theorem 6 is based on the self- adjointness and positive definiteness of the space difference operator A in L_{2h} [14] and on the following stability results.

Theorem 6 ([5]) *For the solution of the source identification difference problem (7), the following stability inequalities hold:*

$$\left\|\{u_k\}_0^N\right\|_{C_\tau(H)}+\left\|A^{-1}p\right\|_H$$

$$\leq M(\delta,\sigma,\beta,l)\left[\|\xi\|_H+\|\psi\|_H+\left\|\{f_k\}_1^N\right\|_{C_\tau(H)}\right],$$

$$\left\|\left\{\frac{u_k-u_{k-1}}{\tau}\right\}_1^N\right\|_{C_\tau(H)}+\left\|\{Au_k\}_0^N\right\|_{C_\tau(H)}+\|p\|_H$$

$$\leq M(\delta,\sigma,\beta,l)\left[\|A\xi\|_H+\|A\psi\|_H+\|f_1\|_H+\max_{2\leq k\leq N}\left\|\left\{\frac{1}{\tau}(f_k-f_{k-1})\right\}_2^N\right\|_H\right],$$

where $M(\delta,\sigma,\beta,l)$ *is independent of* ξ,ψ *and* $f(t)$.

Moreover, the solution of the identification difference problem (6) satisfies the following coercive stability estimate.

Theorem 7 *The solution of the identification difference problem (6) satisfies coercive stability estimate*

$$\left\| \left\{ \frac{u_k^h - u_{k-1}^h}{\tau} \right\}_1^N \right\|_{C_\tau^\alpha (L_{2h})} + \left\| \left\{ u_k^h \right\}_0^N \right\|_{C_\tau^\alpha (W_{2h}^2)} + \left\| p^h \right\|_{L_{2h}}$$

$$\leq M\left(\delta, \sigma, \beta, l\right) \left[\left\| \xi^h \right\|_{W_{2h}^2} + \left\| \psi^h \right\|_{W_{2h}^2} + \left\| \left\{ f_k^h \right\}_1^N \right\|_{C_\tau^\alpha (L_{2h})} \right],$$

where $M(\delta, \sigma, \beta, l)$ *does not depend on* $\tau, h, f_k^h,\ 1 \leq k \leq N,\ \xi^h(x)$ *and* $\psi^h(x)$.

The proof of Theorem 7 is based on the self-adjointness and positive definiteness of the space difference operator A in L_{2h} [14] and on the following coercive stability results.

Theorem 8 *For the solution of the source identification difference problem (7) the following coercive stability inequality holds:*

$$\left\| \left\{ \frac{u_k - u_{k-1}}{\tau} \right\}_1^N \right\|_{C_\tau^\alpha (H)} + \left\| \left\{ A u_k \right\}_0^N \right\|_{C_\tau^\alpha (H)} + \left\| p \right\|_H$$

$$\leq M\left(\delta, \sigma, \beta, l\right) \left[\left\| A\xi \right\|_H + \left\| A\psi \right\|_H + \left\| \left\{ f_k \right\}_1^N \right\|_{C_\tau^\alpha (H)} \right],$$

where $M(\delta, \sigma, \beta, l)$ *does not depend on* $\tau, h, f_k,\ 1 \leq k \leq N,\ \xi$ *and* ψ.

4 Numerical Results

A lot of problems are not analytically solvable and for obtaining approximate solution of a given problem, numerical methods play an important role. In this section, for the approximate solution of a source identification problem, a first order of accuracy difference scheme is constructed.

We consider the identification problem with the Neumann condition

$$\begin{cases} u_t(t,x) - u_{xx}(t,x) - \frac{1}{2} u_{x,x}(t,-x) + u(t,x) \\ = p(x) - \sin x + \cos t \cos x + \frac{5}{2} \sin t \cos x, \quad x \in (-\pi, \pi), t \in (0, \pi), \\ u(0,x) = 0, u(\pi, x) = 0, x \in [-\pi, \pi], \\ u_x(t, -\pi) = u_x(t, \pi) = 0, t \in [0, \pi] \end{cases} \tag{8}$$

for parabolic equation with involution. The exact solution is

$$(u(t,x), p(x)) = (\sin t \cos x, \sin x), -\pi \leq x \leq \pi, 0 \leq t \leq \pi.$$

For the numerical solution of source identification problem (8), we construct a difference scheme. First, the set $[0, \pi]_\tau \times [-\pi, \pi]_h$ of all grid points is defined by

$$[0,\pi]_\tau \times [-\pi,\pi]_h = \{(t_k, x_n) : t_k = k\tau, 0 \le k \le N,$$

$$N\tau = \pi, x_n = nh, -M \le n \le M, Mh = \pi\}.$$

Then, we present the first order of accuracy difference scheme in t

$$\begin{cases} \tau^{-1}\left(u_n^k - u_n^{k-1}\right) - h^{-2}\left(u_{n+1}^k - 2u_n^k + u_{n-1}^k\right) \\ -\frac{1}{2}h^{-2}\left(u_{-n+1}^k - 2u_{-n}^k + u_{-n-1}^k\right) + u_n^k = p_n - \sin x_n \\ +\cos t_k \cos x_n + \frac{5}{2}\sin t_k \cos x_n, 1 \le k \le N, -M+1 \le n \le M-1, \\ u_n^0 = 0, u_n^N = 0, -M \le n \le M, \\ u_{-M}^k - u_{-M+1}^k = u_M^k - u_{M-1}^k = 0, 0 \le k \le N. \end{cases}$$

To get the approximate solution, in the first step, we obtain $\left\{\left\{w_n^k\right\}_0^N\right\}_{n=-M}^M$ as solution of nonlocal boundary value problem

$$\begin{cases} \tau^{-1}\left(w_n^k - w_n^{k-1}\right) - h^{-2}\left(w_{n+1}^k - 2w_n^k + w_{n-1}^k\right) \\ -\frac{1}{2}h^{-2}\left(w_{-n+1}^k - 2w_{-n}^k + w_{-n-1}^k\right) + w_n^k \\ = -\sin x_n + \cos t_k \cos x_n + \frac{5}{2}\sin t_k \cos x_n, 1 \le k \le N, -M+1 \le n \le M-1, \\ w_n^0 - w_n^N = 0, -M \le n \le M, \\ w_{-M}^k - w_{-M+1}^k = w_M^k - w_{M-1}^k = 0, 0 \le k \le N, \end{cases} \tag{9}$$

where w_n^k denotes the numerical approximation of $w(t,x)$ at (t_k, x_n). The solution of difference scheme (9) is obtained after rewriting it in the matrix form

$$\begin{cases} Aw_{n+1} + Bw_n + Aw_{n-1} + Cw_{-n+1} + Dw_{-n} + Cw_{-n-1} = f_n, \\ Aw_{-n+1} + Bw_{-n} + Aw_{-n-1} + Cw_{n+1} + Dw_n + Cw_{n-1} = f_{-n}, \end{cases}, \tag{10}$$

$$1 \le n \le M-1, \begin{pmatrix} w_M \\ w_{-M} \end{pmatrix} = \begin{pmatrix} w_{M-1} \\ w_{-M+1} \end{pmatrix},$$

where w_s for $s = n,\ n \pm 1$, and f_n are $(N+1) \times 1$ column matrices, and $(N+1) \times (N+1)$ square matrices A, B, C, D are defined as follows:

$$A = \begin{bmatrix} 0 & 0 & 0 & . & 0 & 0 \\ 0 & \kappa & 0 & . & 0 & 0 \\ 0 & 0 & \kappa & . & 0 & 0 \\ . & . & . & . & . & . \\ 0 & 0 & 0 & . & \kappa & 0 \\ 0 & 0 & 0 & . & 0 & \kappa \end{bmatrix}, \quad B = \begin{bmatrix} 1 & 0 & 0 & . & 0 & 0 & -1 \\ \lambda & \mu & 0 & . & 0 & 0 & 0 \\ 0 & \lambda & \mu & . & 0 & 0 & 0 \\ . & . & \ddots & & \ddots & . & . \\ 0 & 0 & 0 & . & \lambda & \mu & 0 \\ 0 & 0 & 0 & . & 0 & \lambda & \mu \end{bmatrix},$$

$$C = \frac{1}{2}A, \ D = -A.$$

Here, $\kappa = -\frac{1}{h^2}$, $\lambda = -\frac{1}{\tau}$ and $\mu = \frac{1}{\tau} + \frac{2}{h^2} + 1$. Grouping the above expression (10) as

$$\begin{cases} Aw_{n+1} + Cw_{-n-1} + Bw_n + Dw_{-n} + Aw_{n-1} + Cw_{-n+1} = f_n, \\ Cw_{n+1} + Aw_{-n-1} + Dw_n + Bw_{-n} + Cw_{n-1} + Aw_{-n+1} = f_{-n}, \end{cases},$$

and defining $Z_n = \begin{pmatrix} w_n \\ w_{-n} \end{pmatrix}$ and $\phi_n = \begin{pmatrix} f_n \\ f_{-n} \end{pmatrix}$, the system can be written as

$$\begin{cases} \begin{pmatrix} A & C \\ C & A \end{pmatrix} Z_{n+1} + \begin{pmatrix} B & D \\ D & B \end{pmatrix} Z_n + \begin{pmatrix} A & C \\ C & A \end{pmatrix} Z_{n-1} = \phi_n, \ 1 \le n \le M-1, \\ Z_M = Z_{M-1}. \end{cases} \tag{11}$$

For solving the system (11), we use the Gauss elimination method. Thus, let's define

$$Z_n = \alpha_{n+1} Z_{n+1} + \beta_{n+1}, n = M-1, \ldots, 1, \tag{12}$$

where α_n $(1 \le n \le M)$ are $(2N+2) \times (2N+2)$ square matrices and β_n $(1 \le n \le M)$ are $(2N+2) \times 1$ column vectors, calculated as,

$$\begin{cases} \alpha_{n+1} = -\left(P\alpha_n + Q\right)^{-1} P, \\ \beta_{n+1} = \left(P\alpha_n + Q\right)^{-1} \left(I\phi_n - P\beta_n\right), \\ n = 1, \ldots, M-1, \end{cases} \tag{13}$$

where $P = \begin{pmatrix} A & C \\ C & A \end{pmatrix}$ and $Q = \begin{pmatrix} B & D \\ D & B \end{pmatrix}$ and I is $(2N+2) \times (2N+2)$ identity matrix.

First, we evaluate α_n and β_n $(1 \le n \le M)$. Since,

$$\phi_0 = \begin{pmatrix} f_0 \\ f_0 \end{pmatrix} = \begin{pmatrix} Aw_1 + Cw_{-1} \\ Cw_1 + Aw_{-1} \end{pmatrix} + \begin{pmatrix} Bw_0 + Dw_0 \\ Dw_0 + Bw_0 \end{pmatrix} + \begin{pmatrix} Aw_{-1} + Cw_1 \\ Cw_{-1} + Aw_1 \end{pmatrix},$$

we get

$$Z_0 = \begin{pmatrix} w_0 \\ w_0 \end{pmatrix} = \begin{pmatrix} B & D \\ D & B \end{pmatrix}^{-1} \left\{ -\begin{pmatrix} A+C & A+C \\ A+C & A+C \end{pmatrix} Z_1 + \phi_0 \right\}$$

and

$$\alpha_1 = -\begin{pmatrix} B & D \\ D & B \end{pmatrix}^{-1} \begin{pmatrix} A+C & A+C \\ A+C & A+C \end{pmatrix},$$

$$\beta_1 = \begin{pmatrix} B & D \\ D & B \end{pmatrix}^{-1} \phi_0.$$

We use the iteration (13) to obtain α_n and β_n $(1 \le n \le M)$ values.

Since $Z_M = Z_{M-1,}$ using the formula (12) we first get

$$Z_M = Z_{M-1} = \left(I - \alpha_M\right)^{-1} \beta_M.$$

Table 1 Error analysis

Errors	$\|E_p\|_\infty$	$\|E_u\|_\infty$
$N = M = 20$	0.1329	0.0760
$N = M = 40$	0.0664	0.0370
$N = M = 80$	0.0332	0.0182
$N = M = 160$	0.0166	0.0091

After then, by the formula (12), all $Z_n = \begin{pmatrix} w_n \\ w_{-n} \end{pmatrix}$ values are obtained recursively.

To get p_n, we use the formula ([5, Equation 8])

$$p_n = \frac{w_{n+1}^N - 2w_n^N + w_{n-1}^N}{h^2} + \frac{1}{2}\frac{w_{-n+1}^N - 2w_{-n}^N + w_{-n-1}^N}{h^2} - w_n^N$$

for $-M + 1 \leq n \leq M - 1$.

By the formula (for details, see [5])

$$u_n^k = w_n^k - w_n^N, n = -M, -M + 1, \ldots, M, k = 0, \ldots, N.$$

we obtain the numerical solution $\left\{ \left\{ u_n^k \right\}_{k=0}^N \right\}_{n=-M}^M$.

Let $u\,(t, x)$, $p(x)$ represent the exact solution and u_n^k represent the numerical solutions at (t_k, x_n) and p_n represent the numerical solutions at x_n. Then, the maximum error between the exact solution and numerical solution is computed by the following formulas:

$$\begin{cases} \|E_u\|_\infty = \max\limits_{0 \leq k \leq N, -M \leq n \leq M} \left| u\,(t_k, x_n) - u_n^k \right|, \\ \|E_p\|_\infty = \max\limits_{-M < n < M} \left| p\,(x_n) - p_n \right|. \end{cases}$$

In Table 1, numerical results for various N and M values are given.

5 Conclusion

In this work, we considered the source identification problem for parabolic equation with involution and Neumann condition. We proved the well-posedness theorem for the differential equation that governs the source identification parabolic problem. Also, we gave the stability estimates for the difference scheme of this source identification parabolic problem. We constructed a test problem in order to support the theoretical results. To obtain the numerical results, modified Gauss elimination method is used and the results are summarized in the above table. As expected, errors are proportional to the step size, i.e., as we double the number of steps, the error halves. So, this result supports our theoretical results.

References

1. Choulli, M., Yamamoto, M.: Generic well-posedness of a linear inverse parabolic problem with respect to diffusion parameters. J. Inverse III-Pose. P. **7**(3), 241–254 (1999)
2. Ashyralyev, A., Agirseven, D.: On source identification problem for a delay parabolic equation. Nonlinear Anal-Model. **19**(3), 335–349 (2014)
3. Ashyralyev, A., Ashyralyyev, Ch.: On the problem of determining the parameter of an elliptic equation in a Banach space. Nonlinear Anal-Model. **19**(3), 350–366 (2014)
4. Ashyralyev, A., Erdogan, A.S.: On the second order implicit difference schemes for a right hand side identification problem. Appl. Math. Comput. **226**, 212–228 (2014)
5. Ashyralyev, A., Erdogan, A.S., Demirdag, O.: On the determination of the right-hand side in a parabolic equation. Appl. Numer. Math. **62**(11), 1672–1683 (2012)
6. Ashyralyyev, C.: High order approximation of the inverse elliptic problem with Dirichlet-Neumann conditions. Filomat. **28**(5), 947–962 (2014)
7. Blasio, G.Di., Lorenzi, A.: Identification problems for parabolic delay differential equations with measurement on the boundary. J. Inverse Ill-Pose. P. **15**(7), 709–734 (2007)
8. Jator, S.: Block unification scheme for elliptic, telegraph, and Sine-Gordon partial differential equations. Am. J. Comput. Math. **5**(2), 175–185 (2015)
9. Ashyralyev, A., Sobolevskii, P.E.: New Difference Schemes for Partial Differential Equations. Birkhäuser, Basel, Boston, Berlin (2004)
10. Ashyralyev, A., Ashyralyyeva, M.A.: On source identification problem for a hyperbolic-parabolic equation. Contemp. Anal. Appl. Math. **3**(1), 88–103 (2015)
11. Ashyralyyeva, M.A., Ashyralyyev, A.: Stable difference scheme for the solution of the source identification problem for hyperbolic-parabolic equations. AIP Conf. Proc. **1676**, Article Number: 020024 (2015)
12. Ashyralyyeva, M.A., Ashyraliyev, M.: On the numerical solution of identification hyperbolic-parabolic problems with the Neumann boundary condition. Bull. Karaganda Univ.-Math. **91**(3), 69–74 (2018)
13. Ashyralyyeva, M.A., Ashyraliyev, M.: Numerical solutions of source identification problem for hyperbolic-parabolic equations. AIP Conf. Proc. **1997**, Article Number: 020048 (2018)
14. Ashyralyev, A., Sarsenbi, A.: Well-posedness of an elliptic equation with involution. Electron. J. Differ. Eq. **2015**, 1–8 (2015)
15. Ashyralyev, A., Sarsenbi, A.: Well-Posedness of a parabolic equation with involution. Numer. Func. Anal. Opt. **38**(10), 1295–1304 (2017)
16. Ashyralyev, A., Sarsenbi, A.: Stability of a hyperbolic equation with the involution. In: Kalmenov, TSh, Nursultanov, E.D., Ruzhansky, M.V., Sadybekov, M.A. (eds.) Functional Analysis in Interdisciplinary Applications. Springer Proceedings in Mathematics & Statistics, vol. 216, pp. 204–212 (2016)
17. Ashyralyev, A., Karabaeva, B., Sarsenbi, A.: Stable difference scheme for the solution of an elliptic equation with involution. AIP Conf. Proc. **1759**(1), Article Number: 020111 (2016)
18. Cabada, A., Tojo, F.: Differential Equations with Involutions. Atlantis Press (2015)

Mathematical Methods in Applied Sciences

The Structure of Fractional Spaces Generated by a Two-Dimensional Difference Neutron Transport Operator and Its Applications

Allaberen Ashyralyev and Abdulgafur Taskin

Abstract In this study, the structure of fractional spaces generated by the two-dimensional difference neutron transport operator $A_{r,h}$ defined by formula

$$A_{r,h}u^{r,h} = \left\{ \omega_1 \frac{u_m^n - u_m^{n-1}}{r} + \omega_2 \frac{u_m^{n-1} - u_{m-1}^{n-1}}{h} \right\}_{n,m=0,\pm1,\pm2,\ldots}$$

is investigated. The positivity of $A_{r,h}$ in $C\left(\mathbb{R}^2_{(r,h)}\right)$ is established. The structure of fractional spaces generated by the operator $A_{r,h}$ is studied. It is established that for any $0 < \alpha < 1$ the norms in the spaces $E_\alpha\left(C\left(\mathbb{R}^2_{(r,h)}\right), A_{r,h}\right)$ and $C^\alpha\left(\mathbb{R}^2_{(r,h)}\right)$ are equivalent. The positivity of the difference neutron transport operator in Hölder space $C^\alpha\left(\mathbb{R}^2_{(r,h)}\right)$ is proved.

Keywords Neutron transport operator · Fractional space · Positivity · Hölder space

1 Introduction

The neutron transport theory has a critical importance in nuclear engineering. It plays an important role in the design and safety of nuclear power stations. The most

---

A. Ashyralyev
Department of Mathematics, Near East University, Nicosia TRNC, Mersin 10, Turkey
e-mail: allaberen.ashyralyev@neu.edu.tr; aashyr@yahoo.com

Peoples Friendship University of Russia (RUDN University) , 6 Miklukho-Maklaya St, Moscow 117198, Russian Federation

Institute of Mathematics and Mathematical Modeling, 050010 Almaty, Kazakhstan

A. Taskin (✉)
ENKA Schools Istinye, 34460 Istanbul, Turkey
e-mail: gafur.taskin@enka.k12.tr; gafurtaskin@hotmail.com

A. Ashyralyev et al. (eds.), *Functional Analysis in Interdisciplinary Applications—II*, Springer Proceedings in Mathematics & Statistics 351,
https://doi.org/10.1007/978-3-030-69292-6_18

important equation in neutron transport theory is the neutron transport equation. We use the neutron transport equation in many physical applications, such as neutron transport, radiative transfer high frequency waves in heterogeneous, and random media and in many application of nuclear physics. The neutron transport equation describes the distribution of neutrons in terms of their positions in space and time, their energies, and their travel directions (see, [1–3], and the references therein).

The importance of the positivity property of the differential operators in a Banach space in the study of various properties for partial differential equations is well known (see, [4–9], and the references therein). Several authors have investigated the positivity of a wider class of differential and difference operators in Banach spaces (see, [4, 10–19], and the references therein).

An operator A densely defined in a Banach space E with the domain $D(A)$ is called positive in E, if its spectrum σ_A lies in the interior of the sector of angle φ, $0 < \varphi < \pi$, symmetric with respect to the real axis, and moreover on the edges of this sector $S_1(\varphi) = \{\rho e^{i\varphi} : 0 \leq \rho \leq \infty\}$ and $S_2(\varphi) = \{\rho e^{-i\varphi} : 0 \leq \rho \leq \infty\}$, and outside of the sector the resolvent $(\lambda I - A)^{-1}$ is subject to the bound [4]

$$\left\|(A - \lambda I)^{-1}\right\|_{E \to E} \leq \frac{M}{1 + |\lambda|}.$$

Throughout the present paper, M denotes positive constants, which may differ in time and, thus, is not a subject of precision. However, we will use $M(\alpha, \omega_1, \omega_2, \ldots)$ to stress the fact that the constant depends only on $\alpha, \omega_1, \omega_2, \ldots$

The infimum of all such angles φ is called the spectral angle of the positive operator A and is denoted by $\varphi(A) = \varphi(A, E)$. The operator A is said to be strongly positive in a Banach space E, if $\varphi(A, E) < \frac{\pi}{2}$.

The structure of fractional spaces generated by positive differential and difference operators and its applications to partial differential equations has been investigated by many researchers [12–15, 20–24]. Finally, a survey of results in fractional spaces generated by positive operators and their applications to partial differential equations was given in [25].

Nevertheless, structure of fractional spaces generated by difference transport operators and its applications to partial differential equations has not been investigated sufficiently. In paper [15], the structure of fractional spaces generated by the two-dimensional neutron transport operator is studied.

In this paper, the structure of fractional spaces generated by the two-dimensional difference neutron transport operator is studied. The positivity of the difference neutron transport operator in a Hölder space $C^\alpha\left(\mathbb{R}^2_{(r,h)}\right)$ is proved. In applications, new theorems on the stability of Cauchy problem for the difference neutron transport equation in a Hölder space are provided. Finally, some of these statements were formulated in [26, 27] without proof.

2 The Difference Neutron Transport Operator

We consider the resolvent equation

$$\lambda u_{m-1}^{n-1} - A_{r,h}u^{r,h} = f\left(x_{n-1}, y_{m-1}, \bar{\omega}\right), (x_n, y_m) \in \mathbb{R}^2_{(r,h)} \tag{1}$$

for the two-dimensional difference neutron transport operator $A_{r,h}$ defined by the formula

$$A_{r,h}u^{r,h} = \left\{\omega_1 \frac{u_m^n - u_m^{n-1}}{r} + \omega_2 \frac{u_m^{n-1} - u_{m-1}^{n-1}}{h}\right\}_{n,m=0,\pm1,\pm2,...}. \tag{2}$$

Here, the coefficients ω_1 and ω_2 are directional cosines of neutrons with respect to the x and y axes. $0 < r < r_o$ and $0 < h < h_o$ as the set of all points of the Euclidean space $\mathbb{R}^2_{(r,h)}$ whose coordinates are given by $x_n = nr$ and $y_m = mh$, $-\infty < n, m < \infty$. The numbers r and h are the steps of the grid spaces. Let $u^{r,h} = \left\{u_m^n\right\}_{n,m=0,\pm1,\pm2,...}$ be a grid function defined on $\mathbb{R}^2_{(r,h)}$. Then, for the solution of (1) the following formula holds:

$$\begin{aligned} &\left(\lambda I - A_{r,h}\right)^{-1} f\left(x_n, y_m, \bar{\omega}\right) \\ &= \frac{\Delta s}{\sqrt{\omega_1^2+\omega_2^2}} \sum_{i=1}^{\infty} R^i f\left(x_n + \frac{\omega_1 s_{i-1}}{\sqrt{\omega_1^2+\omega_2^2}}, y_m + \frac{\omega_2 s_{i-1}}{\sqrt{\omega_1^2+\omega_2^2}}, \bar{\omega}\right), \end{aligned} \tag{3}$$

where $\cos\varphi = \frac{\omega_1}{\sqrt{\omega_1^2+\omega_2^2}}$, $\sin\varphi = \frac{\omega_2}{\sqrt{\omega_1^2+\omega_2^2}}$, $x_i = s_i \cos\varphi + x_0$, $y_i = s_i \sin\varphi + y_0$,
$R = \left(I + \frac{\lambda \Delta s}{\sqrt{\omega_1^2+\omega_2^2}}\right)^{-1}$, $x_i - x_{i-1} = \frac{\omega_1 \Delta s}{\sqrt{\omega_1^2+\omega_2^2}} = r$, $y_i - y_{i-1} = \frac{\omega_2 \Delta s}{\sqrt{\omega_1^2+\omega_2^2}} = h$,
$\Delta s = s_i - s_{i-1}$, $1 \le i < \infty$, $x_n = nr$, $y_m = mh$, $-\infty < n, m < \infty$, for the solution of resolvent of the neutron transport operator $A_{r,h}$ defined by (2).

Let us denote $B_{r,h} = -A_{r,h}$. Then, using formula (3), we get

$$\begin{aligned} u_m^n &= \left(\lambda I + B_{r,h}\right)^{-1} f\left(x_n, y_m, \bar{\omega}\right) \\ &= \frac{\Delta s}{\sqrt{\omega_1^2+\omega_2^2}} \sum_{i=1}^{\infty} R^i f\left(x_n + \frac{\omega_1 s_{i-1}}{\sqrt{\omega_1^2+\omega_2^2}}, y_m + \frac{\omega_2 s_{i-1}}{\sqrt{\omega_1^2+\omega_2^2}}, \bar{\omega}\right), \end{aligned} \tag{4}$$

where $n, m = 0, \pm1, \pm2, \ldots$

3 Positivity of the Difference Neutron Transport Operator $B_{r,h}$ in $C\left(\mathbb{R}^2_{(r,h)}\right)$ and the Fractional Space $E_\alpha\left(C\left(\mathbb{R}^2_{(r,h)}\right), B_{r,h}\right)$

Recall that the space $C_{r,h} = C\left(\mathbb{R}^2_{(r,h)}\right)$ is the Banach space of bounded grid functions $f^{r,h} = \left\{f\left(x_n, y_m, \bar{\omega}\right)\right\}_{n,m=0,\pm1,\pm2,\dots}$ defined on $\mathbb{R}^2_{(r,h)}$ equipped with the norm

$$\|f^{r,h}\|_{C\left(\mathbb{R}^2_{(r,h)}\right)} = \sup_{(x_n,y_m)\in\mathbb{R}^2_{(r,h)}} \left|f\left(x_n, y_m, \bar{\omega}\right)\right|. \tag{5}$$

Theorem 1 *For all $\lambda > 0$, the resolvent $\left(\lambda I + B_{r,h}\right)^{-1}$ satisfies the following estimate:*

$$\left\|\left(\lambda I + B_{r,h}\right)^{-1}\right\|_{C\left(\mathbb{R}^2_{(r,h)}\right)\to C\left(\mathbb{R}^2_{(r,h)}\right)} \le \frac{M}{\lambda}. \tag{6}$$

Proof Using formula (4) and triangle inequality, we obtain

$$\left|\left(\lambda I + B_{r,h}\right)^{-1} f\left(x_n, y_m, \bar{\omega}\right)\right|$$

$$\le \sup_{(x_n,y_m)\in\mathbb{R}^2_{(r,h)}} \left|f\left(x_n, y_m, \bar{\omega}\right)\right| \frac{\Delta s}{\sqrt{\omega_1^2+\omega_2^2}} \sum_{i=1}^{\infty} |R|^i$$

for any $(x_n, y_m) \in \mathbb{R}^2_{(r,h)}$ and $\lambda > 0$. Then, we have $|R| = \frac{1}{1+\frac{\lambda\Delta s}{\sqrt{\omega_1^2+\omega_2^2}}}$ for any $\lambda > 0$.

From that it follows $\sum\limits_{i=1}^{\infty} |R|^i = \frac{1}{\frac{\lambda\Delta s}{\sqrt{\omega_1^2+\omega_2^2}}}$.

Therefore,

$$\left|\left(\lambda I + B_{r,h}\right)^{-1} f\left(x_n, y_m, \bar{\omega}\right)\right|$$

$$\le \sup_{(x_n,y_m)\in\mathbb{R}^2_{(r,h)}} \left|f\left(x_n, y_m, \bar{\omega}\right)\right| \frac{1}{\lambda}$$

for any $(x_n, y_m) \in \mathbb{R}^2_{(r,h)}$ and $\lambda > 0$. Thus, we have

$$\left\|\left(\lambda I + B_{r,h}\right)^{-1} f^{r,h}\right\|_{C\left(\mathbb{R}^2_{(r,h)}\right)} \le \tfrac{1}{\lambda} \|f^{r,h}\|_{C\left(\mathbb{R}^2_{(r,h)}\right)} \tag{7}$$

for any $\lambda > 0$. From that it follows

$$\left\| \left(\lambda I + B_{r,h} \right)^{-1} \right\|_{C\left(\mathbb{R}^2_{(r,h)}\right) \to C\left(\mathbb{R}^2_{(r,h)}\right)} \leq \frac{1}{\lambda}. \tag{8}$$

This completes the proof of Theorem 1.

Let us study the structure of the fractional space $E_\alpha \left(C \left(\mathbb{R}^2_{(r,h)} \right), \ B_{r,h} \right)$, $0 < \alpha < 1$, of all grid functions $f^{r,h} = \left\{ f \left(x_n, y_m, \bar{\omega} \right) \right\}_{n,m=0,\pm 1,\pm 2,\ldots}$ defined on $\mathbb{R}^2_{(r,h)}$ for which the following norm is finite (see [25])

$$\begin{aligned} &\left\| f^{r,h} \right\|_{E_\alpha\left(C\left(\mathbb{R}^2_{(r,h)}\right), \ B_{r,h}\right)} = \left\| f^{r,h} \right\|_{C\left(\mathbb{R}^2_{(r,h)}\right)} \\ &+ \sup_{\lambda > 0} \sup_{(x_n, y_m) \in \mathbb{R}^2_{(r,h)}} \left| \lambda^\alpha B_{r,h} \left(\lambda I + B_{r,h} \right)^{-1} f^{r,h} \left(x_n, y_m, \bar{\omega} \right) \right|. \end{aligned} \tag{9}$$

Recall that the Hölder space $C^\alpha \left(\mathbb{R}^2_{(r,h)} \right)$, $0 < \alpha < 1$ of all grid functions $f^{r,h} = \left\{ f \left(x_n, y_m, \bar{\omega} \right) \right\}_{n,m=0,\pm 1,\pm 2,\ldots}$ defined on $\mathbb{R}^2_{(r,h)}$, for which the following norm is finite (see [11])

$$\begin{aligned} &\left\| f^{r,h} \right\|_{C^\alpha\left(\mathbb{R}^2_{(r,h)}\right)} = \left\| f^{r,h} \right\|_{C\left(\mathbb{R}^2_{(r,h)}\right)} \\ &+ \sup_{(x_n, y_m), (x_n + \omega_1 r, y_m + \omega_2 h) \in \mathbb{R}^2_{(r,h)}} \frac{\left| f \left(x_n, y_m, \bar{\omega} \right) - f \left(x_n + \omega_1 r, y_m + \omega_2 h, \bar{w} \right) \right|}{\left(\sqrt{\omega_1^2 r^2 + \omega_2^2 h^2} \right)^\alpha}. \end{aligned} \tag{10}$$

Applying the definition of $E_\alpha \left(C \left(\mathbb{R}^2_{(r,h)} \right), \ B_{r,h} \right)$, we get the following estimate:

$$\begin{aligned} &\left\| \left(\lambda I + B_{r,h} \right)^{-1} \right\|_{E_\alpha\left(C\left(\mathbb{R}^2_{(r,h)}\right), \ B_{r,h}\right) \to E_\alpha\left(C\left(\mathbb{R}^2_{(r,h)}\right), \ B_{r,h}\right)} \\ &\leq \left\| \left(\lambda I + B_{r,h} \right)^{-1} \right\|_{C\left(\mathbb{R}^2_{(r,h)}\right) \to C\left(\mathbb{R}^2_{(r,h)}\right)}. \end{aligned} \tag{11}$$

From (9) and Theorem 1 it follows the following theorem:

Theorem 2 *Let* $0 < \alpha < 1$. *Then,* $B_{r,h}$ *is a positive operator in* $E_\alpha \left(C \left(\mathbb{R}^2_{(r,h)} \right), \ B_{r,h} \right)$.

The proof of this statement is based on the following theorem:

Theorem 3 *Let* $0 < \alpha < 1$. *Then, the spaces* $E_\alpha \left(C \left(\mathbb{R}^2_{(r,h)} \right), \ B_{r,h} \right)$ *and* $C^\alpha \left(\mathbb{R}^2_{(r,h)} \right)$ *coincide and their respective norms are equivalent uniformly in* r *and* h.

Proof Let us prove that

$$\left|\lambda^{\alpha} B_{r,h}\left(\lambda I+B_{r,h}\right)^{-1} f\left(x_n, y_m, \bar{\omega}\right)\right| \leq M(\alpha)\left\|f^{r,h}\right\|_{C^{\alpha}\left(\mathbb{R}^2_{(r,h)}\right)} \tag{12}$$

for $(x_n, y_m) \in \mathbb{R}^2_{(r,h)}$ and $\lambda > 0$. For any $\lambda > 0$, we have the equality

$$B_{r,h}\left(\lambda I+B_{r,h}\right)^{-1} f\left(x_n, y_m, \bar{\omega}\right)=\lambda\left[\frac{1}{\lambda}-\left(\lambda I+B_{r,h}\right)^{-1}\right] f\left(x_n, y_m, \bar{\omega}\right). \tag{13}$$

Applying formulas (4) and (13), we get

$$\begin{aligned} &\lambda^{\alpha} B_{r,h}\left(\lambda I+B_{r,h}\right)^{-1} f\left(x_n, y_m, \bar{\omega}\right) \\ &=\frac{\lambda^{1+\alpha}}{\sqrt{\omega_1^2+\omega_2^2}} \sum_{i=2}^{\infty} s_{i-1}^{\alpha} R^i \Delta s \frac{\left[f\left(x_n, y_m, \bar{\omega}\right)-f\left(x_n+\frac{\omega_1 s_{i-1}}{\sqrt{\omega_1^2+\omega_2^2}}, y_m+\frac{\omega_2 s_{i-1}}{\sqrt{\omega_1^2+\omega_2^2}}, \bar{\omega}\right)\right]}{s_{i-1}^{\alpha}}, \end{aligned} \tag{14}$$

where $\left(\sqrt{\left(\frac{\omega_1 s_{i-1}}{\sqrt{\omega_1^2+\omega_2^2}}\right)^2+\left(\frac{\omega_2 s_{i-1}}{\sqrt{\omega_1^2+\omega_2^2}}\right)^2}\right)^{\alpha}=s_{i-1}^{\alpha}$.

Using formula (14) and the triangle inequality, we obtain

$$\begin{aligned} &\left|\lambda^{\alpha} B_{r,h}\left(\lambda I+B_{r,h}\right)^{-1} f\left(x_n, y_m, \bar{\omega}\right)\right| \\ &\leq\left\|f^{r,h}\right\|_{C^{\alpha}\left(\mathbb{R}^2_{(r,h)}\right)} \frac{\lambda^{1+\alpha}}{\sqrt{\omega_1^2+\omega_2^2}} \sum_{i=2}^{\infty} s_{i-1}^{\alpha} R^i \Delta s \leq J\left\|f^{r,h}\right\|_{C^{\alpha}\left(\mathbb{R}^2_{(r,h)}\right)}, \end{aligned}$$

where $J=\frac{\lambda^{1+\alpha}}{\sqrt{\omega_1^2+\omega_2^2}} \sum_{i=2}^{\infty} s_{i-1}^{\alpha} R^i \Delta s$.

Now, let us estimate J. We have

$$J=\frac{\lambda}{\sqrt{\omega_1^2+\omega_2^2}} \sum_{i=2}^{\infty} \frac{(\lambda(i-1) \Delta s)^{\alpha}}{\left(1+\frac{\lambda \Delta s}{\sqrt{\omega_1^2+\omega_2^2}}\right)^{i-\left[\frac{i}{2}\right]}} \frac{\Delta s}{\left(1+\frac{\lambda \Delta s}{\sqrt{\omega_1^2+\omega_2^2}}\right)^{\left[\frac{i}{2}\right]}}.$$

Applying the Bernoulli inequality $(1+x)^{-\alpha} \leq \frac{1}{1+\alpha x}$, we obtain

$$\frac{1}{\left(1+\frac{\lambda \Delta s}{\sqrt{\omega_1^2+\omega_2^2}}\right)^{i-\left[\frac{i}{2}\right]}} \leq \frac{1}{\left[1+\left(i-\left[\frac{i}{2}\right]\right) \frac{\lambda \Delta s}{\sqrt{\omega_1^2+\omega_2^2}}\right]}.$$

Therefore,

$$\frac{(\lambda (i-1)\Delta s)^{\alpha}}{\left(1+\frac{\lambda \Delta s}{\sqrt{\omega_1^2+\omega_2^2}}\right)^{i-\left[\frac{i}{2}\right]}} \leq \frac{\frac{(\lambda(i-1)\Delta s)^{\alpha}}{\left(\sqrt{\omega_1^2+\omega_2^2}\right)^{\alpha}}\left(\sqrt{\omega_1^2+\omega_2^2}\right)^{\alpha}}{1+\left(i-\left[\frac{i}{2}\right]\right)\frac{\lambda \Delta s}{\sqrt{\omega_1^2+\omega_2^2}}}.$$

Since $i-\left[\frac{i}{2}\right] \geq \frac{1}{2}(i-1)$, we have that

$$\frac{\frac{(\lambda(i-1)\Delta s)^{\alpha}}{\left(\sqrt{\omega_1^2+\omega_2^2}\right)^{\alpha}}\left(\sqrt{\omega_1^2+\omega_2^2}\right)^{\alpha}}{1+\left(i-\left[\frac{i}{2}\right]\right)\frac{\lambda \Delta s}{\sqrt{\omega_1^2+\omega_2^2}}} \leq \left(\sqrt{\omega_1^2+\omega_2^2}\right)^{\alpha},$$

$$\frac{\lambda}{\sqrt{\omega_1^2+\omega_2^2}}\sum_{i=2}^{\infty}\frac{\Delta s}{\left(1+\frac{\lambda \Delta s}{\sqrt{\omega_1^2+\omega_2^2}}\right)^{\left[\frac{i}{2}\right]}} \leq \frac{2\lambda}{\sqrt{\omega_1^2+\omega_2^2}}\sum_{m=1}^{\infty}\frac{\Delta s}{\left(1+\frac{\lambda \Delta s}{\sqrt{\omega_1^2+\omega_2^2}}\right)^{m}} = 2.$$

Then, $J \leq 2$. Thus, for any $\lambda > 0$ and $(x_n, y_m) \in \mathbb{R}^2$, we established the inequality

$$\left|\lambda^{\alpha} B_{r,h}\left(\lambda I + B_{r,h}\right)^{-1} f\left(x_n, y_m, \bar{\omega}\right)\right| \leq M \left\| f^{r,h}\right\|_{C^{\alpha}\left(\mathbb{R}^2_{(r,h)}\right)}.$$

This means that

$$E_{\alpha}\left(C\left(\mathbb{R}^2_{(r,h)}\right), B_{r,h}\right) \subset C^{\alpha}\left(\mathbb{R}^2_{(r,h)}\right). \tag{15}$$

Now, let us prove the reverse inclusion. For any positive operator $B_{r,h}$, we can write

$$\int_0^{\infty}\left(\lambda I + B_{r,h}\right)^{-2} f\left(x_n, y_m, \bar{\omega}\right) d\lambda = B_{r,h}^{-1} f\left(x_n, y_m, \bar{\omega}\right).$$

Thus, we have that,

$$f\left(x_n, y_m, \bar{\omega}\right) = \int_0^{\infty}\left(\lambda I + B_{r,h}\right)^{-1} B_{r,h}\left(\lambda I + B_{r,h}\right)^{-1} f\left(x_n, y_m, \bar{\omega}\right) d\lambda.$$

From identity (4) it follows

$$f\left(x_n, y_m, \bar{\omega}\right) - f\left(x_n + \frac{\omega_1 \Delta s}{\sqrt{\omega_1^2+\omega_2^2}}, y_m + \frac{\omega_2 \Delta s}{\sqrt{\omega_1^2+\omega_2^2}}, \bar{\omega}\right)$$

$$= \int_0^{\infty}\left\{\frac{\Delta s}{\sqrt{\omega_1^2+\omega_2^2}}\sum_{i=2}^{\infty} R^i \tilde{f}\left(x_n + \frac{\omega_1 s_{i-1}}{\sqrt{\omega_1^2+\omega_2^2}}, y_m + \frac{\omega_2 s_{i-1}}{\sqrt{\omega_1^2+\omega_2^2}}, \bar{\omega}\right)\right.$$

$$-\frac{\Delta s}{\sqrt{\omega_1^2+\omega_2^2}}\sum_{i=2}^{\infty}R^i\tilde{f}\left(x_n+\frac{\omega_1(\Delta s+s_{i-1})}{\sqrt{\omega_1^2+\omega_2^2}},y_m+\frac{\omega_2(\Delta s+s_{i-1})}{\sqrt{\omega_1^2+\omega_2^2}},\bar{\omega}\right)\Bigg\}d\lambda.$$

Here, $\tilde{f}\left(x_n,y_m,\bar{\omega}\right)=B_{r,h}\left(\lambda I+B_{r,h}\right)^{-1}f\left(x_n,y_m,\bar{\omega}\right)$. Since $\Delta s=s_i-s_{i-1}$, we get

$$f\left(x_n,y_m,\bar{\omega}\right)-f\left(x_n+\frac{\omega_1\Delta s}{\sqrt{\omega_1^2+\omega_2^2}},y_m+\frac{\omega_2\Delta s}{\sqrt{\omega_1^2+\omega_2^2}},\bar{\omega}\right)$$

$$=\int_0^{\infty}\frac{\Delta s}{\sqrt{\omega_1^2+\omega_2^2}}R\tilde{f}\left(x_n,y_m,\bar{\omega}\right)d\lambda$$

$$+\frac{\Delta s}{\sqrt{\omega_1^2+\omega_2^2}}\int_0^{\infty}\sum_{i=2}^{\infty}R^i\tilde{f}\left(x_n+\frac{\omega_1 s_{i-1}}{\sqrt{\omega_1^2+\omega_2^2}},y_m+\frac{\omega_2 s_{i-1}}{\sqrt{\omega_1^2+\omega_2^2}},\bar{\omega}\right)d\lambda$$

$$-\frac{\Delta s}{\sqrt{\omega_1^2+\omega_2^2}}\int_0^{\infty}\sum_{i=2}^{\infty}R^{i-1}\tilde{f}\left(x_n+\frac{\omega_1 s_{i-1}}{\sqrt{\omega_1^2+\omega_2^2}},y_m+\frac{\omega_2 s_{i-1}}{\sqrt{\omega_1^2+\omega_2^2}},\bar{\omega}\right)d\lambda$$

$$=\int_0^{\infty}\frac{\Delta s}{\sqrt{\omega_1^2+\omega_2^2}}R\tilde{f}\left(x_n,y_m,\bar{\omega}\right)d\lambda$$

$$+\frac{\Delta s}{\sqrt{\omega_1^2+\omega_2^2}}\int_0^{\infty}\sum_{i=2}^{\infty}R^i\frac{(-1)\lambda\Delta s}{\sqrt{\omega_1^2+\omega_2^2}}\tilde{f}\left(x_n+\frac{\omega_1 s_{i-1}}{\sqrt{\omega_1^2+\omega_2^2}},y_m+\frac{\omega_2 s_{i-1}}{\sqrt{\omega_1^2+\omega_2^2}},\bar{\omega}\right)d\lambda=T_1+T_2.$$

Here,

$$T_1=\int_0^{\infty}\frac{\Delta s\lambda^{-\alpha}}{\sqrt{\omega_1^2+\omega_2^2}}R\lambda^{\alpha}B_{r,h}\left(\lambda I+B_{r,h}\right)^{-1}f\left(x_n,y_m,\bar{\omega}\right)d\lambda$$

and

$$T_2=\frac{\Delta s}{\sqrt{\omega_1^2+\omega_2^2}}\int_0^{\infty}\sum_{i=2}^{\infty}R^i\frac{(-1)\lambda^{1-\alpha}\Delta s}{\sqrt{\omega_1^2+\omega_2^2}}\lambda^{\alpha}B_{r,h}\left(\lambda I+B_{r,h}\right)^{-1}$$

$$\times f\left(x_n+\frac{\omega_1 s_{i-1}}{\sqrt{\omega_1^2+\omega_2^2}},y_m+\frac{\omega_2 s_{i-1}}{\sqrt{\omega_1^2+\omega_2^2}},\bar{\omega}\right)d\lambda.$$

Let us estimate T_1 and T_2, separately. First, let us estimate T_1. Using the triangle inequality, we obtain

$$|T_1| \leq \int_0^\infty \frac{\Delta s \lambda^{-\alpha}}{\sqrt{\omega_1^2+\omega_2^2}} \frac{1}{1+\frac{\lambda \Delta s}{\sqrt{\omega_1^2+\omega_2^2}}} d\lambda \sup_{\lambda>0} \lambda^{\alpha} \left| B_{r,h} \left(\lambda I + B_{r,h}\right)^{-1} f\left(x_n, y_m, \bar{\omega}\right) \right|.$$

The change of variable $\frac{\lambda \Delta s}{\sqrt{\omega_1^2+\omega_2^2}} = y$ yields

$$\int_0^\infty \frac{\Delta s \lambda^{-\alpha}}{\sqrt{\omega_1^2+\omega_2^2}} \frac{1}{1+\frac{\lambda \Delta s}{\sqrt{\omega_1^2+\omega_2^2}}} d\lambda = \left(\frac{\Delta s}{\sqrt{\omega_1^2+\omega_2^2}}\right)^{\alpha} \int_0^\infty \frac{dy}{(1+y)\, y^{\alpha}}.$$

From

$$\int_0^\infty \frac{dy}{(1+y)\, y^{\alpha}} \leq \int_0^1 \frac{dy}{y^{\alpha}} + \int_1^\infty \frac{dy}{y^{\alpha+1}} = \frac{1}{\alpha(1-\alpha)}$$

it follows that

$$|T_1| \leq \frac{1}{\alpha(1-\alpha)} \left(\frac{\Delta s}{\sqrt{\omega_1^2+\omega_2^2}}\right)^{\alpha} \left\| f^{r,h} \right\|_{E_\alpha\left(\mathbb{R}^2_{(r,h)}\right)}. \tag{16}$$

Now, let us estimate T_2. Using triangle inequality, we obtain

$$|T_2| \leq \frac{\Delta s}{\sqrt{\omega_1^2+\omega_2^2}} \int_0^\infty \sum_{i=2}^\infty R^i \frac{\lambda^{1-\alpha} \Delta s}{\sqrt{\omega_1^2+\omega_2^2}} d\lambda \sup_{\lambda>0} \lambda^{\alpha} \left| B_{r,h} \left(\lambda I + B_{r,h}\right)^{-1} f\left(x_n, y_m, \bar{\omega}\right) \right|$$

$$\leq \frac{\Delta s}{\sqrt{\omega_1^2+\omega_2^2}} \int_0^\infty \frac{1}{1+\frac{\lambda \Delta s}{\sqrt{\omega_1^2+\omega_2^2}}} \lambda^{-\alpha} d\lambda \left\| f^{r,h} \right\|_{E_\alpha\left(\mathbb{R}^2_{(r,h)}\right)}.$$

The change of variable $\frac{\lambda \Delta s}{\sqrt{\omega_1^2+\omega_2^2}} = y$ yields $\frac{\Delta s}{\sqrt{\omega_1^2+\omega_2^2}} d\lambda = dy$ and making this substitution, we obtain

$$|T_2| \leq \left(\frac{\Delta s}{\sqrt{\omega_1^2+\omega_2^2}}\right)^{\alpha-1} \int_0^\infty \frac{dy}{(1+y)\, y^{\alpha}} \leq \left(\frac{\Delta s}{\sqrt{\omega_1^2+\omega_2^2}}\right)^{\alpha-1} \frac{1}{\alpha(1-\alpha)}. \tag{17}$$

Finally, by combining estimates (16) and (17), we obtain

$$\frac{\left|f\left(x_n,y_m,\bar{\omega}\right)-f\left(x_n+\frac{\omega_1\Delta s}{\sqrt{\omega_1^2+\omega_2^2}},y_m+\frac{\omega_2\Delta s}{\sqrt{\omega_1^2+\omega_2^2}},\bar{\omega}\right)\right|}{\left(\frac{\Delta s}{\sqrt{\omega_1^2+\omega_2^2}}\right)^{\alpha}}\leq\frac{2}{\alpha\left(1-\alpha\right)}\left\|f^{r,h}\right\|_{E_\alpha\left(\mathbb{R}^2_{(r,h)}\right)},$$

for any $(x_n, y_m) \in \mathbb{R}^2_{(r,h)}$ and $\lambda > 0$. Thus, for any $(x_n, y_m) \in \mathbb{R}^2_{(r,h)}$, we get the inequality

$$\left\|f^{r,h}\right\|_{C^\alpha\left(\mathbb{R}^2_{(r,h)}\right)}\leq\frac{2}{\alpha\left(1-\alpha\right)}\left\|f^{r,h}\right\|_{E_\alpha\left(C\left(\mathbb{R}^2_{(r,h)}\right),\ B_{r,h}\right)}.$$

Hence, we have proved that

$$C^\alpha\left(\mathbb{R}^2_{(r,h)}\right)\subset E_\alpha\left(C\left(\mathbb{R}^2_{(r,h)}\right),B_{r,h}\right). \tag{18}$$

This completes the proof of Theorem 3.

Theorem 4 *The operator $B_{r,h}$ is a positive operator in $C^\alpha\left(\mathbb{R}^2_{(r,h)}\right)$, $0 < \alpha < 1$, and the following estimate for $\lambda > 0$ holds:*

$$\left\|\left(\lambda I+B_{r,h}\right)^{-1}\right\|_{C^\alpha\left(\mathbb{R}^2_{(r,h)}\right)\to C^\alpha\left(\mathbb{R}^2_{(r,h)}\right)}\leq\frac{M(\alpha)}{\lambda}.$$

Proof Applying Theorem 3, we can write

$$\left\|\left(\lambda I+B_{r,h}\right)^{-1}\right\|_{C^\alpha\left(\mathbb{R}^2_{(r,h)}\right)\to C^\alpha\left(\mathbb{R}^2_{(r,h)}\right)}\leq M(\alpha)\left\|\left(\lambda I+B_{r,h}\right)^{-1}\right\|_{E^\alpha\left(\mathbb{R}^2_{(r,h)}\right)\to E^\alpha\left(\mathbb{R}^2_{(r,h)}\right)}.$$

Then, from estimate (11) it follows

$$\left\|\left(\lambda I+B_{r,h}\right)^{-1}\right\|_{C^\alpha\left(\mathbb{R}^2_{(r,h)}\right)\to C^\alpha\left(\mathbb{R}^2_{(r,h)}\right)}\leq M(\alpha)\left\|\left(\lambda I+B_{r,h}\right)^{-1}\right\|_{C\left(\mathbb{R}^2_{(r,h)}\right)\to C\left(\mathbb{R}^2_{(r,h)}\right)}.$$

Finally, using last estimate and Theorem 1 we complete the proof of Theorem 4.

4 Applications

In this section we consider the application of results of sections 3 and 4. We consider the difference scheme

$$\begin{cases}\frac{u^k_{n,m}-u^{k-1}_{n,m}}{\tau}-\omega_1\frac{u^k_{n,m}-u^k_{n,m-1}}{r}-\omega_2\frac{u^k_{n-1,m}-u^k_{n-1,m-1}}{h}=f\left(t_k,x_n,y_m\right),\\ t_k=k\tau,\ 1\leq k\leq N,\ N\tau=T,\ \left(x_n,y_m\right)\in\mathbb{R}^2_{(r,h)}\\ u(0,x_n,y_m)=\varphi\left(x_n,y_m\right),\ \ \left(x_n,y_m\right)\in\mathbb{R}^2_{(r,h)}\end{cases} \tag{19}$$

for the numerical solution of initial value problem

$$\begin{cases} \frac{\partial u(t,x,y)}{\partial t} - \omega_1 \frac{\partial u(t,x,y)}{\partial x} - \omega_2 \frac{\partial u(t,x,y)}{\partial y} = f(t,x,y), \\ (x,y) \in \mathbb{R}^2, \ 0 \le t \le T, \\ u(0,x,y) = \varphi(x,y), \ \ (x,y) \in \mathbb{R}^2. \end{cases} \tag{20}$$

Theorem 5 *Let* $0 < \alpha < 1$. *Then, the solution of difference scheme (19) satisfies the following stability inequality:*

$$\max_{0\le k\le N} \left\| \left(u^k\right)^{r,h} \right\|_{C^\alpha\left(\mathbb{R}^2_{(r,h)}\right)} \le M(\alpha) \left[\left\|\varphi^{r,h}\right\|_{C^\alpha\left(\mathbb{R}^2_{(r,h)}\right)} + \max_{1\le k\le N} \left\| \left(f^k\right)^{r,h} \right\|_{C^\alpha\left(\mathbb{R}^2_{(r,h)}\right)} \right].$$

The proof of Theorem 5 is based on Theorem 1 on the positivity of the difference neutron transport operator $B_{(r,h)} = -A_{(r,h)}$defined by formula (2) and on Theorem 3 on the structure of fractional space $E_\alpha = E_\alpha\left(\mathbb{R}^2_{(r,h)}, B_{(r,h)}\right)$ and on the following theorem on stability of the difference scheme:

$$\begin{cases} \frac{u^k - u^{k-1}}{\tau} + Bu^k = f_k = f(t_k), \\ t_k = k\tau, \ 1 \le k \le N, \ N\tau = T, \ u_0 = \varphi \end{cases} \tag{21}$$

for the approximate solution of abstract initial value problem (20).

Theorem 6 *Let* $B_{(r,h)}$ *be a positive operator in a Banach space* $E_\alpha\left(C\left(\mathbb{R}^2_{(r,h)}\right), B_{r,h}\right)$. *Then, for the solution of difference scheme (19) the following stability inequality holds:*

$$\max_{0\le k\le N} \left\|u^k\right\|_{E_\alpha} \le M \left[\|\varphi\|_{E_\alpha} + \max_{1\le k\le N} \left\|f^k\right\|_{E_\alpha} \right].$$

Acknowledgements The publication has been prepared with the support of the "RUDN University Program 5-100".

References

1. Lewis, E., Miller, W.: Computational Methods of Neutron Transport. American Nuclear Society, USA (1993)
2. Marchuk, G.I., Lebedev, V.I.: Numerical Methods in the Theory of Neutron Transport. Taylor and Francis, USA (1986)
3. Kharroubi, M.M.: Mathematical Topics in Neutron Transport Theory, New Aspects. World Scientific, Singapore and River Edge, N.J. (1997)
4. Ashyralyev, A., Sobolevskii, P.E.: Well-Posedness of Parabolic Difference Equations. Operator Theory: Advances and Applications, vol. 69. Birkhäuser, Basel, Boston, Berlin (1994)
5. Krasnosel'skii, M.A., Zabreyko, P.P., Pustylnik, E.I., Sobolevskii, P.E.: Integral Operators in Spaces of Summable Functions. Springer, Netherlands (1976)

6. Krein, S.G.: Linear Differential Equations in a Banach Space. Translations of Mathematical Monographs. Am. Math. Soc., Providence, Rhode Island (1971)
7. Fattorini, H.O.: Second Order Linear Differential Equations in Banach Spaces. Elsevier Science Publishing Company, Providence, North-Holland (1985)
8. Triebel, H.: Interpolation Theory, Function Spaces, Differential Operators. North-Holland, Amsterdam, New York, Oxford (1978)
9. Ashyralyev, A., Akturk, S.: Positivity of a one-dimensional difference operator in the half-line and its applications. Appl. Comput. Math. **14**(2), 204–220 (2015)
10. Ashyralyev, A., Tetikoglu, F.S.: A note on fractional spaces generated by the positive operator with periodic conditions and applications. Bound. Value Probl. **2015**, 31 (2015). https://doi.org/10.1186/s13661-015-0293-9
11. Ashyralyev, A., Sozen, Y.: Well-posedness of parabolic differential and difference equations. Comput. Math. Appl. **60**(3), 792–802 (2010). https://doi.org/10.1016/j.camwa.2010.05.026
12. Ashyralyev, A., Tetikoglu, F.S.: Structure of fractional spaces generated by the difference operator and its applications. Numer. Funct. Anal. Optim. **38**(10), 1325–1340 (2017). https://doi.org/10.1080/01630563.2017.1316999
13. Ashyralyev, A., Nalbant, N., Sozen, Y.: Structure of fractional spaces generated by second order difference operators. J. Frankl. Inst. Eng. Appl. Math. **351**(2), 713–731 (2014). https://doi.org/10.1016/j.jfranklin.2013.07.00
14. Ashyralyev, A., Akturk, S., Sozen, Y.: The structure of fractional spaces generated by a two-dimensional elliptic differential operator and its applications. Bound Value Probl. **3**(2014), 1–17 (2014). https://doi.org/10.1186/1687-2770-2014-3
15. Ashyralyev, A., Taskin, A.: The structure of fractional spaces generated by a two-dimensional neutron transport operator and its applications. Adv. Oper. Theory. **4**(1), 140–155 (2019). https://doi.org/10.15352/aot.1711-1261
16. Lebedova, V.I., Sobolevskii, P.E.: Spectral properties of the transfer operator with constant coefficients in $L_p(\mathbb{R}^n)$ ($1 \leq p < \infty$) spaces. Voronezh. Gosud. Univ. (Russian) Deposited VINITI, **2958**(83), 54 pp (1983)
17. Lebedova, V.I.: Spectral properties of the transfer operator of neutron in $C(\Omega, C(\mathbb{R}^n))$ spaces. Qual. Approx. Methods Solving Oper. Equ. Yaroslavil, (Russian) **9**, 44–51 (1984)
18. Zhukova, V.I., Gamolya, L.N.: Investigation of spectral properties of the transfer operator. Dalnovostochniy Matematicheskiy Zhurnal, (Russian) **5(1)**, 158–164 (2004)
19. Zhukova, V.I.: Spectral properties of the transfer operator. Trudy Vsyesoyuznoy Nauchno-Prakticheskoy Konferensii. Chita **5**(1), 170–174 (2000)
20. Sobolevskii, P.E.: Some properties of the solutions of differential equations in fractional spaces. Trudy Nauchn.-Issled. Inst. Mat. Voronezh. Gos. Univ., (Russian) **74**, 68–76 (1975)
21. Bazarov, M.A.: On the structure of fractional spaces. In: Proceeding of the XXVII All-Union Scientific Student Conference. The Student and Scientific-Technological Progress. Novosibirsk. Gos. Univ., (Russian). Novosibirsk, pp. 3–17 (1989)
22. Ashyralyev, A., Taskin, A.: Structure of fractional spaces generated by the two dimensional neutron transport operator. AIP Conf. Proc. **1759**, 0200661–0200665 (2016). https://doi.org/10.1063/1.4959680
23. Ashyraliyev, M.: On Gronwall's type integral inequalities with singular kernels. Filomat. **31**(4), 1041–1049 (2017). https://doi.org/10.2298/FIL1704041A
24. Ashyralyev, A., Hamad, A.: A note on fractional powers of strongly positive operators and their applications. Fract. Calc. Appl. Anal. **22**, 302–325 (2019). https://doi.org/10.1515/fca-2019-0020
25. Ashyralyev, A.: A survey of results in the theory of fractional spaces generated by positive operators. TWMS J. Pure Appl. Math. **6**(2), 129–157 (2015)
26. Ashyralyev, A., Taskin, A.: Stable difference schemes for the neutron transport equation. AIP Conf. Proc. **1389**, 570–573 (2011). https://doi.org/10.1063/1.3636794
27. Ashyralyev, A., Taskin, A.: Structures of the fractional spaces generated by the difference neutron transport operator. AIP Conf. Proc. **1676**, 020051 (2015). https://doi.org/10.1063/1.4930477

Uncertainty Type Principles for Radial Derivatives

Dina Shilibekova

Abstract In this paper, we provide uncertainty type principles for radial derivatives on an open bounded set $\Omega \subset \mathbb{R}^n$. We obtain several inequalities of uncertainty type principles of the form

$$\int_{\Omega} |f(x)|^p \phi(|x|)\, dx \leq p \left(\int_{\Omega} \frac{|\mathscr{R}_{|x|} f(x)|^p \; |\tilde{\phi}(|x|)|^p}{|x|^{n-p}} dx \right)^{\frac{1}{p}} \left(\int_{\Omega} \frac{|f(x)|^p}{|x|^n} dx \right)^{\frac{p-1}{p}},$$

for $1 < p < +\infty$, acting on functions with the radial derivative $\mathscr{R}_{|x|}$. As byproduct, we present versions of the Hardy and higher order Steklov inequalities for the radial derivatives.

Keywords Radial derivatives · Uncertainty type principle · Hardy inequalities · Steklov inequalities · Sharp remainder terms

1 Introduction

Heisenberg's uncertainty principle serves as a main basis for many theorems in quantum mechanics. In terms of physics, it is well known inequality that shows relationship between position of a particle and its momentum proven by Kennard and the mathematical interpretation is extensively studied as spectral properties of differential operators which was first found in Fefferman's work [5]. Mathematical formulation of uncertainty principle which is derived from the Hardy inequality [1] is

$$\left(\int_{\mathbb{R}} |f(x)|^2 dx \right)^2 \leq 4 \left(\int_{\mathbb{R}} |x|^2 |f(x)|^2 dx \right) \left(\int_{\mathbb{R}} (f'(x))^2 dx \right), \tag{1}$$

D. Shilibekova (✉)
Department of Mathematics, School of Science and Humanities, Nazarbayev University, 53 Kabanbay Batyr Ave, Nur-Sultan 010000, Kazakhstan
e-mail: dina.shilibekova@nu.edu.kz

A. Ashyralyev et al. (eds.), *Functional Analysis in Interdisciplinary Applications—II*, Springer Proceedings in Mathematics & Statistics 351,
https://doi.org/10.1007/978-3-030-69292-6_19

which contributed to the emergence of an open problem stated in Maz'ya's paper [2] that asks to find a sharp constant C if f is a vector field

$$\left(\int_{\mathbb{R}^n} |f|^2 dx\right)^2 \leq C \left(\int_{\mathbb{R}^n} |x|^2 |f|^2 dx\right) \left(\int_{\mathbb{R}^n} |\nabla f|^2 dx\right). \tag{2}$$

Motivated by this open problem, in this paper, we consider generalized uncertainty type principles for radial derivatives. Thus, in Sect. 2 we present the main result of this paper. As byproduct, we discuss radial derivatives of Hardy and higher order Steklov inequalities.

2 Main Results

Theorem 1 *Let $\Omega \subset \mathbb{R}^n$ be an open bounded set. Then for any integrable radial function ϕ, and for any $f \in C_0^1(\Omega)$ we have the following generalized uncertainty type principle*

$$\int_\Omega |f(x)|^p \, \phi(|x|) \, dx \leq p \left(\int_\Omega \frac{|\mathscr{R}_{|x|} f(x)|^p \, |\tilde{\phi}(|x|)|^p}{|x|^{n-p}} dx\right)^{\frac{1}{p}} \left(\int_\Omega \frac{|f(x)|^p}{|x|^n} dx\right)^{\frac{p-1}{p}}, \tag{3}$$

for $1 < p < +\infty$ and here, $\tilde{\phi}(|x|) = \int_0^{|x|} \phi(r) \, r^{n-1} \, dr$.

Proof Let $\Omega \subset B(0, R)$ and B be a ball with sufficiently large radius R. Using integration by parts and having $|x| = r$ in Euclidean case,

$$\begin{aligned}
\int_\Omega |f(x)|^p \, \phi(|x|) \, dx &= \int_{B(0,R)} |f(x)|^p \, \phi(|x|) \, dx \\
&= \int_S \int_0^R \phi(r) |f(r,\omega)|^p \, r^{n-1} \, dr d\omega \\
&= \int_S \left(\tilde{\phi}(r) \, |f(r,\omega)|^p \Big|_0^R - \int_0^R \tilde{\phi}(r) \frac{\partial(|f(r,\omega)|^p)}{\partial r} dr \right) d\omega \\
&\leq p \int_{B(0,R)} |f(x)|^{p-1} \left|\mathscr{R}_{|x|} f(x)\right| \frac{|\tilde{\phi}(|x|)|}{|x|^{n-1}} dx \\
&= p \int_{B(0,R)} \frac{\left|\mathscr{R}_{|x|} f(x)\right| \, |\tilde{\phi}(|x|)|}{|x|^{\frac{n-p}{p}}} \frac{|f(x)|^{p-1}}{|x|^{\frac{n(p-1)}{p}}} dx,
\end{aligned}$$

where in the penultimate line, we have used the Schwartz inequality. Now, by applying Hölder's inequality, we arrive at

$$\int_{\Omega} |f(x)|^p \, \phi(|x|) \, dx \leq p \left(\int_{\Omega} \frac{|\mathcal{R}_{|x|} f(x)|^p \, \left|\tilde{\phi}(|x|)\right|^p}{|x|^{n-p}} \, dx \right)^{\frac{1}{p}} \left(\int_{\Omega} \frac{|f(x)|^p}{|x|^n} \, dx \right)^{\frac{p-1}{p}},$$

for $1 < p < +\infty$. It completes the proof.

Formula (3) can be indeed called uncertainty type principle. If we raise both sides of the inequality to the power of p, then the inequality becomes

$$\left(\int_{\Omega} |f(x)|^p \, \phi(|x|) \, dx \right)^p \leq p^p \left(\int_{\Omega} \frac{|\mathcal{R}_{|x|} f(x)|^p \, |\tilde{\phi}(|x|)|^p}{|x|^{n-p}} \, dx \right) \left(\int_{\Omega} \frac{|f(x)|^p}{|x|^n} \, dx \right)^{p-1},$$

for $1 < p < +\infty$.

To demonstrate consequences of Theorem 1 we consider several examples and obtain some interesting insights. For that let us set $|x| = r$, since $|x|$ is a standard Euclidean distance in $\mathbb{R}^n$. Then we have

$$\frac{d\tilde{\phi}(r)}{dr} = \phi(r) \, r^{n-1}. \tag{4}$$

Example 1 Assume that $\tilde{\phi}(r) = r^{\frac{n-p}{p}}$. From the first order differential equation (4) we can find $\phi(r)$:

$$\phi(r) \, r^{n-1} = \left(\frac{n-p}{p} \right) r^{\frac{n-p}{p} - 1}$$
$$\phi(r) = \left(\frac{n-p}{p} \right) r^{\frac{n-p-np}{p}}.$$

Let us consider several cases for p:

- Let $p = 2$, then we have

$$\phi(r) = \left(\frac{n}{2} - 1 \right) r^{-\frac{n}{2} - 1}.$$

In this case, we obtain

$$\int_{\Omega} |f(x)|^2 \, |x|^{-\frac{n}{2} - 1} \, dx \leq \frac{4}{n-2} \left(\int_{\Omega} |\mathcal{R}_{|x|} f(x)|^2 \, dx \right)^{\frac{1}{2}} \left(\int_{\Omega} \frac{|f(x)|^2}{|x|^n} \, dx \right)^{\frac{1}{2}}.$$

- If $p = 4$, then

$$\phi(r) = \left(\frac{n}{4} - 1 \right) r^{-\frac{3n}{4} - 1}.$$

Therefore, we get

$$\int_{\Omega} |f(x)|^4 |x|^{-\frac{3n}{4}-1} dx \leq \frac{16}{n-4} \left(\int_{\Omega} \left|\mathscr{R}_{|x|} f(x)\right|^4 dx \right)^{\frac{1}{4}} \left(\int_{\Omega} \frac{|f(x)|^4}{|x|^n} dx \right)^{\frac{3}{4}}.$$

The case of $p = n$ is trivial.

Thus, the general formula for the specific function $\tilde{\phi}(r) = r^{\frac{n-p}{p}}$ is

$$\int_{\Omega} |f(x)|^p |x|^{\frac{n(1-p)}{p}-1} dx \leq \frac{p^2}{n-p} \left(\int_{\Omega} \left|\mathscr{R}_{|x|} f(x)\right|^4 dx \right)^{\frac{1}{p}} \left(\int_{\Omega} \frac{|f(x)|^p}{|x|^n} dx \right)^{\frac{p-1}{p}}, \tag{5}$$

when $p \neq n$.

Example 2 Assume that $\phi(r) = \frac{1}{r^{n+\alpha}}$ and $\alpha < 0$, then the first order differential equation (4) can be solved for $\tilde{\phi}(|x|)$:

$$\tilde{\phi}(|x|) = \int_0^{|x|} \frac{1}{r^{n+\alpha}} r^{n-1} dr$$

$$\tilde{\phi}(|x|) = \int_0^{|x|} \frac{1}{r^{n+\alpha}} r^{n-1} dr = \int_0^{|x|} r^{-1-\alpha} dr = -\frac{1}{\alpha |x|^{\alpha}}.$$

Let us set $\alpha = -1$, then

$$\tilde{\phi}(|x|) = \int_0^{|x|} r^{-1+1} dr = \int_0^{|x|} dr = |x|,$$

which gives the following expression

$$\int_{\Omega} \frac{|f(x)|^p}{|x|^{n-1}} dx \leq p \left(\int_{\Omega} \frac{\left|\mathscr{R}_{|x|} f(x)\right|^p}{|x|^{n-2p}} dx \right)^{\frac{1}{p}} \left(\int_{\Omega} \frac{|f(x)|^p}{|x|^n} dx \right)^{\frac{p-1}{p}}.$$

The following proposition is an extension of a specific type of critical Hardy inequality proved in [6] for homogeneous groups with arbitrary quasi-norms. We show it for $\mathbb{R}^n$. However, in our case logarithmic term does not appear.

Proposition 1 *Let $\Omega \subset \mathbb{R}^n$ be an open bounded set. Then for all real-valued functions $f \in C_0^1(\Omega)$ and for any $\alpha < 0$ we have*

$$\int_{\Omega} \frac{|f(x)|^n}{|x|^{n+\alpha}} dx \leq \left(\frac{n}{|\alpha|} \right)^n \int_{\Omega} \frac{\left|\mathscr{R}_{|x|} f(x)\right|^n}{|x|^{\alpha}} dx, \tag{6}$$

where $\left(\frac{n}{|\alpha|} \right)^n$ is optimal.

Proof By setting $\phi(|x|) = \frac{1}{|x|^{n+\alpha}}$ in (3), and using the penultimate line's result in the proof of (3) we get

$$\int_{\Omega} \frac{|f(x)|^p}{|x|^{n+\alpha}} dx \leq \frac{p}{|\alpha|} \int_{\Omega} \frac{\left|\mathcal{R}_{|x|} f(x)\right| |f(x)|^{p-1}}{|x|^{n+\alpha-1}} dx$$
$$= \frac{p}{|\alpha|} \int_{\Omega} \frac{\left|\mathcal{R}_{|x|} f(x)\right|}{|x|^{\frac{n+\alpha-p}{p}}} \frac{|f(x)|^{p-1}}{|x|^{\frac{(n+\alpha)(p-1)}{p}}} dx.$$

By using Hölder's inequality, we obtain

$$\int_{\Omega} \frac{|f(x)|^p}{|x|^{n+\alpha}} dx \leq \frac{p}{|\alpha|} \left(\int_{\Omega} \frac{\left|\mathcal{R}_{|x|} f(x)\right|^p}{|x|^{n+\alpha-p}} dx \right)^{\frac{1}{p}} \left(\int_{\Omega} \frac{|f(x)|^p}{|x|^{n+\alpha}} \right)^{\frac{p-1}{p}} dx.$$

This gives (6).

Now let us show that the constant $\left(\frac{n}{|\alpha|}\right)^n$ is optimal. This is done by checking the equality condition in Hölder's inequality. Let us consider

$$h(x) = |x|.$$

A straightforward computation implies

$$\left(\frac{\left|\mathcal{R}_{|x|} h(x)\right|}{|x|^{\frac{n+\alpha-p}{p}}} \right)^p = \left(\frac{|h(x)|^{p-1}}{|x|^{\frac{(n+\alpha)(p-1)}{p}}} \right)^{\frac{p}{p-1}}.$$

Since, this satisfies the equality condition in Hölder's inequality, the constant $\left(\frac{n}{|\alpha|}\right)^n$ is optimal.

Now we discuss the sharp remainder terms and general formula for higher order Steklov inequalities with respect to the radial derivative $\mathcal{R}_{|x|}$ in $\mathbb{R}^n$. Using representation formulae for higher-order Steklov inequality for vector fields we will obtain similar results for the radial derivative $\mathcal{R}_{|x|}$ instead of a gradient applied in Theorem 2.1 in [4].

Theorem 2 *Let $\Omega \subset \mathbb{R}^n$ be an open bounded set and $\phi \in \Omega$ be any positive smooth function such that $-\mathcal{R}^2_{|x|}\phi = \lambda\phi$. We denote by $\mathcal{R}_{|x|}$ a radial derivative and λ is some positive constant. Then for any $u \in C_0^\infty(\Omega)$ we have*

$$\left|\mathcal{R}^{2m}_{|x|} u\right|^2 - \lambda^{2m}|u|^2 = \sum_{j=0}^{m-1} \lambda^{2(m-1-j)} \left(\left|\mathcal{R}^{2j+2}_{|x|} u + \lambda \mathcal{R}^{2j}_{|x|} u\right|^2 + 2\lambda \left|\mathcal{R}^{2j+1}_{|x|} u - \frac{\mathcal{R}_{|x|}\phi}{\phi} \mathcal{R}^{2j}_{|x|} u\right|^2 \right)$$

$$+\sum_{j=0}^{m-1} 2\lambda^{2(m-1-j)+1}\mathscr{R}_{|x|}\left(\frac{\mathscr{R}_{|x|}\phi}{\phi}|\mathscr{R}_{|x|}^{2j}u|^2 - \mathscr{R}_{|x|}^{2j}u\mathscr{R}_{|x|}^{2j+1}u\right), \tag{7}$$

where $m = 1, 2, \ldots,$ *and*

$$\left|\mathscr{R}_{|x|}^{2m+1}u\right|^2 - \lambda^{2m+1}|u|^2 = \left|\mathscr{R}_{|x|}^{2m+1}u - \frac{\mathscr{R}_{|x|}\phi}{\phi}\mathscr{R}_{|x|}^{2m}u\right|^2$$

$$+\sum_{j=0}^{m-1}\lambda^{2(m-j)-1}\left(\left|\mathscr{R}_{|x|}^{2j+2}u + \lambda\mathscr{R}_{|x|}^{2j}u\right|^2 + 2\lambda\left|\mathscr{R}_{|x|}^{2j+1}u - \frac{\mathscr{R}_{|x|}\phi}{\phi}\mathscr{R}_{|x|}^{2j}u\right|^2\right)$$

$$+2\sum_{j=0}^{m-1}\lambda^{2(m-j)}\mathscr{R}_{|x|}\left(\frac{\mathscr{R}_{|x|}\phi}{\phi}|\mathscr{R}_{|x|}^{2j}u|^2 - \mathscr{R}_{|x|}^{2j}u\mathscr{R}_{|x|}^{2j+1}u\right) + \mathscr{R}_{|x|}\left(\frac{\mathscr{R}_{|x|}\phi}{\phi}\left(\mathscr{R}_{|x|}^{2m}u\right)^2\right), \tag{8}$$

where $m = 0, 1, 2, \ldots$.

The proof of Theorem 2 is similar to the proof of Theorem 2.1 [4], because $\mathscr{R}_{|x|}$ acts in the same way on radial functions as a gradient operator on vector fields. This Theorem 2 is an important tool for obtaining the sharp remainder terms for higher order Steklov inequalities for several operators such as polyharmonic and hypoelliptic operators [4]. Using Theorem 2 for $\mathbb{R}$, our operator $\mathscr{R}_{|x|}$ becomes ∂_x, so that the following Theorem 3 is a consequence of Theorem 2 for $\mathbb{R}$. Moreover, the following expression confirms that the generalized difference in [3] is indeed nonnegative.

Theorem 3 *We have the sharp remainder of the higher order Steklov inequality*

$$\int_0^1 \left|\partial_x^{2m}u\right|^2 dx - \pi^{4m}\int_0^1 |u|^2 dx$$

$$= \sum_{j=0}^{m-1}\pi^{4(m-1-j)}\left(\int_0^1\left|\partial_x^{2j+2}u + \pi^2\partial_x^{2j}u\right|^2 dx + 2\pi^2\int_0^1\left|\partial_x^{2j+1}u - \frac{\pi\cos(\pi x)}{\sin(\pi x)}\partial_x^{2j}u\right|^2 dx\right) \geq 0, \tag{9}$$

where $m = 1, 2, \ldots,$ *and*

$$\int_0^1 \left|\partial_x^{2m+1}u\right|^2 dx - \pi^{4m+2}\int_0^1 |u|^2 dx = \int_0^1\left|\partial_x^{2m+1}u - \frac{\pi\cos(\pi x)}{\sin(\pi x)}\partial_x^{2m}u\right|^2 dx$$

$$+\sum_{j=0}^{m-1}\pi^{4(m-j)-2}\left(\int_0^1\left|\partial_x^{2j+2}u + \pi^2\partial_x^{2j}u\right|^2 dx + 2\pi^2\int_0^1\left|\partial_x^{2j+1}u - \frac{\pi\cos(\pi x)}{\sin(\pi x)}\partial_x^{2j}u\right|^2 dx\right) \geq 0, \tag{10}$$

where $m = 0, 1, 2 \ldots$, *for all* $u \in C_0^\infty(0,1)$. *The equality cases hold if and only if* u *is proportional to* $\sin(\pi x)$.

Proof Since $\partial_x(\sin(\pi x)) = \pi \cos(\pi x)$ and $\partial_x^2(\sin(\pi x)) = -\pi^2 \sin(\pi x)$, the constant $\lambda = \pi^2$. Integrating over $(0, 1)$ both sides of the equations in (7) and (8) and knowing the fact that u vanishes on the boundary of $(0, 1)$, we arrive at (9) and (10). Having $\lambda > 0$, the equality case in (9) holds if and only if

$$0 = \left|\partial_x^{2j+2} u + \pi^2 \partial_x^{2j} u\right|^2, \; j = 0, ..., m-1,$$

and

$$0 = \left|\partial_x^{2j+1} u - \frac{\pi \cos(\pi x)}{\sin(\pi x)} \partial_x^{2j} u\right|^2 = \left|\partial_x \left(\frac{\partial_x^{2j} u}{\sin(\pi x)}\right)\right|^2 \sin^2(\pi x), \; j = 0, ..., m-1,$$

which means $\frac{u}{\sin(\pi x)} = const$. Which in turn, requires u to be proportional to $\sin(\pi x)$ to have the equality case in (9).

Similarly, the equality case in (10) holds if and only if

$$0 = \left|\partial_x^{2j+2} u + \pi^2 \partial_x^{2j} u\right|^2, \; j = 0, ..., m-1,$$

and

$$0 = \left|\partial_x^{2j+1} u - \frac{\pi \cos(\pi x)}{\sin(\pi x)} \partial_x^{2j} u\right|^2 = \left|\partial_x \left(\frac{\partial_x^{2j} u}{\sin(\pi x)}\right)\right|^2 \sin^2(\pi x), \; j = 0, ..., m,$$

which means $\frac{u}{\sin(\pi x)} = const$. Consequently, this implies that u has to be proportional to $\sin(\pi x)$ to have the equality case in (10). It completes the proof.

Acknowledgements The author was supported by the Nazarbayev University grant 240919FD3901.

References

1. Balinsky, A., Evans, D., Lewis, R.: The Analysis and Geometry of Hardy's Inequality. Universitext, Springer International, 263 pp (2015)
2. Maz'ya, V.: Seventy five (thousand) unsolved problems in analysis and partial differential equations. Integr. Equ. Oper. Theory **90**, 44 pp (2018). https://doi.org/10.1007/s00020-018-2460-8
3. Ozawa, T., Suragan, D.: Poincaré inequalities with exact missing terms on homogeneous groups. To appear in JMSJ (2020)
4. Ozawa, T., Suragan, D.: Representation formulae for the higher order Steklov and L^{2^m}-Friedrichs inequalities (2020). arXiv:2005.04482v1

5. Ruzhansky, M., Suragan, D.: Uncertainty relations on nilpotent Lie groups. Proc. R. Soc. A **473**, 12 pp (2017). https://doi.org/10.1098/rspa.2017.0082
6. Ruzhansky, M., Suragan, D.: Critical Hardy inequalities. Annales Academi æ Scientiarum Fennic æ. Mathematica **44**, 1159–1174 (2019). https://doi.org/10.5186/aasfm.2019.4467

Boundary Conditions of Volume Hyperbolic Potential in a Domain with Curvilinear Boundary

Makhmud A. Sadybekov and Bauyrzhan O. Derbissaly

Abstract A one-dimensional volume hyperbolic potential in a domain with curvilinear boundaries is studied. As a kernel of the hyperbolic potential the fundamental solution of the Cauchy problem is chosen. It is well-known that in this case the volume hyperbolic potential satisfies homogeneous initial conditions. The boundary conditions to which the hyperbolic potential satisfies at lateral boundaries of the domain are constructed. It is shown that the formulated initial-boundary value problem has the unique classical solution.

Keywords Hyperbolic equation · Initial-boundary value problem · Boundary condition · Hyperbolic potential

1 Introduction

In [2], the Riemann-Green method is used to give general solutions of Cauchy problems for a hyperbolic equation in an arbitrary domain. Riemann first engaged in such tasks in the twentieth century [20]. After Riemann, Darboux made a great progress in this area. In these papers, the foundations were laid for representation of solutions of hyperbolic equations in an integral form.

The volume elliptic potential is widely used in solving classical problems of Dirichlet, Neumann and other boundary value problems in domains of an arbitrary form. But, at the same time, the boundary conditions and the spectral problems of the volume potential have not been researched till the recent time. That is, despite

M. A. Sadybekov (✉) · B. O. Derbissaly
Institute of Mathematics and Mathematical Modeling, 125 Pushkin st., Almaty 050010, Kazakhstan
e-mail: sadybekov@math.kz

B. O. Derbissaly
e-mail: derbissaly@math.kz

B. O. Derbissaly
Al-Farabi Kazakh National University, Al-Farabi ave. 71, 050040 Almaty, Kazakhstan

A. Ashyralyev et al. (eds.), *Functional Analysis in Interdisciplinary Applications—II*, Springer Proceedings in Mathematics & Statistics 351,
https://doi.org/10.1007/978-3-030-69292-6_20

the deep research of the general theory of the volume potential, till the recent time the Newton volume potential

$$u_{NP}(x) = \int_{\Omega} \varepsilon(x-y) f(y) dy$$

has not been considered as an independent operator being a solution of some boundary value problem. The scientists as Engquist B. and Majda A. [3], Givoli D. [4–6], Li J.R., Greengard L. [7], Hagstrom T. [8], Tsynkov S.V. [27], Wu X. and Zhang J. [28] used the foundations of the theory of the boundary value problems for different kinds of the volume potentials for solving various problems of the mathematical physics and numerical calculations.

In T. Sh. Kal'menov and D. Suragan's paper the boundary conditions of the volume potential u_{NP} for the case of multidimensional Laplace operator were built for the first time [11]. New non-local boundary conditions, which uniquely define the Newton volume potential, have the form

$$\frac{u(x)}{2} - \int_{\partial\Omega} \left(\frac{\partial \varepsilon(x-y)}{\partial n_y} u(y) - \varepsilon(x-y) \frac{\partial u(y)}{\partial n_y} \right) dS_y = 0, \quad x \in \partial\Omega.$$

Despite the complexity of these boundary conditions, they were quite convenient to use. Using these boundary conditions, all eigenvalues and eigenfunctions were constructed for the volume potential in a two-dimensional circle and a three-dimensional ball considered in [12].

The trace of the Newton potential on a boundary surface appeared in Kac's work [9], where he called it as "the principle of not feeling the boundary" and he made the subsequent spectral analysis. This was further expanded in Kac's book [10] (see also Saito [26]) with several further applications to the spectral theory and the asymptotics of the Weyl eigenvalue counting function. For the general background details on potential theory of the time-fractional diffusion equation we refer to [1, 15, 16].

It is shown in [13] that self-adjoint differential operators are generated by boundary conditions. Further the boundary conditions were built for the non-self-adjoint operators. In [14] the initial-boundary value problem for the wave equation in the domain with rectilinear boundaries is considered. In [25] a generalized heat potential for the degenerate (heat) diffusion equation, which satisfies the initial condition with respect to the time variable is considered. In this work the boundary condition for this potential is found. The nonlocal initial boundary value problem for the time-fractional diffusion equation for the Kohn Laplacian and its powers on the Heisenberg group have been recently investigated by Ruzhanksy and Suragan in [22] as well as in [23] for general stratified Lie groups.

The study of the correctness of non-local problems for hyperbolic equations with integral conditions is recently an urgent problem. One of the first works in this direction was an article by L.S. Pulkina [18] in which the existence and uniqueness of generalized solution for a second-order hyperbolic equation with integral conditions in a rectangle are proved. In [17] a boundary-value problem for one-dimensional hyperbolic equation with nonlocal initial data in integral form was considered. In a recent paper [19], a problem was considered for a hyperbolic equation with standard initial data and a non-local integral condition of the second kind, which degenerates and turns into the first kind.

In this paper we investigate the problem of constructing boundary conditions for a one-dimensional hyperbolic volume potential in a domain with curvilinear boundary. We show that the solution of boundary value problem is uniquely determined by the volume potential.

2 Formulation of Problem

Let $Q \subset R^2$ be a finite domain bounded at the sides by the curves $x = \alpha_1(t)$ and $x = \beta_1(t)$, and bounded above and below by the segments $t = 0$, $0 < x < 1$ and $t = T$, $x_0 < x < x_1$. Here $T > 0$, $\alpha_1(0) = 0$, $\beta_1(0) = 1$, $\alpha_1(T) = x_0$, $\beta_1(T) = x_1$, $\alpha_1(t) < \beta_1(t)$.

We consider the following hyperbolic equation

$$Lu \equiv \frac{\partial^2 u(x,t)}{\partial t^2} - \frac{\partial^2 u(x,t)}{\partial x^2} + a_1(x,t)\frac{\partial u(x,t)}{\partial x}$$

$$+ b_1(x,t)\frac{\partial u(x,t)}{\partial x} + c_1(x,t)u(x,t) = f_1(x,t),\ (x,t) \in Q, \tag{1}$$

with the initial conditions

$$u(x,0) = \frac{\partial u}{\partial t}(x,0) = 0,\ 0 \le x \le 1, \tag{2}$$

where $a_1, b_1, c_1 \in C^1\left(\overline{Q}\right)$. Additionally, assume that

$$|\alpha_1'(t)| < 1,\ \ |\beta_1'(t)| < 1. \tag{3}$$

It is known that for $T > 1/2$ the solution of the hyperbolic equation (1) in Q is restored under the initial conditions (2) not uniquely. For the uniqueness it is necessary to use boundary conditions. We set a task to construct boundary conditions under which (together with the initial conditions) the solution of Eq. (1) in Q will be uniquely defined in the form of a volume hyperbolic potential (see Eq. (4)). In the case, when $\alpha(t) \equiv 0$, $\beta(t) \equiv 1$ and $a_1, b_1, c_1 = 0$ this problem was considered in [14]. When the domain Q remains the same and $a_1, b_1, c_1 = 0$, this case was considered in our paper [24].

Let $Q_{x,t}$ be a part of Q: $Q_{x,t} = \{(x_1, t_1) \in Q : |x - x_1| < t - t_1\}$. In Q we have the volume hyperbolic potential

$$u(x,t) = -\int_{\Omega_{x,t}} R_1(x,t;x_1,t_1) f_1(x_1,t_1)dx_1dt_1, \tag{4}$$

where $R_1(x,t;x_1,t_1)$ is the Riemann-Green function [2], which satisfies the conjugate homogeneous equation

$$L_1^* R_1 \equiv \frac{\partial^2}{\partial x_1 \partial t_1} R(x,t;x_1,t_1) - \frac{\partial}{\partial x_1}\left(a_1(x_1,t_1)R(x,t;x_1,t_1)\right)$$

$$-\frac{\partial}{\partial t_1}\left(R(x,t;x_1,t_1)b_1(x_1,t_1)\right) + c_1(x_1,t_1)R(x,t;x_1,t_1) = 0,\ (x,t) \in \Omega,$$

and the following characteristic equations

$$\frac{\partial R_1(x,t;x_1,t_1)}{\partial x_1} - b_1(x_1,t_1)R_1(x,t;x_1,t_1) = 0, \quad \text{when } x_1 = x;$$

$$\frac{\partial R_1(x,t;x_1,t_1)}{\partial t_1} - a_1(x_1,t_1)R_1(x,t;x_1,t_1) = 0, \quad \text{when } t_1 = t;$$

$$R_1(x,t;x,t) = 1.$$

In the characteristic coordinates $\xi = x + t, \eta = x - t$ Eq. (1) has the form

$$\frac{\partial^2 u(\xi,\eta)}{\partial \xi \partial \eta} + a\frac{\partial u(\xi,\eta)}{\partial \xi} + b\frac{\partial u(\xi,\eta)}{\partial \eta} + cu(\xi,\eta) = f(\xi,\eta),\ (\xi,\eta) \in \Omega \tag{5}$$

and initial conditions (2) have the form

$$u = \frac{\partial u}{\partial \xi} - \frac{\partial u}{\partial \eta} = 0, \text{ at } \xi = \eta, 0 \le \eta \le 1, \tag{6}$$

where $a(\xi,\eta), b(\xi,\eta), c(\xi,\eta) \in C^1\left(\overline{\Omega}\right)$ and

$$a(\xi,\eta) = \frac{1}{4}\left(a_1\left(\frac{\xi+\eta}{2}, \frac{\xi-\eta}{2}\right) + b_1\left(\frac{\xi+\eta}{2}, \frac{\xi-\eta}{2}\right)\right),$$

$$b(\xi,\eta) = \frac{1}{4}\left(a_1\left(\frac{\xi+\eta}{2}, \frac{\xi-\eta}{2}\right) - b_1\left(\frac{\xi+\eta}{2}, \frac{\xi-\eta}{2}\right)\right),$$

$$c(\xi,\eta) = \frac{1}{4}c_1\left(\frac{\xi+\eta}{2}, \frac{\xi-\eta}{2}\right), f(\xi,\eta) = \frac{1}{4}f_1\left(\frac{\xi+\eta}{2}, \frac{\xi-\eta}{2}\right).$$

Here $\Omega \subset R^2$ is a domain bounded at the sides by the curves $\xi = \alpha(\eta)$ and $\xi = \beta(\eta)$, and bounded from above and below by the segment $\xi - \eta = 0$ and $\xi - \eta = 2T$. Here

$\alpha(0) = 0$, $\beta(1) = 1$, $\alpha(\eta) < \beta(\eta)$. From (3) we have

$$-\infty < \alpha'(\eta) < 0, \tag{7}$$

$$-\infty < \beta'(\eta) < 0. \tag{8}$$

3 Construction of Boundary Conditions

By $\Omega_{\xi,\eta}$ we denote a part of Ω: $\Omega_{\xi,\eta} = \{(\xi_1, \eta_1) \in \Omega : \xi_1 < \xi, \eta_1 > \eta\}$. Then the volume potential (4) can be written in the form

$$u(\xi, \eta) = -\int_{\Omega_{\xi,\eta}} R(\xi, \eta; \xi_1 \eta_1) f(\xi_1, \eta_1) d\xi_1 d\eta_1, \tag{9}$$

where $R(\xi, \eta; \xi_1, \eta_1)$ is the Riemann-Green function [2], which satisfies the following equations

$$L^* R \equiv \frac{\partial^2}{\partial \xi_1 \partial \eta_1} R(\xi, \eta; \xi_1, \eta_1) - \frac{\partial}{\partial \xi_1} (a(\xi_1, \eta_1) R(\xi, \eta; \xi_1, \eta_1))$$

$$- \frac{\partial}{\partial \eta_1} (b(\xi_1, \eta_1) R(\xi, \eta; \xi_1, \eta_1)) + c(\xi_1, \eta_1) R(\xi, \eta; \xi_1, \eta_1) = 0, \ (\xi, \eta) \in \Omega; \tag{10}$$

$$\frac{\partial R(\xi, \eta; \xi_1, \eta_1)}{\partial \xi_1} - b(\xi_1, \eta_1) R(\xi, \eta; \xi_1, \eta_1) = 0, \text{ when } \xi_1 = \xi; \tag{11}$$

$$\frac{\partial R(\xi, \eta; \xi_1, \eta_1)}{\partial \eta_1} - a(\xi_1, \eta_1) R(\xi, \eta; \xi_1, \eta_1) = 0, \text{ when } \eta_1 = \eta; \tag{12}$$

$$R(\xi, \eta; \xi, \eta) = 1. \tag{13}$$

Evidently that for any $f(\xi, \eta) \in C^1(\overline{\Omega})$ the volume potential (9) gives a classical solution of the inhomogeneous hyperbolic equation (5) from the class $u(\xi, \eta) \in C^2(\overline{\Omega})$. Our task is to construct homogeneous boundary conditions on the lateral boundaries $\xi = \alpha(\eta)$ and $\xi = \beta(\eta)$, which the volume potential (9) satisfies for all $f(\xi, \eta)$.

We consider separately various cases of placing $\Omega_{\xi,\eta}$ inside Ω.

Case I

Firstly, we consider a case, when $0 < \eta < \xi < 1$, $(\xi, \eta) \in \Omega$. In this case $\Omega_{\xi,\eta}$ is a triangle that is bounded from above by $\xi_1 = \xi$, is bounded from below by $\eta_1 = \eta$, and is bounded from the right by $\xi_1 = \eta_1$. In this case the domain $\Omega_{\xi,\eta}$ nowhere touches the lateral boundaries Ω.

Therefore there is no need to construct the boundary conditions for the hyperbolic volume potential. By direct calculation it is easy to see that the volume potential (9) satisfies the homogeneous initial conditions (6).

CASE II

The case, when $\xi < 1, \eta < 0, (\xi, \eta) \in \Omega$. Let $\xi = \alpha(\eta)$, then $\Omega_{\alpha(\eta),\eta} = \{(\alpha(\eta), \eta) \in \Omega : \eta_1 < \xi_1 < \alpha(\eta),$ at $\eta_1 > 0; \alpha(\eta_1) < \xi_1 < \alpha(\eta),$ at $\eta_1 < 0\}$ is a curvilinear triangle, which is bounded from the right by $\xi_1 = \alpha(\eta)$, is bounded from below by $\xi_1 = \alpha(\eta_1)$, and is bounded from above by $\xi_1 = \eta_1$).

Hereinafter we will use the Green's theorem in a plane [21]: *Let C be a positively oriented, piecewise smooth, simple closed curve in a plane, and let D be a domain bounded by C. If L and M are functions of* (ξ_1, η_1) *defined on an open domain containing D and have continuous partial derivatives there, then*

$$\oint_C (L\, d\xi_1 + M\, d\eta_1) = \int_D \left(\frac{\partial M}{\partial \xi_1} - \frac{\partial L}{\partial \eta_1} \right) d\xi_1\, d\eta_1,$$

where the left-hand side is a line integral and the right-hand side is a surface integral, and the path of integration along C is anticlockwise.

Applying Green's theorem in a plane, from (9) we get the following chain of equalities:

$$u\,(\alpha(\eta), \eta) = -\int_{\Omega_{\alpha(\eta),\eta}} R(\alpha(\eta), \eta; \xi_1 \eta_1) f(\xi_1, \eta_1) d\xi_1 d\eta_1$$

$$-\int_{\Omega_{\alpha(\eta),\eta}} \left(RLu - uL^*R\right) d\xi_1 d\eta_1$$

$$= -\oint_{\partial\Omega_{\alpha(\eta),\eta}} \left(-bRu + \frac{1}{2}\frac{\partial R}{\partial \xi_1} u - \frac{1}{2} R \frac{\partial u}{\partial \xi_1} \right) d\xi_1 + \left(aRu - \frac{1}{2}\frac{\partial R}{\partial \eta_1} u + \frac{1}{2} R \frac{\partial u}{\partial \eta_1} \right) d\eta_1.$$

Calculating the obtained line integrals, taking into account the initial conditions (6), and conditions (10)–(12), we have

$$I_\alpha u \equiv \int_0^\eta \frac{\partial u(\alpha(\eta_1), \eta_1)}{\partial \eta_1} R(\alpha(\eta), \eta; \alpha(\eta_1), \eta_1) d\eta_1$$

$$+ \int_0^\eta u(\alpha(\eta_1), \eta_1) R(\alpha(\eta), \eta; \alpha(\eta_1), \eta_1) \left(a(\alpha(\eta_1), \eta_1) - b(\alpha(\eta_1), \eta_1) \alpha'(\eta_1) \right) d\eta_1$$

$$- \int_0^\eta u(\alpha(\eta_1), \eta_1) \frac{\partial R(\alpha(\eta), \eta; \alpha(\eta_1), \eta_1)}{\partial \xi_1} d\eta_1 = 0. \tag{14}$$

Note that (14) is a condition on the boundary $\xi = \alpha(\eta)$ connecting the values of the function u and its derivative on this boundary.

CASE III

Consider a case, when $0 < \eta, 1 < \xi, (\xi, \eta) \in \Omega$. Let $\xi = \beta(\eta)$, then $\Omega_{\beta(\eta),\eta} = \{(\beta(\eta), \eta) \in \Omega : \eta_1 < \xi_1 < \beta(\eta_1) \text{ and } \eta_1 > \eta\}$ is a curvilinear triangle, which is bounded from the right by $\xi_1 = \beta(\eta_1)$, is bounded from below by $\eta_1 = \eta$, and is bounded from the left by $\xi_1 = \eta_1$.

Analogously, as in Case II, applying the Green's theorem, from (9) we have the boundary condition

$$I_\beta u \equiv \int_1^\eta \frac{\partial u(\beta(\eta_1), \eta_1)}{\partial \xi_1} R(\beta(\eta), \eta; \beta(\eta_1), \eta_1)\beta'(\eta_1)d\eta_1$$

$$-\int_1^\eta u(\beta(\eta_1), \eta_1)R(\beta(\eta), \eta; \beta(\eta_1), \eta_1)\left(a(\beta(\eta_1), \eta_1) - b(\beta(\eta_1), \eta_1)\beta'(\eta_1)\right)d\eta_1$$

$$+\int_1^\eta u(\beta(\eta_1), \eta_1)\frac{\partial R(\beta(\eta), \eta; \beta(\eta_1), \eta_1)}{\partial \eta_1}d\eta_1 = 0. \tag{15}$$

Note that (15) is a condition on the boundary $\xi = \beta(\eta)$ connecting the values of the function u and its derivative on this boundary.

CASE IV

Consider a domain, when $\eta < 0$ and $1 < \xi$. In this case the domain $\Omega_{\xi,\eta}$ is a curvilinear pentagon, which is bounded from above by $\xi_1 = \beta(\eta_1)$ and $\eta_1 = \xi_1$; is bounded from below by $\xi_1 = \alpha(\eta_1)$ and $\eta_1 = \eta$; is bounded from the right by $\xi_1 = \xi$.

We apply the Green's theorem in a plane for the volume hyperbolic potential

$$u(\xi, \eta) = -\int_{\Omega_{\xi,\eta}} R(\xi, \eta; \xi_1\eta_1)f(\xi_1, \eta_1)d\xi_1 d\eta_1$$

$$-\int_{\Omega_{\xi,\eta}} \left(RLu - uL^*R\right)d\xi_1 d\eta_1$$

$$= -\oint_{\partial\Omega_{\xi,\eta}} \left(-bRu + \frac{1}{2}\frac{\partial R}{\partial \xi_1}u - \frac{1}{2}R\frac{\partial u}{\partial \xi_1}\right)d\xi_1 + \left(aRu - \frac{1}{2}\frac{\partial R}{\partial \eta_1}u + \frac{1}{2}R\frac{\partial u}{\partial \eta_1}\right)d\eta_1$$

$$= \int_{AE} \left(-bRu + \frac{1}{2}\frac{\partial R}{\partial \xi_1}u - \frac{1}{2}R\frac{\partial u}{\partial \xi_1}\right)d\xi_1 + \left(aRu - \frac{1}{2}\frac{\partial R}{\partial \eta_1}u + \frac{1}{2}R\frac{\partial u}{\partial \eta_1}\right)d\eta_1$$

$$+\int_{ED} \left(-bRu + \frac{1}{2}\frac{\partial R}{\partial \xi_1}u - \frac{1}{2}R\frac{\partial u}{\partial \xi_1}\right)d\xi_1 + \left(aRu - \frac{1}{2}\frac{\partial R}{\partial \eta_1}u + \frac{1}{2}R\frac{\partial u}{\partial \eta_1}\right)d\eta_1$$

$$+\int_{DC} \left(-bRu + \frac{1}{2}\frac{\partial R}{\partial \xi_1}u - \frac{1}{2}R\frac{\partial u}{\partial \xi_1}\right)d\xi_1 + \left(aRu - \frac{1}{2}\frac{\partial R}{\partial \eta_1}u + \frac{1}{2}R\frac{\partial u}{\partial \eta_1}\right)d\eta_1$$

$$+\int_{CB}\left(-bRu+\frac{1}{2}\frac{\partial R}{\partial \xi_1}u-\frac{1}{2}R\frac{\partial u}{\partial \xi_1}\right)d\xi_1+\left(aRu-\frac{1}{2}\frac{\partial R}{\partial \eta_1}u+\frac{1}{2}R\frac{\partial u}{\partial \eta_1}\right)d\eta_1$$

$$+\int_{BA}\left(-bRu+\frac{1}{2}\frac{\partial R}{\partial \xi_1}u-\frac{1}{2}R\frac{\partial u}{\partial \xi_1}\right)d\xi_1+\left(aRu-\frac{1}{2}\frac{\partial R}{\partial \eta_1}u+\frac{1}{2}R\frac{\partial u}{\partial \eta_1}\right)d\eta_1.$$

Then we obtain the identity

$$\int_1^{\eta_2}\left(\frac{1}{2}\frac{\partial R(\xi,\eta;\beta(\eta_1),\eta_1)}{\partial \xi_1}u(\beta(\eta_1),\eta_1)-\frac{1}{2}\frac{\partial u(\beta(\eta_1),\eta_1)}{\partial \xi_1}R(\xi,\eta;\beta(\eta_1),\eta_1)\right)\beta'(\eta_1)d\eta_1$$

$$-\int_1^{\eta_2} b(\beta(\eta_1),\eta_1)R(\xi,\eta;\beta(\eta_1),\eta_1)u(\beta(\eta_1),\eta_1)\beta'(\eta_1)d\eta_1$$

$$+\int_1^{\eta_2}\left(\frac{1}{2}\frac{\partial u(\beta(\eta_1),\eta_1)}{\partial \eta_1}R(\xi,\eta;\beta(\eta_1),\eta_1)-\frac{1}{2}\frac{\partial R(\xi,\eta;\beta(\eta_1),\eta_1)}{\partial \eta_1}u(\beta(\eta_1),\eta_1)\right)d\eta_1$$

$$+\int_1^{\eta_2} a(\beta(\eta_1),\eta_1)R(\xi,\eta;\beta(\eta_1),\eta_1)u(\beta(\eta_1),\eta_1)d\eta_1$$

$$-\int_0^{\eta}\left(\frac{1}{2}\frac{\partial R(\xi,\eta;\alpha(\eta_1),\eta_1)}{\partial \xi_1}u(\alpha(\eta_1),\eta_1)-\frac{1}{2}\frac{\partial u(\alpha(\eta_1),\eta_1)}{\partial \xi_1}R(\xi,\eta;\alpha(\eta_1),\eta_1)\alpha'(\eta_1)\right)d\eta_1$$

$$-\int_0^{\eta} b(\alpha(\eta_1),\eta_1)R(\xi,\eta;\alpha(\eta_1),\eta_1)u(\alpha(\eta_1),\eta_1)\alpha'(\eta_1)d\eta_1$$

$$-\int_0^{\eta}\left(\frac{1}{2}\frac{\partial u(\alpha(\eta_1),\eta_1)}{\partial \eta_1}R(\xi,\eta;\alpha(\eta_1),\eta_1)+\frac{1}{2}\frac{\partial R(\xi,\eta;\alpha(\eta_1),\eta_1)}{\partial \eta_1}u(\alpha(\eta_1),\eta_1)\right)d\eta_1$$

$$+\int_0^{\eta} a(\alpha(\eta_1),\eta_1)R(\xi,\eta,\alpha(\eta_1),\eta_1)u(\alpha(\eta_1),\eta_1)d\eta_1$$

$$-\frac{1}{2}R(\xi,\eta;\xi,\eta_2)u(\xi,\eta_2)-\frac{1}{2}R(\xi,\eta;\xi_0,\eta)u(\xi_0,\eta)=0, \tag{16}$$

where $(\beta(\eta_2),\eta_2)$ is a point of crossing $\xi_1=\beta(\eta_1)$ and $\xi_1=\xi$; and (ξ_0,η) is a point of crossing $\xi_1=\alpha(\eta_1)$ and $\eta_1=\eta$.

In (16) equating firstly $\xi=\alpha(\eta)$ and then $\xi=\beta(\eta)$, we get two identities

$$J_\alpha u \equiv I_\alpha u \tag{17}$$

$$-\int_1^{\eta_2}\left(\frac{\partial R(\alpha(\eta),\eta;\beta(\eta_1),\eta_1)}{\partial \eta_1}u(\beta(\eta_1),\eta_1)-\frac{\partial u(\beta(\eta_1),\eta_1)}{\partial \xi_1}R(\alpha(\eta),\eta;\beta(\eta_1),\eta_1)\right)d\eta_1$$

$$+\int_1^{\eta_2}\left(a(\beta(\eta_1),\eta_1)-b(\beta(\eta_1),\eta_1)\beta'(\eta_1)\right)R(\alpha(\eta),\eta;\beta(\eta_1),\eta_1)u(\beta(\eta_1),\eta_1)d\eta_1=0,$$

$$J_\beta u\equiv I_\beta u \tag{18}$$

$$+\int_0^{\eta}\left(\frac{\partial R(\beta(\eta),\eta;\alpha(\eta_1),\eta_1)}{\partial\eta_1}u(\alpha(\eta_1),\eta_1)-\frac{\partial u(\alpha(\eta_1),\eta_1)}{\partial\eta_1}R(\beta(\eta),\eta;\alpha(\eta_1),\eta_1)\right)d\eta_1$$

$$+\int_0^{\eta}\left(a(\alpha(\eta_1)),\eta_1)-b(\alpha(\eta_1),\eta_1)\alpha'(\eta_1)\right)R(\beta(\eta),\eta;\alpha(\eta_1),\eta_1)u(\alpha(\eta_1),\eta_1)d\eta_1=0.$$

Note that (17) and (18) are conditions connecting the values of the function u and its derivative on the boundaries $\xi=\alpha(\eta)$ and $\xi=\beta(\eta)$.

Thus, the following Lemma is proved:

Lemma 1 *The volume hyperbolic potential (9) satisfies the hyperbolic equation (5), the homogeneous initial conditions (6) and the boundary conditions:*

$$\begin{cases} I_\alpha u=0 \;\; at\; \alpha(\eta)\le 1,\\ J_\alpha u=0 \;\; at\; \alpha(\eta)\ge 1,\\ I_\beta u=0 \;\; at\; \eta\ge 0,\\ J_\alpha u=0 \;\; at\; \eta\le 0.\end{cases} \tag{19}$$

Corollary 1 *The volume hyperbolic potential (9) is the solution of the initial- boundary problem (5), (6), (19).*

4 Uniqueness of Solution of Problem (5), (6), (19)

The constructed in Sect. 3 boundary conditions (19) will uniquely define the volume hyperbolic potential (9) if the initial-boundary problem (5), (6), (19) has no other solutions except (9).

Lemma 2 *The solution of the initial-boundary problem (5), (6), (19) is unique.*

Proof. As usual, by $u_1(\xi;\eta)$ and $u_2(\xi;\eta)$ we denote two solutions of the initial-boundary problem (5), (6), (19). Then their difference $u(\xi;\eta)=u_1(\xi;\eta)-u_2(\xi;\eta)$ satisfies the homogeneous hyperbolic equation

$$\frac{\partial^2 u(\xi,\eta)}{\partial\xi\partial\eta}+a(\xi,\eta)\frac{\partial u(\xi,\eta)}{\partial\xi}+b(\xi,\eta)\frac{\partial u(\xi,\eta)}{\partial\eta}+c(\xi,\eta)u(\xi,\eta)=0,\ (\xi,\eta)\in\Omega, \tag{20}$$

the homogeneous initial conditions (6) and the boundary conditions (19). We apply the Green's theorem in a plane to the integral

$$0 = -\int_{\Omega_{\xi,\eta}} R(\xi,\eta;\xi_1\eta_1)\cdot 0\cdot d\xi_1 d\eta_1 = -\int_{\Omega_{\xi,\eta}} \left(RLu - uL^*R\right) d\xi_1 d\eta_1$$

$$= \int_1^{\eta_2} \left(\frac{1}{2}\frac{\partial R(\xi,\eta;\beta(\eta_1),\eta_1)}{\partial \xi_1} u(\beta(\eta_1),\eta_1) - \frac{1}{2}\frac{\partial u(\beta(\eta_1),\eta_1)}{\partial \xi_1} R(\xi,\eta;\beta(\eta_1),\eta_1)\right)\beta'(\eta_1)d\eta_1$$

$$-\int_1^{\eta_2} b(\beta(\eta_1),\eta_1)R(\xi,\eta;\beta(\eta_1),\eta_1)u(\beta(\eta_1),\eta_1)\beta'(\eta_1)d\eta_1$$

$$+\int_1^{\eta_2} \left(\frac{1}{2}\frac{\partial u(\beta(\eta_1),\eta_1)}{\partial \eta_1} R(\xi,\eta;\beta(\eta_1),\eta_1) - \frac{1}{2}\frac{\partial R(\xi,\eta;\beta(\eta_1),\eta_1)}{\partial \eta_1} u(\beta(\eta_1),\eta_1)\right) d\eta_1$$

$$+\int_1^{\eta_2} a(\beta(\eta_1),\eta_1)R(\xi,\eta;\beta(\eta_1),\eta_1)u(\beta(\eta_1),\eta_1)d\eta_1$$

$$-\int_0^{\eta} \left(\frac{1}{2}\frac{\partial R(\xi,\eta;\alpha(\eta_1),\eta_1)}{\partial \xi_1} u(\alpha(\eta_1),\eta_1) - \frac{1}{2}\frac{\partial u(\alpha(\eta_1),\eta_1)}{\partial \xi_1} R(\xi,\eta;\alpha(\eta_1),\eta_1)\alpha'(\eta_1)\right) d\eta_1$$

$$-\int_0^{\eta} b(\alpha(\eta_1),\eta_1)R(\xi,\eta;\alpha(\eta_1),\eta_1)u(\alpha(\eta_1),\eta_1)\alpha'(\eta_1)d\eta_1$$

$$-\int_0^{\eta} \left(\frac{1}{2}\frac{\partial u(\alpha(\eta_1),\eta_1)}{\partial \eta_1} R(\xi,\eta;\alpha(\eta_1),\eta_1) + \frac{1}{2}\frac{\partial R(\xi,\eta;\alpha(\eta_1),\eta_1)}{\partial \eta_1} u(\alpha(\eta_1),\eta_1)\right) d\eta_1$$

$$+\int_0^{\eta} a(\alpha(\eta_1),\eta_1)R(\xi,\eta,\alpha(\eta_1),\eta_1)u(\alpha(\eta_1),\eta_1)d\eta_1$$

$$-\frac{1}{2}R(\xi,\eta;\xi,\eta_2)u(\xi,\eta_2) - \frac{1}{2}R(\xi,\eta;\xi_0,\eta)u(\xi_0,\eta) + u(\xi,\eta) = 0. \tag{21}$$

In (21) equating firstly $\xi = \alpha(\eta)$ and then $\xi = \beta(\eta)$, we get two identities:

$$-J_\alpha u + u(\alpha(\eta),\eta) = 0, \tag{22}$$

and

$$-J_\beta u + u(\beta(\eta),\eta) = 0. \tag{23}$$

Taking into account homogenous boundary conditions (19), from (22), (23) we obtain that

$$u(\alpha(\eta),\eta) = 0,\ \eta < \eta_0, \tag{24}$$

$$u(\beta(\eta),\eta) = 0,\ \eta < 0, \tag{25}$$

where $\alpha(\eta_0) = 1$. Similarly to the case II and case III we have

$$u(\alpha(\eta), \eta) = 0, \eta_0 < \eta, \tag{26}$$

$$u(\beta(\eta), \eta) = 0, \eta > 0. \tag{27}$$

Thus, the function $u(\xi, \eta)$ satisfies the homogeneous hyperbolic equation (5), the homogeneous initial conditions (6) and the boundary conditions (24)–(27), it is the solution of the homogeneous first initial-boundary value problem. By virtue of the uniqueness of its solution we have $u(\xi, \eta) = 0$ at $(\xi, \eta) \in \Omega$. Consequently, $u_1(\xi, \eta) = u_2(\xi, \eta)$. Lemma 2 is proved.

5 Main Result Statement

Definition 1 As a *classical solution* of the initial-boundary problem (5), (6), (19) we call a function $u(\xi, \eta)$ from the class $u(\xi, \eta) \in C^2\left(\overline{\Omega}\right)$ satisfying Eq. (5) and the initial conditions (6) and the boundary conditions (19).

Combining the results of Lemmas 1 and 2, we obtain the main result of the paper.

Theorem 1 *Let $f(\xi, \eta) \in C^1\left(\overline{\Omega}\right)$. The volume hyperbolic potential (9) satisfies the hyperbolic equation (5), homogeneous initial conditions (6), boundary conditions (19).*
Conversely, for any $f(\xi, \eta) \in C^1\left(\overline{\Omega}\right)$ the initial boundary problem (5), (6), (19) has the unique classical solution $u(\xi, \eta) \in C^2\left(\overline{\Omega}\right)$ and this solution is presented in the form of the hyperbolic potential (9).

Corollary 2 *The boundary conditions (19) together with the initial conditions (6) uniquely determine the volume hyperbolic potential (9), i.e. are boundary conditions of the hyperbolic potential (19).*

6 The Case of Wave Potential

In this section we consider a special case, when $a, b, c = 0$. In this case

$$R(\xi, \eta; \xi_1, \eta_1) = 1.$$

In the case II, from Eq. (14), substituting $\xi = \alpha(\eta)$ and differentiating with respect to η, and taking into account the initial conditions (6), we have the following condition:

$$\frac{\partial u(\alpha(\eta), \eta_1)}{\partial \eta} = 0, \eta_0 < \eta < 0. \tag{28}$$

In the case III, from Eq. (15) substituting $\xi = \beta(\eta)$ and differentiating with respect to η, we have the following condition:

$$\frac{\partial u(\beta(\eta),\eta)}{\partial \xi}=0, 0<\eta<1. \tag{29}$$

For the case IV in the special case from (16) we have the next identity

$$-\int_1^{\eta_2}\frac{\partial u(\beta(\eta_1),\eta_1)}{\partial \xi_1}\beta^{'}(\eta_1)d\eta_1+\int_1^{\eta_2}\frac{\partial u(\beta(\eta_1),\eta_1)}{\partial \eta_1}d\eta_1$$

$$+\int_0^{\eta}\frac{\partial u(\alpha(\eta_1),\eta_1)}{\partial \xi_1}\alpha^{'}(\eta_1)d\eta_1-\int_0^{\eta}\frac{\partial u(\alpha(\eta_1),\eta_1)}{\partial \eta_1}d\eta_1-u(\xi,\eta_2)-u(\xi_0,\eta)=0. \tag{30}$$

Firstly in (30) equating $\xi=\alpha(\eta)$ and then $\xi=\beta(\eta)$, and differentiating with respect to η, and taking into account the initial conditions (6), we have the next conditions:

$$-\alpha^{'}(\eta)\frac{\partial u(\beta(\eta_2),\eta_2)}{\partial \xi}=\frac{\partial u(\alpha(\eta),\eta)}{\partial \eta}, \eta_0<\eta<0, \tag{31}$$

$$-\frac{\partial u(\alpha(\eta),\eta)}{\partial \eta}=\frac{\partial u(\beta(\eta),\eta)}{\partial \xi}\beta'(\eta), 0<\eta<1. \tag{32}$$

These boundary conditions look more clearly in variables (x,t). In coordinates (x,t) the volume hyperbolic potential is written in the form

$$u(x,t)=-\int_{\Omega_{x,t}} f_1(x_1,t_1)dx_1dt_1, \tag{33}$$

hyperbolic equation (1) has the form

$$\frac{\partial^2 u(x,t)}{\partial t^2}-\frac{\partial^2 u(x,t)}{\partial x^2}=f_1(x,t), (x,t)\in Q, \tag{34}$$

and the initial conditions (2) has the form

$$u(x,0)=\frac{\partial u}{\partial t}(x,0)=0,\ 0\le x\le 1, \tag{35}$$

For the case II, when $x=\alpha_1(t)$ from (28) we have

$$\left(\frac{\partial u}{\partial x}-\frac{\partial u}{\partial t}\right)(\alpha_1(t),t)=0, 0<t<t_1. \tag{36}$$

For the case III, when $x=\beta_1(t)$ from (29) we have

$$\left(\frac{\partial u}{\partial x}+\frac{\partial u}{\partial t}\right)(\beta_1(t),t)=0, 0<t<t_1. \tag{37}$$

For the case IV from (31), (32), when $x = \alpha_1(t)$ and $x = \beta_1(t)$ we have the next boundary conditions on the left-hand and right-hand sides of the domain $Q_{x,t}$:

$$\left(\frac{\partial u}{\partial x} - \frac{\partial u}{\partial t}\right)(\alpha_1(t), t)$$

$$= \frac{1+\alpha_1'(t)}{1-\alpha_1'(t)}\left(\frac{\partial u}{\partial x} + \frac{\partial u}{\partial t}\right)(\beta_1(t_2(t)), t_2(t)), t_1 < t < T, \tag{38}$$

$$\left(\frac{\partial u}{\partial x} + \frac{\partial u}{\partial t}\right)(\beta_1(t), t)$$

$$= \frac{1-\beta_1'(t)}{1+\beta_1'(t)}\left(\frac{\partial u}{\partial x} - \frac{\partial u}{\partial t}\right)(\alpha_1(t_0(t)), t_0(t)), t_1 < t < T, \tag{39}$$

where $(\alpha_1(t_0(t)), t_0(t))$ is a point of crossing of the boundary curve $x_1 = \alpha_1(t_1)$ and of the characteristics $x_1 = t_1 - t + \beta_1(t)$; $(\beta_1(t_2(t)), t_2(t))$ is a point of crossing of the boundary curve $x_1 = \beta_1(t_1)$ and of the characteristics $x_1 = t + \alpha_1(t) - t_1$.

The identity (38) holds for $t + \alpha_1(t) > 1$, and the identity (39) holds for $\beta_1(t) - t < 0$.

Both obtained identities (38), (39) connect with each other the traces of variables on the left-hand and right-hand boundaries of the domain $Q_{x,t}$. Herewith, since $t > t_2(t)$ and $t > t_0(t)$, then the points in which values are taken in the left-hand parts of this identities, are "above" than the points, in which values are taken in the right-hand parts of the identities.

Lemma 3 *The volume wave potential (28) satisfies the wave equation (34), the homogeneous initial conditions (35), the boundary condition on the left-hand boundary of the domain*

$$\left(\frac{\partial u}{\partial x} - \frac{\partial u}{\partial t}\right)(\alpha_1(t), t) = 0, \quad at\ 0 \le t \le T, \tag{40}$$

and the boundary condition on the right-hand boundary of the domain

$$\left(\frac{\partial u}{\partial x} + \frac{\partial u}{\partial t}\right)(\beta_1(t), t) = 0, \quad at\ 0 \le t \le T. \tag{41}$$

Corollary 3 *The volume wave potential (33) is the solution of the initial boundary value problem (34), (35), (40), (41).*

Boundary conditions (40), (41) have the following physical interpretation. It is well known that the general solution of the homogeneous equation (34), that is, the equation

$$\frac{\partial^2 u}{\partial x^2} - \frac{\partial^2 u}{\partial t^2} = 0 \tag{42}$$

is a superposition of two waves

$$u(x,t) = \phi(x+t) + \psi(x-t),$$

one of which ($\phi(x+t)$) extends to the left, and the second ($\psi(x-t)$) extends to the right.

It is easy to see that the boundary condition (40) is "transparent" for the wave going to the left, that is, for the wave of the form $\phi(x+t)$. Similarly, the boundary condition (41) is "transparent" for the wave going to the right, that is, for the wave of the form $\psi(x-t)$. These waves occur at some nonzero initial perturbation

$$u(x,0) = \tau(x),\ u_t(x,0) = \upsilon(x), \tag{43}$$

given at $t = 0$ on the segment $0 \le x \le 1$. These waves are given by

$$\phi(x+t) = \frac{1}{2}\tau(x+t) + \frac{1}{2}\int_a^{x+t} \upsilon(s)ds,\ \psi(x-t) = \frac{1}{2}\tau(x-t) + \frac{1}{2}\int_a^{x-t} \upsilon(s)ds.$$

Thus, if we consider the wave process of oscillation of an infinite string described by Eq. (34) at $-\infty < x < +\infty, t > 0$, with locally inhomogeneous initial conditions (43) (that is, in the case when $supp\{\tau(x)\} \subset [0, 1]$ and then to study the behavior of the string at the interval $0 \le x \le 1$ it is sufficient to consider the solutions of the Eq. (42) only at $0 \le x \le 1, t > 0$ with boundary conditions (40), (41). $supp\{\upsilon(x)\} \subset [0, 1]$)), then to study the behavior of the string at the interval $0 \le x \le 1$ it is sufficient to consider the solutions of the Eq. (42) only at $0 \le x \le 1, t > 0$ with boundary conditions (40), (41).

Acknowledgements The first author was supported by the MES RK grant AP05133271, and the second author was supported by the MES RK grant AP05133239.

References

1. Aleroev, T.S., Kirane, M., Malik, S.A.: Determination of a source term for a time fractional diffusion equation with an integral type over-determining condition. Electron. J. Differ. Equ. **270**, 1–16 (2013)
2. Copson, E.T.: On the Riemann – Green function. Arch. Ration. Mech. Anal. **1**, 324–348 (1957/58)
3. Engquist, B., Majda, A.: Radiation boundary conditions for acoustic and elastic wave calculations. Commun. Pure Appl. Math. **32**, 313–357 (1979)
4. Givoli, D.: Recent advances in the DtN finite element method for unbounded domains. Arch. Comput. Methods Eng. **6**, 71–116 (1999)
5. Givoli, D.: Numerical Methods for Problems in Infinite Domains. Elsevier, Amsterdam (1992)
6. Givoli, D.: Non-reflecting boundary conditions: a review. J. Comput. Phys. **94**, 1–29 (1991)
7. Greengard, L., Li, J.R.: On the numerical solution of the heat equation I: fast solvers in free space. J. Comput. Phys. **226**, 1891–1901 (2007)

8. Hagstrom, T.: Radiation boundary conditions for the numerical simulation of waves. Acta Numer. **8**, 47–106 (1999)
9. Kac, M.: On some connections between probability theory and differential and integral equations. In: Proceedings of the Second Berkeley Symposium on Mathematical Statistics and Probability, 1950, pp. 189–215, University of California Press, Berkeley and Los Angeles (1951)
10. Kac, M.: Integration in function spaces and some of its applications, Accademia Nazionale dei Lincei, Pisa (1980). Lezioni Fermiane. [Fermi Lectures]
11. Kal'menov, T.Sh., Suragan, D.: A boundary condition and spectral problems for the newton potential. Oper. Theory: Adv. Appl. **216**, 187–210 (2010)
12. Kal'menov, T.Sh., Suragan, D.: To spectral problems for the volume potential. Doklady Math. **80**(2), 646–649 (2009)
13. Kal'menov, T.Sh., Suragan, D.: Boundary conditions for the volume potential for the polyharmonic equation. Differ. Equ. **48**, 604–608 (2012)
14. Kal'menov, T.Sh., Suragan, D.: Initial-boundary value problems for the wave equation. Electron. J. Differ. Equ. **2014**, 1–7 (2014)
15. Kemppainen, J.: Properties of the single layer potential for the time fractional diffusion equation. J. Integral Equ. Appl. **23**(3), 437–455 (2011)
16. Kemppainen, J., Ruotsalainen, K.: Boundary integral solution of the time-fractional diffusion equation. Integr. Equ. Oper. Theory **64**, 239–249 (2009)
17. Kirichenko, S.V., Pul'kina, L.S.: A problem with nonlocal initial data for one-dimensional hyperbolic equation. Russ. Math. **58**(9), 13–21 (2014)
18. Pulkina, L.S.: A non-local problem with integral conditions for hyperbolic equations. Electron. J. Differ. Equ. **1999**(45), 1–6 (1999)
19. Pulkina, L.S.: Nonlocal problems for hyperbolic equations with degenerate integral conditions. Electron. J. Differ. Equ. **2016**(193), 1–12 (2016)
20. Riemann, B.: On the propagation of plane waves of finite amplitude, Works. OGIZ 376–395 (1948)
21. Riley, K.F., Hobson, M.P., Bence, S.J.: Mathematical Methods for Physics and Engineering. Cambridge University Press (2010)
22. Ruzhansky, M., Suragan, D.: Layer potentials, Kac's problem, and refined Hardy inequality on homogeneous Carnot groups. Adv. Math. **308**, 483–528 (2017)
23. Ruzhansky, M., Suragan, D.: On Kac's principle of not feeling the boundary for the Kohn Laplacian on the Heisenberg group. Proc. Amer. Math. Soc. **144**(2), 709–721 (2016)
24. Sadybekov, M., Derbissaly, B.: On an initial-boundary value problem for the wave potential in a domain with a curvilinear boundary. Kazakh Math. J. **18**, 53–66 (2018)
25. Sadybekov, M., Oralsyn, G.: On trace formulae of the generalised heat potential operator. Pseudo-Differ. Oper. Appl. **9**, 143–150 (2018)
26. Saito, N.: Data analysis and representation on a general domain using eigenfunctions of Laplacian. Appl. Comput. Harmon. Anal. **25**(1), 68–97 (2008)
27. Tsynkov, S.V.: Numerical solution of problems on unbounded domains. Appl. Numer. Math. **27**, 465–532 (1998)
28. Wu, X., Zhang, J.: High-order local absorbing boundary conditions for heat equation in unbounded domains. J. Comput. Math. **1**(29), 74–90 (2011)

Basic Theory of Impulsive Quaternion-Valued Linear Systems

Ardak Kashkynbayev and Manat Mustafa

Abstract In this chapter, we consider linear quaternion-valued ordinary differential equations (QDEs) with impulses at fixed times. We consider linear homogeneous and nonhomogeneous impulsive QDEs. Further, we prove the analogue of the Floquet theorem for linear periodic QDEs and discuss some consequences. Moreover, we prove the existence of bounded solutions to nonhomogeneous impulsive QDEs. Finally, we study periodic solutions of nonhomogeneous impulsive QDEs.

Keywords Impulsive differential equations · Quaternions · Floquet theory · Periodic solutions · Stability

1 Introduction and Preliminaries

In the last decades, a new type of differential equations so-called quaternion-valued differential equations (QDEs) were introduced and studied intensively by many authors with applications in various areas of physics and life sciences. The author in [13] have considered many applications of QDEs including quantum mechanics, attitude dynamics, Kalman filter design and spatial rigid body dynamics. Although there are already certain processes in the study of QDEs (see, e.g. [10, 13, 15, 17]), there are many open problems in the theory of QDEs with impulsive perturbations. Impulsive systems play an extremely important role in the applications of many real world problems [16]. Furthermore, by means of impulsive perturbations one can obtain rich dynamics which cannot be seen in regular continuous case. In this regard, many interesting results were obtained in bifurcation theory and chaos [2–8]. Recently, the authors in [12] obtained the existence and uniqueness conditions for the quaternion-valued nonlinear impulsive system. However, linear theory for impul-

A. Kashkynbayev (✉) · M. Mustafa
Department of Mathematics, Nazarbayev University, 010000 Nur-Sultan, Kazakhstan
e-mail: ardak.kashkynbayev@nu.edu.kz

M. Mustafa
e-mail: manat.mustafa@nu.edu.kz

A. Ashyralyev et al. (eds.), *Functional Analysis in Interdisciplinary Applications—II*,
Springer Proceedings in Mathematics & Statistics 351,
https://doi.org/10.1007/978-3-030-69292-6_21

sive QDEs has not been studied yet. Linear systems of QDEs can be seen as a first approximation to nonlinear problems. Therefore, the theory of linear systems are used in the analysis of many nonlinear problems. Motivated by this discussion, the aim of the present study is to extend the basics of linear QDEs to impulsive systems.

Throughout the paper we use the following notations: $\mathbb{R}^n$, $\mathbb{C}^n$, and $\mathbb{H}^n$, $n \in \mathbb{N}$, are the n -dimensional real, complex, and quaternion spaces, respectively. Let us briefly recall the definition of the set $\mathbb{H}$. The quaternion $q \in \mathbb{H}$ is the 4−component number of the from $q = q_1 + q_2 i + q_3 j + q_4 k$, where $q_1, q_2, q_3, q_4 \in \mathbb{R}$ and the imaginary parts i, j and k satisfy $i^2 = j^2 = k^2 = -1$ and $ij = -ji = k$, $ki = -ik = j$, $jk = -kj = i$. The norm of q is defined by the formula $|q| = \sqrt{q_1^2 + q_2^2 + q_3^2 + q_4^2}$. Denote by $\zeta = \{\zeta_l\}$ a sequence of real numbers such that $|\zeta_l| \to \infty$ as $|l| \to \infty$. We say that a function $x(t)$ which is defined on $\mathbb{H}$, belongs to the set $\mathscr{P}C(\mathbb{H}, \zeta)$ if it is left-continuous and continuous except, possibly $t = \zeta_l$, where it has discontinuities of the first kind.

Let $M_{m\times n}(\mathbb{H})$ be a space of $m \times n$ quaternion-valued matrices. The norms for the quaternionic vector $v = (v_1, \ldots, v_n) \in \mathbb{H}^n$ and matrix $A = (a_{ij})_{m\times n} \in M_{m\times n}(\mathbb{H})$ are defined respectively as $||v|| = \sum_{l=1}^{n} |v_l|$ and $||A|| = \sum_{\substack{1\le i\le m\\ 1\le j\le n}} |a_{ij}|$. Then a quaternion λ is called a right (left) eigenvalue of a matrix $P \in M_{n\times n}(\mathbb{H})$ if $Pv = v\lambda$ $(Pv = \lambda v)$. The left and right eigenvalues coincide if $P \in M_{n\times n}(\mathbb{C})$ and, of course, they are in general not equal if $P \in M_{n\times n}(\mathbb{H})$. Moreover, any $P \in M_{n\times n}(\mathbb{H})$ has infinitely many right eigenvalues among which exactly n are complex numbers with nonnegative imaginary parts [9, 14]. These right eigenvalues of P shall be called as the standard eigenvalues of P [18].

Let G_n be the symmetric group on $\{1, \ldots, n\}$. For any $\eta \in G_n$ we write η as the product of disjoint cycles: $\eta = (n_1 a_2 \cdots a_s)(n_2 b_2 \cdots b_h) \cdots (n_r c_2 \cdots c_l)$, where each n_e, $e = 1, \ldots, r$ is the largest number in its cycle and $n = n_1 > n_2 > \cdots > n_r \geq 1$. Let us denote the sign of permutation η by $\sigma(\eta)$. Then, the determinant of a square matrix $P = [p_{lm}]$ is defined as $\det P = \sum_{\eta\in G_n} \sigma(\eta) p_\eta$. Furthermore, we use so-called the double determinant of P as $\mathrm{ddet}P =: \det(P^* P)$, where P^* is the conjugate transpose of P [18]. It turns out that the double determinant is more useful for analysis for QDEs [13]. In particular, the following result will be useful in our proofs.

Lemma 1 ([18]) *Let P and Q be $n \times n$ quaternion-valued matrices. Then the followings hold true.*

(i) P is nonsingular if and only if $ddetP \neq 0$.
(ii) $ddet(PQ) = (ddetP)(ddetQ)$.
(iii) $ddetP = \prod_{r=1}^{n} |\lambda_r|^2$, where λ_r's are the standard eigenvalues of P.

The quaternion-valued functions $f_1(t)$, $f_2(t), \ldots, f_n(t)$ are said to be right linearly dependent on a real interval I if there are quaternionic constants $q_1, q_2, \ldots, q_n$ not all zero such that

$$f_1(t)q_1 + f_2(t)q_2 + \cdots + f_n(t)q_n = 0$$

for all $t \in I$. The functions $f_1(t), f_2(t), \ldots, f_n(t)$ are said to be right linearly independent on I if the equation

$$f_1(t)q_1 + f_2(t)q_2 + \cdots + f_n(t)q_n = 0$$

has only the trivial solution $q_1 = q_2 = \cdots = q_n = 0$.

2 Linear Homogeneous Systems

Let us consider a linear homogeneous impulsive QDE in the form

$$\begin{aligned} x'(t) &= A(t)x(t), \quad t \neq \zeta_l, \\ \Delta x\Big|_{t=\zeta_l} &=: x(\zeta_l+) - x(\zeta_l) = B_l x(\zeta_l). \end{aligned} \tag{1}$$

We assume that the entries of $n \times n$ matrix $A(t)$ are from $\mathscr{P}C(\mathbb{H}, \zeta)$, quaternion-valued matrices B_l are such that $B_l + I$ are invertible for each $l \in \mathbb{Z}$, i.e., $\mathrm{ddet}(B_l + I) \neq 0$, where I is the identity matrix and the real valued sequence $\{\zeta_l\}$ satisfies $|\zeta_l| \to \infty$ as $|l| \to \infty$.

Theorem 1 *Every solution $x(t) = x(t, t_0, x_0)$, $(t_0, x_0) \in \mathbb{R} \times \mathbb{H}^n$, of (1) is unique and continuable on $\mathbb{R}$.*

The proof easily follows from Theorem 3.4 [12] and Theorem 1 [13]. However, it is worth mentioning that the theory of QDEs is not the same as ODEs even for the linear case. In particular, one can show that the set of solutions of (1) does **not** form a linear vector space.

Definition 1 A set of right linearly independent solutions, $\{\phi_1(t), \phi_2(t), \ldots, \phi_n(t)\}$, of (1) is called a fundamental set of solutions of (1) and the $n \times n$ matrix

$$\Phi(t) = [\phi_1(t), \phi_2(t), \ldots, \phi_n(t)]$$

is called a fundamental matrix of (1).

Note that there are infinitely many different fundamental sets of solutions of (1)) and hence, infinitely many different fundamental matrices for (1).

Theorem 2 *A set of solutions, $\{\phi_1(t), \phi_2(t), \ldots, \phi_n(t)\}$, of (1) is a fundamental set of solutions if and only if the double determinant of $\Phi(t)$ is **not** zero all $t \in \mathbb{R}$, where $\Phi(t) = [\phi_1(t), \phi_2(t), \ldots, \phi_n(t)]$.*

Proof A set of solutions $\{\phi_1(t), \phi_2(t), \ldots, \phi_n(t)\}$ is a fundamental set of solutions of (1) if and only if it is right linearly independent, that is, the equation

$$\phi_1(t)q_1 + \phi_2(t)q_2 + \cdots + \phi_n(t)q_n = 0$$

has only the trivial solution for all $t \in \mathbb{R}$. In other words, $\Phi(t)q = 0$ has only the trivial solution for all $t \in \mathbb{R}$, where $q = (q_1, q_2, \dots, q_n)^T$. Thus, $\Phi(t)$ is nonsingular for all $t \in \mathbb{R}$. In the view of the item (i) of Lemma 1 this possible if and only if $\text{ddet}\Phi(t) \neq 0$ for all $t \in \mathbb{R}$.

Theorem 3 *A fundamental matrix $\Phi(t)$ of* (1) *satisfies the system* (1).

Proof It is straightforward to check that if $t \neq \zeta_l$ then

$$\begin{aligned}\Phi'(t) &= [\phi_1'(t), \phi_2'(t), \dots, \phi_n'(t)] = [A(t)\phi_1(t), A(t)\phi_2(t), \dots, A(t)\phi_n(t)] \\ &= A(t)[\phi_1(t), \phi_2(t), \dots, \phi_n(t)] = A(t)\Phi(t).\end{aligned}$$

Further, if $t = \zeta_l$ then

$$\begin{aligned}\Delta\Phi\Big|_{t=\zeta_l} &= \Phi(\zeta_l+) - \Phi(\zeta_l) \\ &= [\phi_1(\zeta_l+), \phi_2(\zeta_l+), \dots, \phi_n(\zeta_l+)] - [\phi_1(\zeta_l), \phi_2(\zeta_l), \dots, \phi_n(\zeta_l)] \\ &= [\phi_1(\zeta_l+) - \phi_1(\zeta_l), \phi_2(\zeta_l+) - \phi_2(\zeta_l), \dots, \phi_n(\zeta_l+) - \phi_n(\zeta_l)] \\ &= [B_l\phi_1(\zeta_l), B_l\phi_2(\zeta_l), \dots, B_l\phi_n(\zeta_l)] \\ &= B_l\Phi(\zeta_l).\end{aligned}$$

Theorem 4 *If $\Phi(t)$ is a fundamental matrix of* (1) *and if Q is any nonsingular constant $n \times n$ matrix, then $\Phi(t)Q$ is also a fundamental matrix of* (1). *Moreover, if $\Psi(t)$ is any other fundamental matrix of* (1), *then there exists a constant $n \times n$ nonsingular matrix P such that $\Psi(t) = \Phi(t)P$.*

Proof One can easily check if $t \neq \zeta_l$ then

$$(\Phi(t)Q)' = \Phi'(t)Q = (A(t)\Phi(t))\,Q = A(t)\,(\Phi(t)Q)\,,$$

and if $t = \zeta_l$ then

$$\Delta\Phi Q\Big|_{t=\zeta_l} = \Phi(\zeta_l+)Q - \Phi(\zeta_l)Q = (B_l + I)\Phi(\zeta_l) - \Phi(\zeta_l)Q = B_l\Phi(\zeta_l)Q.$$

On the other hand, by means of the item (ii) of Lemma 1 we have $\text{ddet}(\Phi(t)Q) = \text{ddet}\Phi(t)\text{ddet}Q \neq 0$ since $\Phi(t)$ is a fundamental matrix and Q is a nonsingular matrix. Thus, by Theorem 2 $\Phi(t)Q$ is a fundamental matrix of (1).

Next, let $\Psi(t)$ be any other fundamental matrix. Note that since $\Phi(t)$ is a fundamental matrix $\Phi(t)$ is nonsingular for all $t \in \mathbb{R}$. Thus, $\Phi^{-1}(t)$ exits and $\Phi(t)\Phi^{-1}(t) = I$ for all $t \in \mathbb{R}$. The by the product rule one has $\Phi'(t)\Phi^{-1}(t) + \Phi(t)\left(\Phi^{-1}(t)\right)' = 0$. Thus,
$\left(\Phi^{-1}(t)\right)' = -\Phi^{-1}(t)\Phi'(t)\Phi^{-1}(t) = -\Phi^{-1}(t)A(t)\Phi(t)\Phi^{-1}(t) = -\Phi^{-1}(t)A(t)$.
Now, consider the derivative of $\Phi^{-1}(t)\Psi(t)$

$$\begin{aligned}\left(\Phi^{-1}(t)\Psi(t)\right)' &= \left(\Phi^{-1}(t)\right)'\Psi(t) + \Phi^{-1}(t)\Psi'(t)\\ &= -\Phi^{-1}(t)A(t)\Psi(t) + \Phi^{-1}(t)A(t)\Psi(t) = 0.\end{aligned}$$

Hence, $\Phi^{-1}(t)\Psi(t) = P$ or $\Psi(t) = \Phi(t)P$.

Definition 2 A fundamental matrix Φ of (1) whose columns are determined by the right linearly independent solutions $\phi_1(t), \phi_2(t), \ldots, \phi_n(t)$ with

$$\phi_1(s) = e_1,\ \phi_2(s) = e_2,\ \ldots,\ \phi_n(s) = e_n,\quad s \in \mathbb{R},$$

is called the state transition matrix Φ for (1), where $\{e_1, e_2, \ldots, e_n\}$ is the standard basis of $\mathbb{R}^n$. Equivalently, if Ψ is any fundamental matrix of (1) , then the matrix Φ determined by $\Phi(t,s) =: \Psi(t)\Psi^{-1}(s)$, for all $t,\ s \in \mathbb{R}$, is said to be the state transition matrix of (1).

Let $X(t,s)$ be the state transition matrix of the linear QDE

$$x'(t) = A(t)x(t). \tag{2}$$

Let us construct a fundamental matrix $\Psi(t)$ of (1) using the transition matrix $X(t,s)$ of (2) and matrices $B_l,\ l \in \mathbb{Z}$. To this end, recall that all solutions of (1) are defined on $\mathbb{H}$ and the solution of the initial value problem is unique by Theorem 1. Next, assume that $\zeta_0 < t_0 \leq \zeta_1$. Then one can obtain that

$$\Psi(t) = \begin{cases}\Psi(t_0), & \text{if } t = t_0\\ X(t,t_0)\Psi(t_0), & \text{if } t_0 \leq t \leq \zeta_1,\end{cases} \tag{3}$$

and $\Psi(t) = X(t,\zeta_m)(I + B_m)X(\zeta_m,\zeta_{m-1})\cdots(I + B_1)X(\zeta_1,t_0)\Psi(t_0)$, if $\zeta_1 \leq \zeta_m < t \leq \zeta_{m+1}$. There are many different cases to compute the state transition matrix of (1). We consider the following cases:

$$\Phi(t,s) = \Psi(t)\Psi^{-1}(s) = \begin{cases}I, & \text{if } t = s\\ X(t,s), & \text{if } t_0 \leq s \leq t \leq \zeta_1,\end{cases} \tag{4}$$

and $\Phi(t,s) = X(t,\zeta_m)(I + B_m)X(\zeta_m,\zeta_{m-1})\cdots(I + B_p)X(\zeta_p,s)$, if $s < \zeta_p \leq \zeta_m < t \leq \zeta_{m+1}$.

Theorem 5 *Let $x(t_0) = x_0$ and $\Phi(t,s)$ denote the state transition matrix for (1) for all $t, s \in \mathbb{R}$, Then*

(i) $\Phi(t,s)$ is the unique solution of the matrix equation

$$\frac{\partial}{\partial t}\Phi(t,s) =: \Phi'(t,s) = A(t)\Phi(t,s),$$

with $\Phi(s,s) = I$.

(ii) $\Phi(t,s)$ *is nonsingular for all* $t, s \in \mathbb{R}$.
(iii) *for any* $t, \tau, s \in \mathbb{R}$ *we have* $\Phi(t,s) = \Phi(t,\tau)\Phi(\tau,s)$.
(iv) *the unique solution* $\phi(t, t_0, x_0)$ *of* (1), *with* $\phi(t_0, t_0, x_0) = x_0$, *is given by* $\phi(t, t_0, x_0) = \Phi(t, t_0)x_0$.

Proof (i) Let $\Psi(t)$ be any fundamental matrix of (1). Then, Definition 2, $\Phi(t,s) = .\Psi(t)\Psi^{-1}(s)$ Thus,

$$\frac{\partial}{\partial t}\Phi(t,s) = \Phi'(t,s) = \Psi'(t)\Psi^{-1}(s) = A(t)\Psi(t)\Psi^{-1}(s) = A(t)\Phi(t,s).$$

Moreover, $\Phi(s,s) = \Psi(s)\Psi^{-1}(s) = I$.

(ii) Employing the item (ii) of Lemma 1 we have $\mathrm{ddet}\Phi(t,s) = \mathrm{ddet}(\Psi(t)\Psi^{-1}(s)) = \mathrm{ddet}(\Psi(t))\mathrm{ddet}(\Psi^{-1}(s)) \neq 0$ since $\Psi(t)$ is a fundamental matrix. Thus, by the item (i) of Lemma 1 $\Phi(t,s)$ is nonsingular for all $t, s \in \mathbb{R}$.

(iii) It is easy to see that $\Phi(t,s) = \Psi(t)\Psi^{-1}(s) = \Psi(t)\Psi^{-1}(\tau)\Psi(\tau)\Psi^{-1}(s) = \Phi(t,\tau)\Phi(\tau,s)$.

(iv) By the uniqueness results in Theorem 1 we know that the system (1) has a unique solution $x(t)$ with $\phi(t_0) = x_0$. Note that $\phi(t_0) = \Phi(t_0, t_0)x_0 = x_0$. Next, differentiating both sides of $\phi(t) = \Phi(t, t_0)x_0$ we get

$$\phi'(t) = \Phi'(t, t_0)x_0 = A(t)\Phi(t, t_0)x_0 = A(t)\phi(t)$$

which shows that $\phi(t)$ is the desired solution.

2.1 Linear Systems with Periodic Coefficients and Floquet Theory

Now, let us consider a linear impulsive system (1) in the form

$$\begin{aligned} x'(t) &= A(t)x(t), \quad t \neq \zeta_l, \\ \Delta x\big|_{t=\zeta_l} &=: x(\zeta_l+) - x(\zeta_l) = B_l x(\zeta_l), \end{aligned} \tag{5}$$

where $A(t)$ is a $n \times n$ quaternion-valued $\omega-$periodic matrix, i.e., $A(t+\omega) = A(t)$ and there exits a natural number p such that $B_{l+p} = B_l$ and $\zeta_{l+p} = \zeta_l + \omega$ for all $l \in \mathbb{Z}$. We shall call the system (5) $(\omega, p)-$periodic. Further, as in (1) the matrices B_l satisfy $\mathrm{ddet}(B_l + I) \neq 0$.

In the sequel, we will need so-called Liouville formula quaternion-valued ODEs. In this regard, we define the Wronskian of (1) by $W(t) = \frac{1}{2}\mathrm{ddet}X(t)$ which also agrees with the definition of the Wronskian defined in [15]. The following result proved in [13] for 2×2 matrices, can be easily generalized for the $n \times n$ case.

Theorem 6 ([13]) *The Wronskian of* (1) *satisfies*

$$W(t) = exp\left(\int_{t_0}^{t} tr(A(s) + A^*(s))ds\right) W(t_0),$$

where $tr\,A(t)$ *is the trace of the matrix* $A(t)$.

The authors in [13] used the row formula for the determinant

$$\det A = \text{rdet}\begin{pmatrix} a_{11} & a_{12} \\ a_{21} & a_{22} \end{pmatrix} =: a_{11}a_{22} - a_{12}a_{21}$$

to prove the Liouville formula. However, the results are also valid if $\text{rdet}A$ is replaced by $\det A = \text{cdet}\begin{pmatrix} a_{11} & a_{12} \\ a_{21} & a_{22} \end{pmatrix} =: a_{11}a_{22} - a_{21}a_{12}$ which corresponds to the definition of the determinant in this chapter.

Theorem 7 *Let* $\Phi(t)$ *be a fundamental matrix solution of* (5). *Then*

i. $\Phi(t+\omega)$ *is also a fundamental matrix.*
ii. *There exist a nontrivial solution* $\phi(t)$ *to* (5) *such that* $\phi(t+\omega) = \phi(t)\mu$, $\mu \in \mathbb{H}$.

Proof *i*. The first item is straightforward. If $t \neq \zeta_l$, then

$$\Phi'(t+\omega) = A(t+\omega)\Phi(t+\omega) = A(t)\Phi(t+\omega).$$

On the other hand, If $t = \zeta_l$, then

$$\begin{aligned} \Delta\Phi(t+\omega)\Big|_{t=\zeta_l} &= \Phi(\zeta_l+\omega+) - \Phi(\zeta_l+\omega) \\ &= [\phi_1(\zeta_l+\omega+) - \phi_1(\zeta_l+\omega), \ldots, \phi_n(\zeta_l+\omega+) - \phi_n(\zeta_l+\omega)] \\ &= [B_l\phi_1(\zeta_l+\omega), B_l\phi_2(\zeta_l+\omega), \ldots, B_l\phi_n(\zeta_l+\omega)] \\ &= B_l\Phi(\zeta_l+\omega). \end{aligned}$$

Moreover, $\text{ddet}\Phi(t+\omega) \neq 0$ since $\text{ddet}\Phi(t) \neq 0$. Thus, $\Phi(t+\omega)$ is a fundamental matrix of (5).

ii. By Theorem 4 there exist an invertible constant quaternion-valued matrix T such that $\Phi(t+\omega) = \Phi(t)T$. Thus, we can find the matrix T by $T = \Phi^{-1}(t_0)\Phi(t_0+\omega)$. Without loss of generality we choose $t_0 = 0$ and $\Phi(0) = I$ and thus $T = \Phi(\omega)$. Let μ be a right eigenvalue of $\Phi(\omega)$ corresponding to an eigenvector v, i.e., $\Phi(\omega)v = v\mu$. Consider the solution $\phi(t) = \Phi(t)v$. Then, it is easy to see that

$$\phi(t+\omega) = \Phi(t+\omega)v = \Phi(t)\Phi(\omega)v = \Phi(t)v\mu = \phi(t)\mu.$$

The proof is complete.

Corollary 1 *The constant* μ *does not depend on the choice of* Φ.

Proof Indeed, let $\Psi(t)$ be a fundamental matrix of (5) different from $\Phi(t)$. Then, by Theorem 4 there exist an invertible constant matrix P such that $\Psi(t) = \Phi(t)P$. Thus,

$\Psi(t+\omega) = \Phi(t+\omega)P = \Phi(t)\Phi(\omega)P = \Psi(t)Q^{-1}\Phi(\omega)P$. Setting $R = P^{-1}\Phi(\omega)P$ and $u = P^{-1}v$ yields $Ru = P^{-1}\Phi(\omega)Pu = P^{-1}\Phi(\omega)v = P^{-1}v\mu = u\mu$. Thus, μ is the right eigenvalue of R associated with the eigenvector u. This completes the proof.

Corollary 2 *The system* (5) *has a nontrivial periodic solution if μ is the lth root of unity, i.e., $\mu^l = 1$, $l \in \mathbb{N}$.*

Proof Theorem 7 (ii) yields $\phi(t+l\omega) = \phi(t+(l-1)\omega)\mu = \cdots = \phi(t)\mu^l = \phi(t)$.

Definition 3 We shall call the matrix $\Phi(\omega)$ as the monodromy matrix and its standard eigenvalues $\mu_1, \ldots, \mu_n$ as the characteristic multipliers. The characteristic exponents or Floquet exponents are defined to be the numbers $\tau_1, \ldots, \tau_n$ such that $\mu_1 = e^{\tau_1\omega}, \ldots, \mu_n = e^{\tau_n\omega}$.

Since τ_m's, $m = 1, \ldots, n$, are defined to an additive multiple of $2\pi\sqrt{-1}/\omega$, i.e., $\mu_m = e^{\tau_m\omega} = e^{(\tau_m + 2\pi\sqrt{-1}/\omega)\omega}$, we require all the imaginary components of $\tau_m\omega$ lie within the interval $[-\pi, \pi]$ for $m = 1, \ldots, n$.

Theorem 8 *Let τ_m be the characteristic exponents corresponding to standard eigenvalues μ_m. Then*

i. There exist $\omega-$periodic functions $v_m(t)$ such that $\phi_m(t) = v_m(t)e^{\tau_m t}$.

ii. $|\mu_1|\cdots|\mu_n| = exp\left(\int_0^\omega Re(tr(A(s)))ds\right)$.

Proof *i.* By Theorem 7 (ii), we know that $\phi_m(t+\omega) = \phi_m(t)\mu_m = \phi_m(t)e^{\tau_m\omega}$. Setting $v_m(t) = \phi_m(t)e^{-\tau_m t}$ yields $v_m(t+\omega) = \phi_m(t+\omega)e^{-\tau_m(t+\omega)} = \phi_m(t)e^{\tau_m\omega}e^{-\tau_m(t+\omega)}$. As $\tau_m\omega$ commutes with $-\tau_m(t+\omega)$ we have $e^{\tau_m\omega}e^{-\tau_m(t+\omega)} = e^{-\tau_m t}$. Thus, $v_m(t+\omega) = \phi_m(t)e^{-\tau_m t} = v_m(t)$.

ii. Note that $W(0) = \text{ddet}\Phi(0) = 1$ since $\Phi(0) = I$. Thus, we conclude the proof by using the item (iii) of Lemma 1 and Theorem 6 with $t_0 = 0$ and $t = \omega$, that is,

$$
\begin{aligned}
|\mu_1|^2\cdots|\mu_n|^2 &= \text{ddet}\Phi(\omega) = W(T) \\
&= \exp\left(\int_0^\omega Re(tr(A(s) + A^*(s)))ds\right) \\
&= \exp\left(\int_0^\omega 2Re(tr(A(s)))ds\right).
\end{aligned}
$$

Thus, it yields that $|\mu_1|\cdots|\mu_n| = \exp\left(\int_0^\omega Re(tr(A(s)))ds\right)$. □

Corollary 3 *Every fundamental matrix $\Psi(t)$ of* (5) *has the form $\Psi(t) = V(t)e^{Rt}$, where $V(t)$ is an $\omega-$periodic $n \times n$ matrix and the diagonal matrix R is such that $R = diag(\tau_1, \ldots, \tau_n)$.*

Proof Let $\Psi(t) = [\psi_1(t) \cdots \psi_n(t)]$ where $\psi_m(t)$ is a solution of (5) for $m = 1, \ldots, n$. Then Theorem 8 (i) yields
$\Psi(t) = [\psi_1(t) \cdots \psi_n(t)] = \left[v_1(t)e^{\tau_1 t} \cdots v_n(t)e^{\tau_n t}\right] = [v_1(t) \cdots v_n(t)] \operatorname{diag}(e^{\tau_1 t}, \ldots, e^{\tau_n t}) = [v_1(t) \cdots v_n(t)]\, e^{Rt}$. We complete the proof by defining $V(t) = [v_1(t) \cdots v_n(t)]$.

Corollary 4 *If the norm of the characteristic multipliers is not greater than one, i.e., $|\mu| \leq 1$, and the the characteristic multipliers with the norm $|\mu| = 1$ have simple divisors then the system (5) is stable. If the the norm of the characteristic multipliers is less than one, i.e., $|\mu| < 1$, then the system (5) is asymptotically stable.*

Proof The proof follows immediately from part Theorem 8 (i) since $|\mu| < 1$ if and only if $Re(\tau) < 0$, $|\mu| > 1$ if and only if $Re(\tau) > 0$, and $|\mu| = 1$ if and only if $Re(\tau) = 0$.

3 Linear Nonhomogeneous Systems

In this part, we consider a linear nonhomogeneous impulsive QDE in the form

$$\begin{aligned} x' &= A(t)x + f(t), \quad t \neq \zeta_l, \\ \Delta x\big|_{t=\zeta_l} &= B_l x + J_l, \end{aligned} \tag{6}$$

where the entries of $n \times n$ matrix $A(t)$ are from $\mathscr{PC}(\mathbb{H}, \zeta)$ and bounded on $\mathbb{H}$, quaternion-valued matrices B_l are such that $\operatorname{ddet}(B_l + I) \neq 0$, and the real valued sequence $\{\zeta_l\}$ satisfies $|\zeta_l| \to \infty$ as $|l| \to \infty$. We assume that there exit positive real numbers $\underline{\zeta}$ and $\overline{\zeta}$ such that $\underline{\zeta} \leq \zeta_{l+1} - \zeta_l \leq \overline{\zeta}$ for all $l \in \mathbb{Z}$. Components of the vector-function $f : \mathbb{R} \to \mathbb{H}^n$ belong to $\mathscr{PC}(\mathbb{H}, \zeta)$ and J_l is the sequence of quaternion-valued vectors. Further, we assume that there exits a positive real number M such that

$$\sup_t ||f|| + \sup_l ||J_l|| < M. \tag{7}$$

One can easily check that the system (6) satisfies all of the conditions of Theorem 3.1 in [12]. Thus, we have the following uniqueness result.

Theorem 9 *Every solution $x(t) = x(t, t_0, x_0)$, $(t_0, x_0) \in \mathbb{R} \times \mathbb{H}^n$, of (6) is unique and continuable on $\mathbb{R}$.*

Proof To construct the general solution of the system (6) we use the method of variation of parameters. That is, we seek to find a solution of the from $x(t) = \Psi(t)y(t)$, where $\Psi(t)$ is a fundamental matrix of (1). Substituting the transformation to the system (6) yields $x'(t) = A(t)\Psi(t)y(t) + f(t)$ if $t \neq \zeta_l$. On the other hand, by the Leibnitz product rule one has $x'(t) = \Psi'(t)y(t) + \Psi(t)y'(t) = A(t)\Psi(t)y(t) + \Psi(t)y'(t)$. Thus, we have $y'(t) = \Psi^{-1}(t)f(t)$ if $t \neq \zeta_l$.
If $t = \zeta_l$ then

$$\begin{aligned}B_l\Psi(\zeta_l)y(\zeta_l)+J_l &= \Psi(\zeta_l+)y(\zeta_l+)-\Psi(\zeta_l)y(\zeta_l)\\ &= \Psi(\zeta_l+)(y(\zeta_l+)-y(\zeta_l))+(\Psi(\zeta_l+)-\Psi(\zeta_l))y(\zeta_l)\\ &= \Psi(\zeta_l+)(y(\zeta_l+)-y(\zeta_l))+B_l\Psi(\zeta_l)y(\zeta_l).\end{aligned}$$

The last equation follows from $\Psi(\zeta_l+)-\Psi(\zeta_l)=B_l\Psi(\zeta_l)$ since $\Psi(t)$ is a fundamental matrix of (1). Thus, we obtain $\Delta y\big|_{t=\zeta_l}=\Psi^{-1}(\zeta_l+)J_l$. Consequently, it follows that $y(t)$ satisfies the following system

$$\begin{aligned}&y'(t)=\Psi^{-1}(t)f(t),\ \ t\neq\zeta_l,\\ &\Delta y\big|_{t=\zeta_l}=\Psi^{-1}(\zeta_l+)J_l.\end{aligned}\tag{8}$$

Next, by Theorem 3.3 from [12] or Theorem 2.4.1 from [1] a solution of (8) with $y(t_0)=y_0$ satisfies

$$y(t)=y_0+\int_{t_0}^{t}\Psi^{-1}(s)f(s)ds+\sum_{t_0\le\zeta_l<t}\Psi^{-1}(\zeta_l+)J_l,\ \ t\ge t_0.$$

Since $x(t)=\Psi(t)y(t)$ and $y_0=\Psi^{-1}(t_0)x(t_0)$ we obtain

$$x(t)=\Psi(t)\Psi^{-1}(t_0)x(t_0)+\int_{t_0}^{t}\Psi(t)\Psi^{-1}(s)f(s)ds+\sum_{t_0\le\zeta_l<t}\Psi(t)\Psi^{-1}(\zeta_l+)J_l,\ \ t\ge t_0.\tag{9}$$

Since $\Psi(t)\Psi^{-1}(s)=\Phi(t,s)$ the last expression is equivalent to

$$x(t,t_0,x_0)=\Phi(t,t_0)x(t_0)+\int_{t_0}^{t}\Phi(t,s)f(s)ds+\sum_{t_0\le\zeta_l<t}\Phi(t,\zeta_l+)J_l,\ \ t\ge t_0.\tag{10}$$

3.1 Bounded Solutions

Let us give the definition of exponential dichotomy for the system (1).

Definition 4 ([11]) The system (1) is said to satisfy exponential dichotomy, if there exists a projection matrix P and positive real numbers K_1, K_2, α_1 and α_2 such that

$$\begin{aligned}&||\Phi(t)P\Phi^{-1}(s)||\le K_1e^{-\alpha_1(t-s)},\ \ t\ge s,\\ &||\Phi(t)(I-P)\Phi^{-1}(s)||\le K_2e^{\alpha_2(t-s)},\ \ t\le s.\end{aligned}\tag{11}$$

Theorem 10 *Assume that for the system* (6) *the condition* (7) *is fulfilled and the system* (1) *satisfy exponential dichotomy. Then, bounded on* $\mathbb{H}$, *function* $x(t)$ *is a unique solution* (6) *if and only if it satisfies the following equation*

$$
\begin{aligned}
x(t) = & \int_{-\infty}^{t} \Phi(t)P\Phi^{-1}(s)f(s)ds - \int_{t}^{\infty} \Phi(t)(I-P)\Phi^{-1}(s)f(s)ds \\
& + \sum_{\zeta_l<t} \Phi(t)P\Phi^{-1}(\zeta_l+)J_l - \sum_{t\leq\zeta_l} \Phi(t)(I-P)\Phi^{-1}(\zeta_l+)J_l.
\end{aligned}
\tag{12}
$$

Proof First of all we show that the function (12) satisfies the system (6). Indeed, if $t \neq \zeta_l$, then

$$
\begin{aligned}
x'(t) = & \int_{-\infty}^{t} \Phi'(t)P\Phi^{-1}(s)f(s)ds + \Phi(t)P\Phi^{-1}(t)f(t) \\
& - \int_{t}^{\infty} \Phi'(t)(I-P)\Phi^{-1}(s)f(s)ds + \Phi(t)(I-P)\Phi^{-1}(t)f(t) \\
& + \sum_{\zeta_l<t} \Phi'(t)P\Phi^{-1}(\zeta_l+)J_l - \sum_{t\leq\zeta_l} \Phi'(t)(I-P)\Phi^{-1}(\zeta_l+)J_l \\
= & A(t)\int_{-\infty}^{t} \Phi(t)P\Phi^{-1}(s)f(s)ds + \Phi(t)P\Phi^{-1}(t)f(t) \\
& - A(t)\int_{t}^{\infty} \Phi(t)(I-P)\Phi^{-1}(s)f(s)ds + f(t) - \Phi(t)P\Phi^{-1}(t)f(t) \\
& + A(t)\sum_{\zeta_l<t} \Phi(t)P\Phi^{-1}(\zeta_l+)J_l - A(t)\sum_{t\leq\zeta_l} \Phi(t)(I-P)\Phi^{-1}(\zeta_l+)J_l \\
= & A(t)x(t) + f(t).
\end{aligned}
$$

Further, for the fixed l, if $t = \zeta_l$, taking into account $\Phi(\zeta_l+) = (B_l + I)\Phi(\zeta_l)$ we have

$$
\begin{aligned}
\Delta x\big|_{t=\zeta_l} = & x(\zeta_l+) - x(\zeta_l) \\
= & \int_{-\infty}^{\zeta_l+} \Phi(\zeta_l+)P\Phi^{-1}(s)f(s)ds - \int_{\zeta_l+}^{\infty} \Phi(\zeta_l+)(I-P)\Phi^{-1}(s)f(s)ds \\
& + \sum_{\zeta_l<\zeta_l+} \Phi(\zeta_l+)P\Phi^{-1}(\zeta_l+)J_l - \sum_{\zeta_l+\leq\zeta_l} \Phi(\zeta_l+)(I-P)\Phi^{-1}(\zeta_l+)J_l \\
& - \int_{-\infty}^{\zeta_l} \Phi(\zeta_l)P\Phi^{-1}(s)f(s)ds + \int_{\zeta_l}^{\infty} \Phi(\zeta_l)(I-P)\Phi^{-1}(s)f(s)ds \\
& - \sum_{\zeta_l<\zeta_l} \Phi(\zeta_l)P\Phi^{-1}(\zeta_l+)J_l + \sum_{\zeta_l\leq\zeta_l} \Phi(\zeta_l)(I-P)\Phi^{-1}(\zeta_l+)J_l \\
= & B_l\int_{-\infty}^{\zeta_l} \Phi(\zeta_l)P\Phi^{-1}(s)f(s)ds - B_l\int_{\zeta_l}^{\infty} \Phi(\zeta_l)(I-P)\Phi^{-1}(s)f(s)ds \\
& + B_l\sum_{\zeta_l<\zeta_l} \Phi(\zeta_l)P\Phi^{-1}(\zeta_l+)J_l - B_l\sum_{\zeta_l\leq\zeta_l} \Phi(\zeta_l)(I-P)\Phi^{-1}(\zeta_l+)J_l + J_l \\
= & B_l x(\zeta_l) + J_l.
\end{aligned}
$$

Thus, $x(t)$ satisfies the system (6). It remains to show that both the integral and infinite sum are convergent uniformly for all t. One can easily check that

$$\begin{aligned}||x(t)|| &\leq \int_{-\infty}^{t} K_1 e^{-\alpha_1(t-s)}||f(s)||ds + \int_{t}^{\infty} K_2 e^{\alpha_2(t-s)}||f(s)||ds \\ &+ \sum_{\zeta_l<t} K_1 e^{-\alpha_1(t-\zeta_l)}||J_l|| + \sum_{t\leq\zeta_l} K_2 e^{\alpha_2(t-\zeta_l)}||J_l|| \\ &\leq \frac{K_1 M}{\alpha_1} + \frac{K_2 M}{\alpha_2} + K_1 M \sum_{i=0}^{\infty} e^{-\alpha_1 \underline{\zeta} i} + K_2 M \sum_{i=0}^{\infty} e^{-\alpha_2 \overline{\zeta} i} \\ &= \frac{K_1 M}{\alpha_1} + \frac{K_2 M}{\alpha_2} + \frac{K_1 M}{1-e^{-\alpha_1 \underline{\zeta}}} + \frac{K_2 M}{1-e^{-\alpha_2 \overline{\zeta}}} < \infty.\end{aligned}$$

Thus, series are convergent. This finalizes the proof.

It is worth mentioning that the Eq. (12) can be represented in the following form.

$$x(t) = \int_{-\infty}^{\infty} G_b(t,s) f(s)ds + \sum_{l=-\infty}^{\infty} G_b(t,\zeta_l+)J_l, \tag{13}$$

where $G_b(t,s)$ is a Green function for the bounded solution given by

$$G_b(t,s) = \begin{cases} \Phi(t)P\Phi^{-1}(s), & t \geq s, \\ \Phi(t)(I-P)\Phi^{-1}(s). & t < s. \end{cases} \tag{14}$$

3.2 Periodic Solutions

Let us consider the system (6) with periodic coefficients.

$$\begin{aligned} x' &= A(t)x + f(t), \ \ t \neq \zeta_l, \\ \Delta x\big|_{t=\zeta_l} &= B_l x + J_l, \end{aligned} \tag{15}$$

where $A(t)$ and $f(t)$ are $\omega-$periodic, i.e., $A(t+\omega) = A(t)$ and $f(t+\omega) = f(t)$, B_l and J_l are $p-$periodic and $\zeta_{l+p} = \zeta_l + \omega$. Further, the matrices B_l satisfy ddet$(I + B_l) \neq 0$.

Let us assume that the linear homogeneous system does not have any nontrivial $\omega-$periodic solution, i.e. the identity is not one of its characteristic multipliers. Thus, $\Phi(\omega)v = v$ has only the trivial solution. In other words, the matrix $\Phi(\omega) - I$ is invertible. Further, without loss of generality, we assume $\zeta_l \neq 0$ for all $l \in \mathbb{Z}$. Then, $\zeta_l \neq \omega$. Let us introduce a Green's function for (15).

$$G_p(t,s)=\begin{cases}\Phi(t)(I-\Phi(\omega))^{-1}\Phi^{-1}(s), & 0\le s\le t\le\omega,\\ \Phi(t+\omega)(I-\Phi(\omega))^{-1}\Phi^{-1}(s). & 0\le t<s\le\omega.\end{cases} \tag{16}$$

Lemma 2 *The Green's function* (16) *satisfies the following properties.*

1. $G'_p(t,s)=A(t)G_p(t,s),\ t\neq s$ *and* $\Delta G_p(t,s)\big|_{t=\zeta_l}=B_lG_p(\zeta_l,s),\ s\neq\zeta_l,\ l\in\mathbb{Z}$.
2. $G_p(s+,s)-G_p(s,s)=I,\ \ s\neq\zeta_l,\ 1\le l\le p$.
3. $G_p(\omega,s)=G_p(0,s),\ \ s\in[0,\omega)$.
4. $G_p(\omega,\omega)-G_p(0,\omega)=I$.

The proof of Lemma 2 is straightforward and left as an exercise to the reader. We shall need the following lemma.

Lemma 3 *The system* (15) *admits a nontrivial* $\omega-$*periodic solution if and only if* $x(0)=x(\omega)$.

Proof If $x(t)$ is $\omega-$periodic solution of (15), then it is clear that $x(0)=x(\omega)$. Next, let $y(t)=x(t+\omega)$ for $t\in\mathbb{R}$. Then, it is easy to see that $y(0)=x(0)$ and

$$\begin{aligned}&y'(t)=x'(t+\omega)=A(t)x(t+\omega)+f(t+\omega)=A(t)y(t)+f(t),\quad t\neq\zeta_l,\\ &\Delta y\big|_{t=\zeta_l}=x(\zeta_l+\omega+)-x(\zeta_l+\omega)=B_lx(\zeta_l+\omega)+J_l=B_ly(\zeta_l)+J_l.\end{aligned}$$

Thus, $y(t)$ is a solution of (15). By the uniqueness result of Theorem 9 we have $x(t)=y(t)=x(t+\omega)$ for $t\in\mathbb{R}$.

Theorem 11 *If the homogeneous system* (5) *does not admit a nontrivial* $\omega-$*periodic solution, then there exits a unique* $\omega-$*periodic solution of the system* (15) *which satisfy the following equation*

$$x(t)=\int_0^\omega G_p(t,s)f(s)ds+\sum_{0<\zeta_l<\omega}G_p(t,\zeta_l+)J_l. \tag{17}$$

Proof One can show that (17) satisfies the system (5) by Lemma 2. The solution $x(t)$ of (5) with $x(0)=x_0$ satisfies (9). That is,

$$x(t)=\Phi(t)x(0)+\int_0^t\Phi(t)\Phi^{-1}(s)f(s)ds+\sum_{0\le\zeta_l<t}\Phi(t)\Phi^{-1}(\zeta_l+)J_l. \tag{18}$$

Thus, by Lemma 3 (15) admits a nontrivial $\omega-$periodic solution if and only if

$$x(0)=x(\omega)=\Phi(\omega)x(0)+\int_0^\omega\Phi(\omega)\Phi^{-1}(s)f(s)ds+\sum_{0\le\zeta_l<\omega}\Phi(\omega)\Phi^{-1}(\zeta_l+)J_l.$$

Thus, since $I-\Phi(\omega)$ invertible x_0 satisfy

$$x(0) = (I - \Phi(\omega))^{-1}\left(\int_0^{\omega} \Phi(\omega)\Phi^{-1}(s)f(s)ds + \sum_{0\le\zeta_l<\omega} \Phi(\omega)\Phi^{-1}(\zeta_l+)J_l\right). \tag{19}$$

Plugging (19) into (18) yields

$$\begin{aligned}
x(t) &= \Phi(t)(I - \Phi(\omega))^{-1}\left(\int_0^{\omega} \Phi(\omega)\Phi^{-1}(s)f(s)ds + \sum_{0\le\zeta_l<\omega} \Phi(\omega)\Phi^{-1}(\zeta_l+)J_l\right)\\
&+ \int_0^{t} \Phi(t)\Phi^{-1}(s)f(s)ds + \sum_{0\le\zeta_l<t} \Phi(t)\Phi^{-1}(\zeta_l+)J_l\\
&= \Phi(t)(I - \Phi(\omega))^{-1}\left(\int_0^{t} \Phi(\omega)\Phi^{-1}(s)f(s)ds + \sum_{0\le\zeta_l<t} \Phi(\omega)\Phi^{-1}(\zeta_l+)J_l\right)\\
&+ \Phi(t)(I - \Phi(\omega))^{-1}\left(\int_t^{\omega} \Phi(\omega)\Phi^{-1}(s)f(s)ds + \sum_{t\le\zeta_l<\omega} \Phi(\omega)\Phi^{-1}(\zeta_l+)J_l\right)\\
&+ \Phi(t)(I - \Phi(\omega))^{-1}\int_0^{t} (I - \Phi(\omega))\Phi^{-1}(s)f(s)ds\\
&+ \Phi(t)(I - \Phi(\omega))^{-1}\sum_{0\le\zeta_l<t} (I - \Phi(\omega))\Phi^{-1}(\zeta_l+)J_l\\
&= \int_0^{t} \Phi(t)(I - \Phi(\omega))^{-1}\Phi^{-1}(s)f(s)ds\\
&+ \sum_{0\le\zeta_l<t} \Phi(t)(I - \Phi(\omega))^{-1}\Phi^{-1}(\zeta_l+)J_l\\
&+ \int_t^{\omega} \Phi(t+\omega)(I - \Phi(\omega))^{-1}\Phi^{-1}(s)f(s)ds\\
&+ \sum_{t\le\zeta_l<\omega} \Phi(t+\omega)(I - \Phi(\omega))^{-1}\Phi^{-1}(\zeta_l+)J_l\\
&= \int_0^{\omega} G_p(t,s)f(s)ds + \sum_{0<\zeta_l<\omega} G_p(t,\zeta_l+)J_l.
\end{aligned}$$

In the foregoing calculations we have used the fact that $(I - \Phi(\omega))^{-1}$ commutes with $\Phi(\omega)$ and $\Phi(t+\omega) = \Phi(t)\Phi(\omega)$. Next, suppose that there exists a $\omega-$periodic solution $y(t)$ of the system (15) different from $x(t)$. Let $z(t) = x(t) - y(t)$. Then,

$$\begin{aligned}
&z'(t) = x'(t) - y'(t) = A(t)x(t) + f(t) - A(t)y(t) - f(t) = A(t)z(t), \quad t \ne \zeta_l,\\
&\Delta z\big|_{t=\zeta_l} = x(\zeta_l+) - y(\zeta_l+) - x(\zeta_l) + y(\zeta_l) = B_l x(\zeta_l) + J_l - B_l y(\zeta_l) - J_l = B_l z(\zeta_l).
\end{aligned}$$

Thus, $z(t)$ is a nontrivial $\omega-$periodic solution of the homogeneous system (5). This is a contradiction. This completes the proof.

Acknowledgements AK was supported by Nazarbayev University Faculty Development Competitive Research Grants N 090118FD5353. MM was supported by Nazarbayev University Faculty Development Competitive Research Grants N 090118FD5342.

References

1. Akhmet, M.: Principle of Discontinuous Dynamical Systems. Springer, Berlin (2010)
2. Akhmet, M., Kashkynbayev, A.: Bifurcation in Autonomous and Nonautonomous Differential Equations with Discontinuities. Springer & Higher Education Press (2017)
3. Akhmet, M.U., Kashkynbayev, A.: Non-autonomous bifurcation in impulsive systems. Electron. J. Qual. Theory Differ. Equ. **2013**(74), 1–23 (2013)
4. Akhmet, M.U., Kashkynbayev, A.: Nonautonomous transcritical and pitchfork bifurcations in impulsive systems. Miskolc Math. Notes **14**, 737–748 (2013)
5. Akhmet, M., Kashkynbayev, A.: Finite-time nonautonomous bifurcation in impulsive systems.Electron. J. Qual. Theory Differ. Equ., Proc. 10'th Coll. Qualitative Theory of Diff. Equ. **2016**(1), 1–13 (2016)
6. Akhmet, M.U., Kashkynbayev, A.: Nonautonomous bifurcations in nonlinear impulsive systems. Differ. Equ. Dyn. Syst. **28**, 177–190 (2020)
7. Akhmet, M.U., Fen, M.O., Kashkynbayev, A.: Persistence of Li-Yorke chaos in systems with relay. Electron. J. Qual. Theory Differ. Equ. **2017**(72), 1–18 (2017)
8. Akhmet, M.U., Feckan, M., Fen, M.O., Kashkynbayev, A.: Perturbed Li-Yorke homoclinic chaos. Electron. J. Qual. Theory Differ. Equ. **2018**(75), 1–18 (2018)
9. Brenner, L.: Matrices of quaternions. Pacific J. Math. **1**, 329–335 (1951)
10. Campos, J., Mawhin, J.: Periodic solutions of quaternionic-values ordinary differential equations. Annali di Mathmatica **185**(5), 109–127 (2006)
11. Coppel, W.A.: Dichotomies in Stability Theory. Springer (1978)
12. Jingwen, Z., Jitao, S.: Existence and uniqueness results for quaternion-valued nonlinear impulsive differential systems. J. Syst. Sci. Complex **31**, 596–607 (2018)
13. Kou, K.I., Xia, Y.H.: Linear quaternion differential equations: basic theory and fundamental results. Stud. Appl. Math. **141**(1), 3–45 (2018)
14. Lee, H.C.: Eigenvalues of canonical forms of matrices with quaternion coefficients. Proc. Roy. Irish Acad. Sect. A **52**, 253–260 (1949)
15. Leo, S., Ducati, G.: Solving simple quaternionic differential equations. J. Math. Phys. **44**, 2224 (2003)
16. Samoilenko, A.M., Perestyuk, N.A.: Impulsive Differential Equations. World Scientific, Singapore (1995)
17. Wilczynski, P.: Quaternionic-valued ordinary differential equations. The Riccati equation. J. Differ. Equ. **247**(7), 2163–2187 (2009)
18. Zhang, F.: Quaternions and matrices of quaternions. Linear Algebra Appl. **251**, 21–57 (1997)

Author Index

A. Ashyralyev et al. (eds.), *Functional Analysis in Interdisciplinary Applications—II*, Springer Proceedings in Mathematics & Statistics 351,
https://doi.org/10.1007/978-3-030-69292-6

Subject Index

A. Ashyralyev et al. (eds.), *Functional Analysis in Interdisciplinary Applications—II*,
Springer Proceedings in Mathematics & Statistics 351,
https://doi.org/10.1007/978-3-030-69292-6

Printed by Printforce, the Netherlands